ELECTRICAL PRINCIPLES

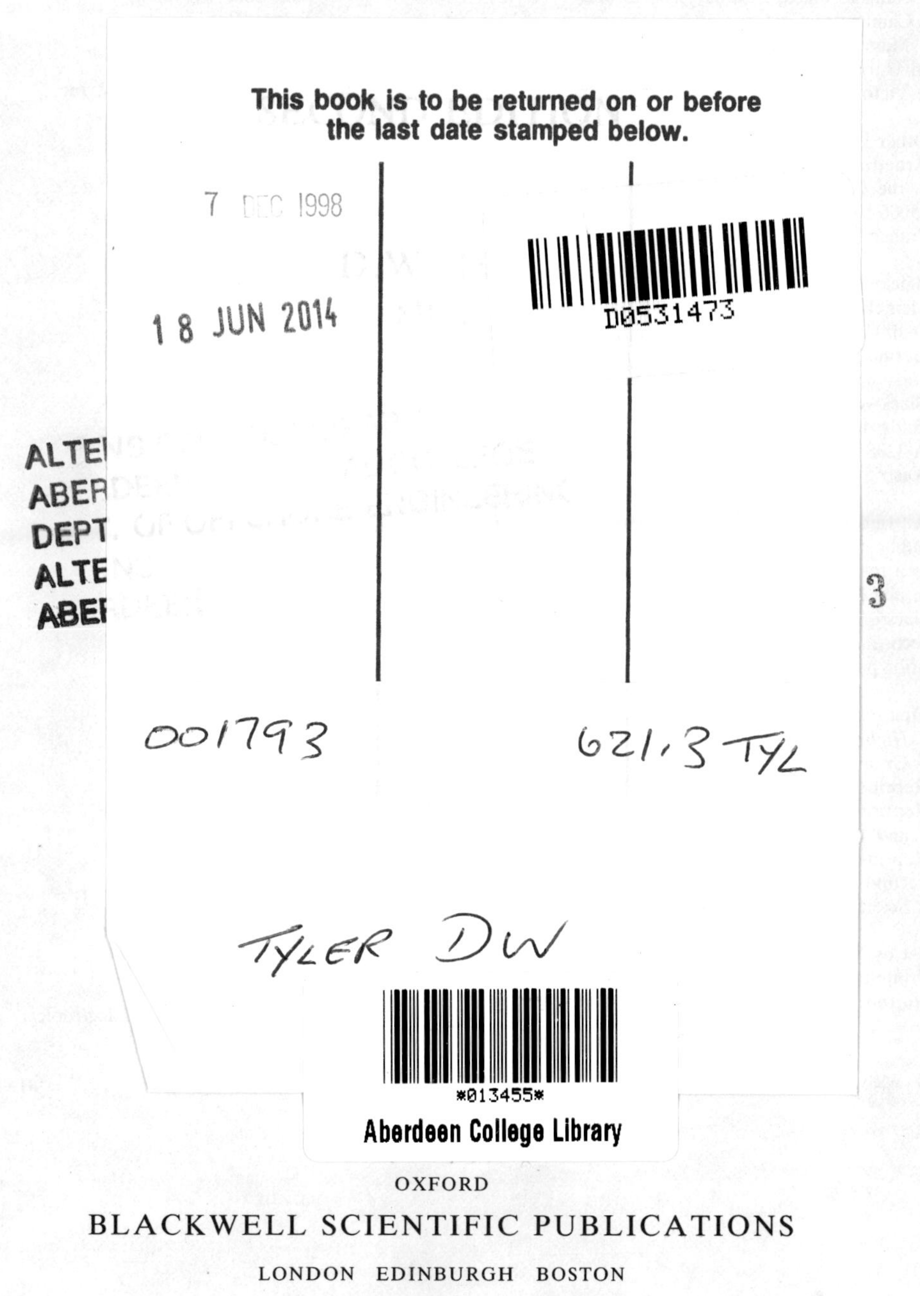

OXFORD
BLACKWELL SCIENTIFIC PUBLICATIONS
LONDON EDINBURGH BOSTON
MELBOURNE PARIS BERLIN VIENNA

Blackwell Scientific Publications
Editorial offices:
Osney Mead, Oxford OX2 0EL
25 John Street, London WC1N 2BL
23 Ainslie Place, Edinburgh EH3 6AJ
3 Cambridge Center, Cambridge,
Massachusetts 02142, USA
54 University Street, Carlton
Victoria 3053, Australia

Other Editorial Offices:
Arnette SA
2, rue Casimir-Delavigne
75006 Paris
France

Blackwell Wissenschaft
Meinekestrasse 4
D-1000 Berlin 15
Germany

Blackwell MZV
Feldgasse 13
A-1238 Wien
Austria

First edition of *Electrical Principles for Higher TEC* published by Granada Publishing Ltd 1982
Reprinted 1984
Reprinted by Collins Professional and Technical Books 1985
Reprinted by BSP Professional Books 1988
Second edition published by Blackwell Scientific Publications 1991

Set by Setrite Typesetters Ltd, H.K.
Printed and bound in Great Britain by Hartnolls, Bodmin, Cornwall

DISTRIBUTORS

Marston Book Services Ltd
PO Box 87
Oxford OX2 ODT
(*Orders*: Tel: 0865 791155
Fax: 0865 791927
Telex: 837515)

USA
Blackwell Scientific Publications, Inc.
3 Cambridge Center
Cambrige, MA 02142
(*Orders*: Tel: 800 759–6102)

Canada
Oxford University Press
70 Wynford Drive
Don Mills
Ontario M3C 1J9
(*Orders*: Tel: 416 441–2941)

Australia
Blackwell Scientific Publications
(Australia) Pty Ltd
54 University Street
Carlton, Victoria 3053
(*Orders*: Tel: 03 347–0300)

British Library
Cataloguing in Publication Data

Tyler, D.W. (David W)
Electrical principles. — 2nd ed.
1. Electrical engineering
I. Title
621.3

ISBN 0–632–02536–0

Library of Congress
Cataloging in Publication Data

Tyler, David W.
Electrical principles/D.W. Tyler. — 2nd ed.
p. cm.
Includes index.
ISBN 0–632–02536–0
1. Electric engineering. 2. Electronics.
I. Title.
TK146.T953 1991
621.3–dc20 90–20611
CIP

Contents

Preface to Second Edition

The contents of the first edition of this book were determined by consultation between the author, the publishers and several individual technical colleges. BTEC have now issued a guide syllabus and the new book has been written to take this into account. It is gratifying to see how much of the original book is still relevant and this reinforces my thoughts expressed in the original preface concerning the rapid rate of change of equipments whereas the principles involved are constant. Existing sections have been reviewed, extra material added together with new chapters as indicated by the guidelines. It has been necessary to exercise an element of choice in the material included since a book to cover every single objective of every section in the guide syllabus would be approaching twice the length of this one. In my view such sections as are not covered are best dealt with by the many specialist works on, for example, electronics and control systems.

I would like to thank all my ex-colleagues for their support and help over the years. They have used the first edition extensively and have made suggestions as to improvements, and on occasion pointed out an incorrect answer, all of these having been incorporated in the new book. As in the Preface to the first edition, "I hope that this book will go at least some way towards a greater appreciation of 'Principles' ".

DWT

Preface to First Edition

This book provides good coverage of the electrical principles required by undergraduates and Higher TEC students at levels IV and V.

A sound grasp of the principles involved in electrical and electronic engineering is of considerable value. To appreciate this one has only to look back and view the extremely rapid development of equipment which has taken place. We have moved from thermionic valves to p-n-p transistors, on to n-p-n transistors, integrated circuits, NMOS, PMOS, CMOS, etc. The speed at which changes take place seems ever accelerating. In power engineering, control, instrumentation and supervision methods have been revolutionised, many of the older electro-mechanical devices having been replaced by integrated electronics and the omnipresent computer.

All such equipment conforms to common principles and a good knowledge of these enables the new technology, as it progresses, to be mastered. For example, the steady-state and transient response of circuits, mutual inductance as a means of obtaining improved bandwidth for power matching and transfer; electro-magnetism and electro-statics, transmission systems and the effects of harmonics.

In many courses the principles (and often the mathematics) seem to be given second rate attention, being sacrificed to a specialist study of particular devices. When these become obsolete, re-training is that much more difficult without a thorough grasp of fundamentals. I hope that this book will go at least some way towards a greater appreciation of 'Principles'.

DWT

Chapter 1
Symbolic Notation

1.1 Phasor manipulation

This section is a brief revisionary introduction to the circuit theory which should have been covered fully in previous work.

Consider the two voltages shown phasorially in Fig. 1.1.

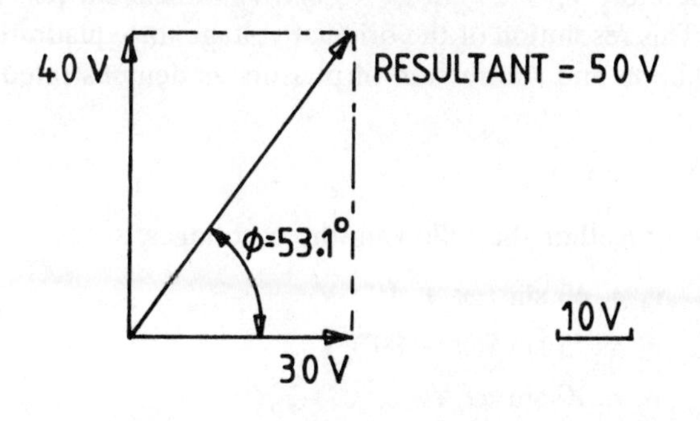

Fig. 1.1

$v_1 = 30 \sin \omega t$ V $\qquad v_2 = 40 \sin (\omega t + 90°)$ V.

The peak values of the two voltages are 30 V and 40 V respectively. v_2 leads v_1 by 90°.

To add these two voltages together the parallelogram is constructed using the peak values of voltage. By scale drawing or by the use of Pythagoras' theorem the resultant V_R = 50 V. The angle between V_1 and V_R = 53.1°.

$$V_R = \sqrt{(30^2 + 40^2)} = 50 \text{ V}; \qquad \tan \phi = \frac{40}{30} = 1.33, \text{ hence } \phi = 53.1°.$$

The resultant voltage is therefore given by the equation

$$v_R = 50 \sin (\omega t + 53.1°) \text{ V}.$$

In this case the two voltages have been added together to find a resultant. It is possible to resolve a single voltage into two component voltages which are at right angles.

Given a voltage $v = 45 \sin (\omega t + 30°)$ V, as shown in Fig. 1.2, this is to be resolved into horizontal and vertical components.

$$\frac{x}{45} = \cos 30°, \qquad \text{hence } x = 45 \cos 30° = 39 \text{ V}.$$

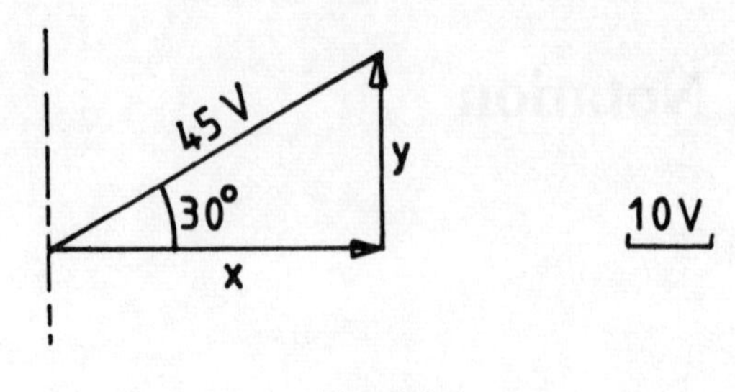

Fig. 1.2

$$\frac{y}{45} = \sin 30°, \qquad \text{hence } y = 45 \sin 30° = 22.5 \text{ V}.$$

Therefore $v_x = 39 \sin \omega t$ V, and $v_y = 22.5 \sin(\omega t + 90°)$ V.

This resolution of the original voltage into quadrature voltages facilitates the addition and subtraction of phasors as demonstrated in worked example 1.1.

WORKED EXAMPLE 1.1

Add together the following three voltages:

$v_1 = 60 \sin(\omega t + 45°)$ V

$v_2 = 75 \sin(\omega t + 80°)$ V

$v_3 = 10 \sin \omega t$ V.

Express the answer in the same form.

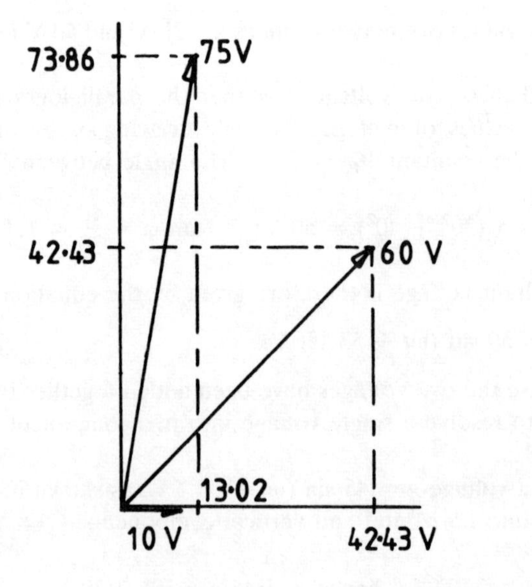

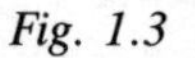

Fig. 1.3

Resolve each of the voltages into horizontal (x) and vertical (y) components.

$V_{x1} = 60 \cos 45° = 42.43$ V $V_{y1} = 60 \sin 45° = 42.43$ V

$V_{x2} = 75 \cos 80° = 13.02$ V $V_{y2} = 75 \sin 80° = 73.86$ V

$V_{x3} = 10$ V. There is no component of V_3 in the y direction.

Adding horizontal components:

$$V_{x(\text{total})} = 42.43 + 13.02 + 10 = 65.45 \text{ V}.$$

Adding vertical components:

$$V_{y(\text{total})} = 42.43 + 73.86 = 116.29 \text{ V}.$$

The resultant voltage has a modulus (size) of $\sqrt{(65.45^2 + 116.29^2)} = 133.4$ V.

$$\tan \phi = \frac{116.29}{65.45} = 1.77, \qquad \text{hence } \phi = 60.63°.$$

Hence $v_R = 133.4 \sin(\omega t + 60.63°)$ V (see Fig. 1.4).

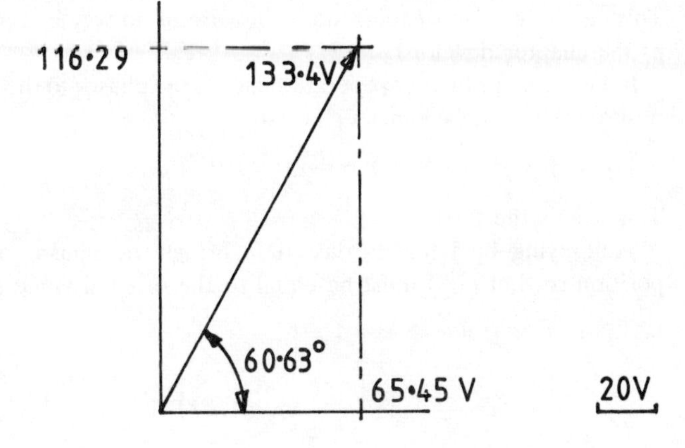

Fig. 1.4

1.2 The j operator

The use of the j operator can reduce considerably the amount of work involved in solving electrical circuits containing inductance and capacitance, especially when these are in parallel or series parallel combination.

By definition, multiplying a phasor by j moves its position 90° anti-clockwise without changing its magnitude.

In Fig. 1.5(a) a phasor three units long is drawn horizontally to the right. It represents an electrical quantity to scale. On a graph this is the usual $+x$ direction. If we write (j3), by definition this is the same phasor turned through 90° anti-clockwise. This is shown in Fig. 1.5(b). A further 90° shift is shown in

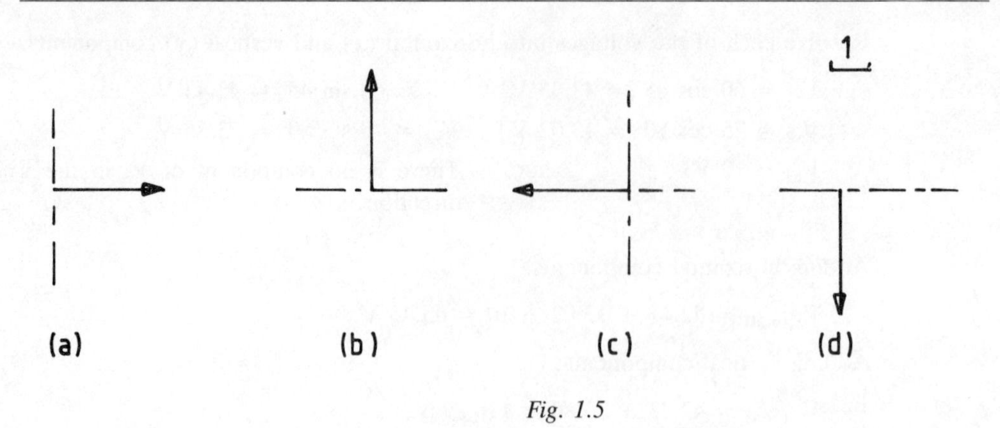

Fig. 1.5

Fig. 1.5(c) and this is described as (j × j3) or ($j^2$3). This is the original phasor reversed, i.e. (−3). This means that

$$j^2 = -1, \qquad \text{hence } j = \sqrt{-1}.$$

Since there is no numerical solution for this quantity (−1 × −1 = 1 and +1 × +1 = 1), the numbers involving j are often called imaginary numbers. This can be very misleading since they are in no way imaginary as will be seen as the chapter develops.

In Fig. 1.5(d) a further j operation moves the phasor to the vertically downward position where it becomes ($j^3$3). Now

$$j^3 = j \times j^2 = j \times -1 = -j.$$

This axis is the (−j) axis.

Multiplying by j for the last time brings the phasor back to the original position so that ($j^4$3) must be equal to the original value (3).

$$j^4 = j^2 \times j^2 = -1 \times -1 = 1.$$

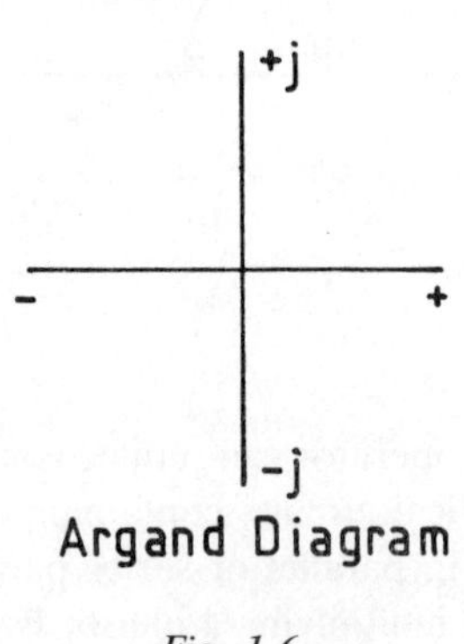

Fig. 1.6

Figure 1.6 combines these axes in one diagram. This is known as the Argand diagram.

Two quantities, OP = (4 + j5) and OQ = (−6 + j3), are shown plotted on an Argand diagram in Fig. 1.7.

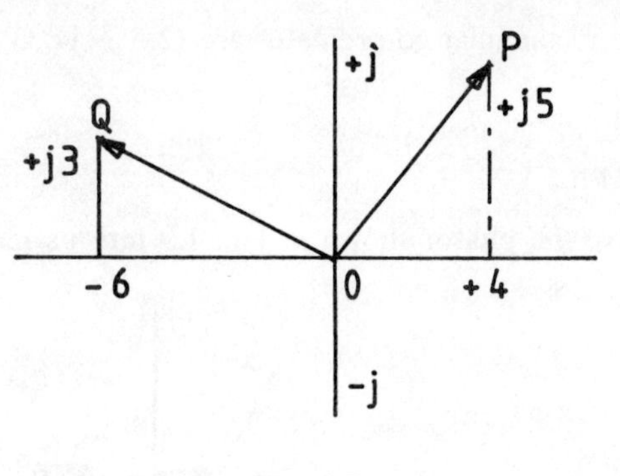

Fig. 1.7

1.3 Rectangular co-ordinates

The position of a phasor representing an electrical quantity can be described by quoting its horizontal and vertical components. Return to Fig. 1.2, in which $v = 45 \sin (\omega t + 30°)$ V. The horizontal component $x = 39$ V; the vertical component $y = 22.5$ V. This voltage may be expressed as

$$(39 + j22.5) \text{ V.}$$

Such a quantity is called a complex number, the values 39 and j22.5 being its rectangular co-ordinates.

WORKED EXAMPLE 1.2

Resolve the phasor shown in Fig. 1.8 into its rectangular co-ordinates.

$$\frac{x}{5} = \cos 60° \qquad x = 5 \cos 60° = 2.5$$

$$\frac{y}{5} = \sin 60° \qquad y = 5 \sin 60° = 4.33$$

The horizontal component is from O, the origin, extending to the right and is in the $+x$ direction. The vertical component is from the origin upwards in the +j direction.

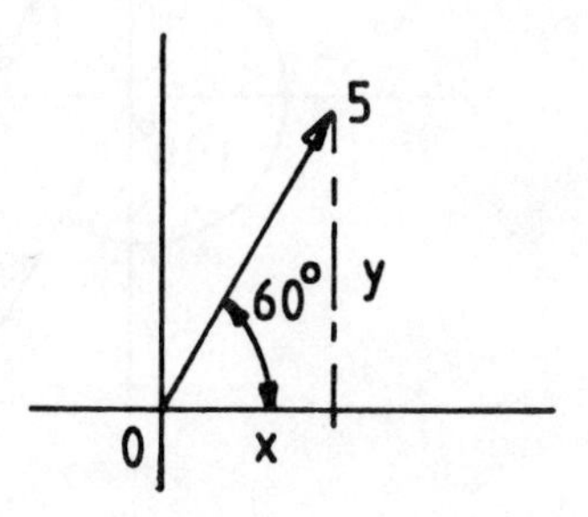

Fig. 1.8

The rectangular co-ordinates are $(2.5 + j4.33)$.

WORKED EXAMPLE 1.3

Resolve the phasor shown in Fig. 1.9 into its rectangular co-ordinates.

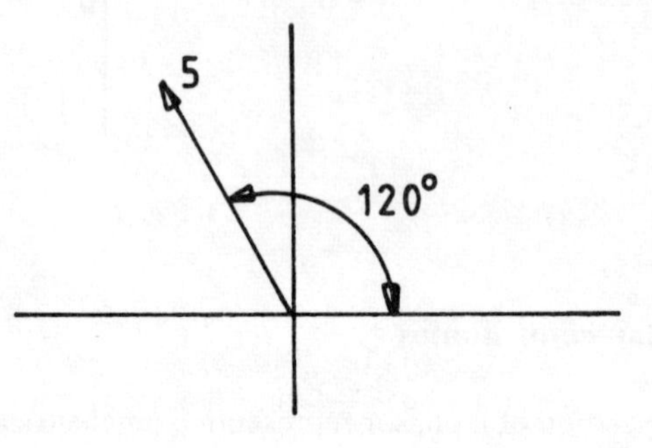

Fig. 1.9

$$\begin{aligned}\text{Horizontal component} &= 5\cos 120^\circ \qquad (\cos 120^\circ = -\cos 60^\circ)\\ &= 5 \times (-0.5)\\ &= -2.5\\ \text{Vertical component} &= 5\sin 120^\circ \qquad (\sin 120^\circ = \sin 60^\circ)\\ &= 4.33\end{aligned}$$

The horizontal component is 2.5 measured to the left of the origin, i.e. in the $-x$ direction. The vertical component is again upwards.

The rectangular co-ordinates are $(-2.5 + j4.33)$.

SELF-ASSESSMENT EXAMPLE 1.4

Resolve the phasor shown in Fig. 1.10 into its rectangular co-ordinates.

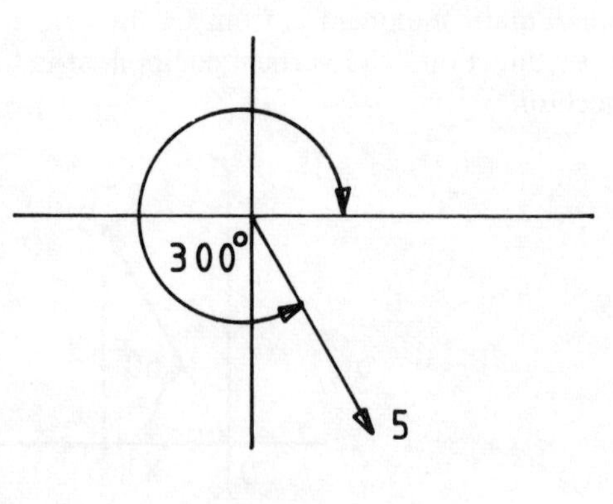

Fig. 1.10

1.4 The polar form

Let z denote a complex number represented by the phasor OP in Fig. 1.11.

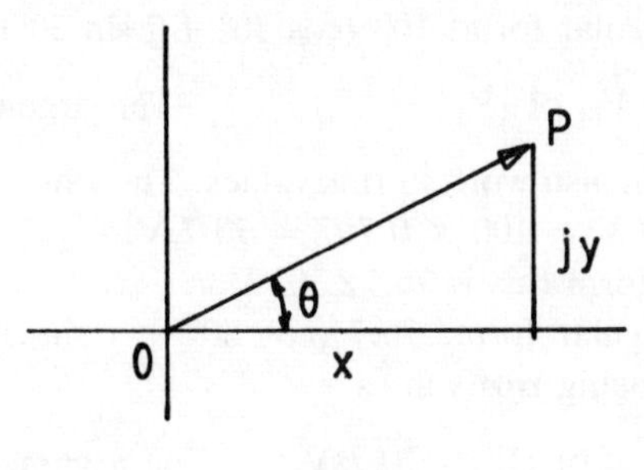

Fig. 1.11

In rectangular form it has been shown that z may be expressed as $(x + jy)$.

By Pythagoras' theorem, $OP = \sqrt{(x^2 + y^2)}$, and $\tan \theta = y/x$, hence $\theta = \arctan (y/x)$.

If the length of OP = r, the phasor may be described as $r\angle\theta$ (read as 'r angle theta'). The quantity r is called the modulus of the complex number; θ is called the argument of the complex number.

Mod z or $|z|$ can be written as an abbreviation for the modulus, i.e. r together with arg z for the angle the phasor makes with the $+x$ direction, where:

$$z = (x + jy) = r\angle\theta,$$
$$\text{mod } z \text{ or } |z| = \sqrt{(x^2 + y^2)},$$
$$\text{and arg } z = \arctan y/x.$$

$r\angle\theta$ is known as the polar form of the complex number.

The phasor in Fig. 1.8 could be described as $5\angle 60°$. The size or modulus of the phasor is 5 and the angle it makes with the $+x$ direction or argument is 60°. The phasor in Fig. 1.9 would be described as $5\angle 120°$, and that in Fig. 1.10 as $5\angle 300°$. The angles may be measured in a clockwise direction when they become negative. The phasor in Fig. 1.10 could equally well be described as $5\angle -60°$.

When using symbolic phasor notation it is sound practice always to draw the triangle concerned roughly to scale so that it is possible to see the angle involved and in which quadrant of the Argand diagram one is working.

1.5 The use of peak and rms values

When expressing a voltage as $v = V_m \sin (\omega t + \phi)$ volts,

v = instantaneous value at time t seconds,
V_m = maximum value of the wave,
ω (= $2\pi f$) is the angular velocity in radians per second.

A phasor drawn to scale may be used to represent this voltage and this in turn may be expressed in either rectangular or polar forms.
For example: $v = 100 \sin(\omega t + 30°)$.
In polar form this is $100\angle 30°$ V.
In rectangular form, $100(\cos 30° + j \sin 30°) = (86.6 + j50)$ V.

$$\text{Mod } V_m \text{ or } |V_m| = 100 \text{ V}. \qquad \text{The argument} = +30°.$$

One can also work in rms values. The rms value of a sinusoid with maximum value 100 V = $100 \times 0.707 = 70.7$ V.
In polar form this is $70.7\angle 30°$ V.
In rectangular form, $70.7(\cos 30° + j \sin 30°) = (61.2 + j35.35)$ V, these voltages being rms values.

$$\text{Mod } V \text{ or } |V| = 70.7 \text{ V}. \qquad \text{The argument} = +30°.$$

Ammeters and voltmeters which are rms indicating are usually employed and Fig. 1.12 shows such meters connected to measure current and voltage in a single-phase circuit. These indicate rms moduli and have no means of resolving angle.

$$\frac{|V|}{|I|} = |Z|.$$

Nothing may be deduced about the phase angle of the load unless either a wattmeter is connected in the circuit to measure power, or other forms of ammeter and voltmeter are used which are capable of indicating phase.

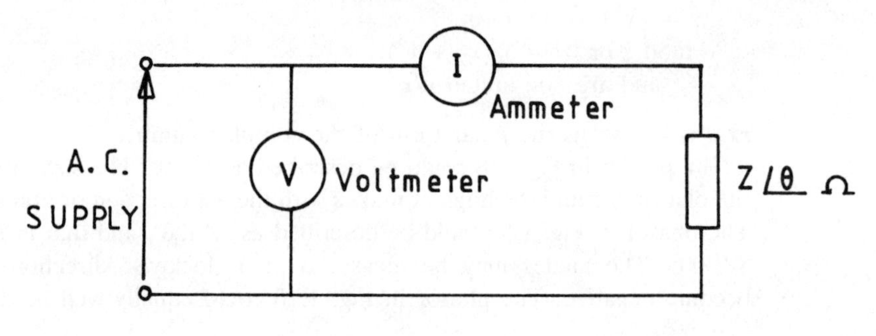

Fig. 1.12

WORKED EXAMPLE 1.5

Express $(1 + j7)$ in the polar form.
Drawing the quantities approximately to scale (Fig. 1.13) shows that the resultant lies in the first quadrant of the Argand diagram. The angle θ lies between 0° and 90°. By Pythagoras' theorem,

$$\text{the resultant} = \sqrt{(1^2 + 7^2)} = 7.07$$

$$\tan\theta = \frac{7}{1}, \qquad \text{hence } \theta = 81.9°.$$

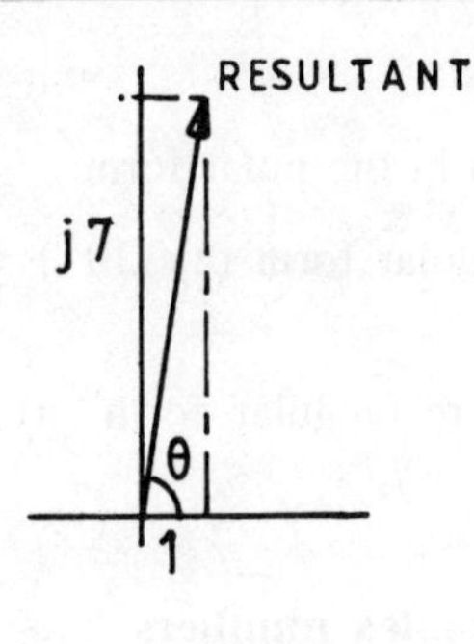

Fig. 1.13

The polar form is $7.07\angle 81.9°$.

WORKED EXAMPLE 1.6

Express $(-7 - j9)$ in the polar form.

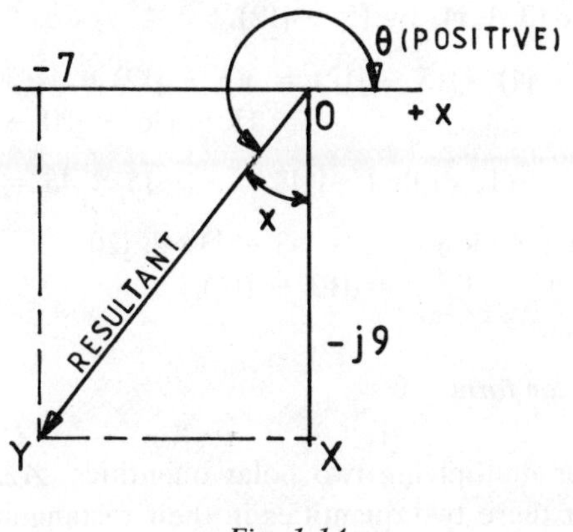

Fig. 1.14

Drawing the rectangular co-ordinates in their correct positions as shown in Fig. 1.14 shows that the resultant lies in the third quadrant and that θ lies between 180° and 270°. In the figure, work in triangle OXY and for the moment ignore the (j) and negative signs. OXY is a triangle with two known sides, OX = 9 and XY = 7.

$$\text{The resultant} = \sqrt{(9^2 + 7^2)} = 11.4$$

To find the angle θ,

$$\tan x = \frac{7}{9} = 0.777, \qquad \text{hence } x = 37.9°.$$

Therefore the resultant phasor lies at 90° + 37.9° = 127.9° from the $+x$ direction (clockwise). Hence θ = 360° − 127.9° = 232.1° (anti-clockwise).

The resultant may be described as (i) $11.4\angle -127.9°$
or (ii) $11.4\angle 232.1°$.

SELF-ASSESSMENT EXAMPLES

1.7 Express $(3 - j4)$ in the polar form.

1.8 Convert to the polar form (a) $(10 + j2)$, (b) $(-4 - j6)$, (c) $(-9 + j4)$, (d) $(5 - j12)$.

1.9 Convert to the rectangular form (a) $28\angle 85°$, (b) $4\angle 220°$, (c) $6\angle 165°$, (d) $10\angle 350°$, (e) $14\angle -75°$.

1.6 Multiplication of complex numbers

1.6.1 Using rectangular co-ordinates

All the rules of algebra apply with the additional requirement that where j^2 occurs it must be remembered that this has a value of -1.

WORKED EXAMPLE 1.10

Multiply $(3 + j4)$ by $(5 - j12)$.

$$\begin{aligned}(3 + j4) \times (5 - j12) &= 3(5 - j12) + j4(5 - j12)\\ &= 15 - j36 + j20 - j^2 48\end{aligned}$$

Now $j^2 = -1$, so that $-j^2 48 = -(-1) \times 48 = +48$.

$$\begin{aligned}\text{The solution is } &15 + 48 - j36 + j20\\ &= (63 - j16).\end{aligned}$$

1.6.2 Using the polar form

Consider multiplying two polar quantities, $A\angle\theta$ and $B\angle\phi$, together. Firstly consider these two quantities in their rectangular form.

$$A\angle\theta = A(\cos\theta + j\sin\theta) \qquad B\angle\phi = B(\cos\phi + j\sin\phi).$$

$$\begin{aligned}\text{Therefore } A\angle\theta \times B\angle\phi &= A(\cos\theta + j\sin\theta) \times B(\cos\phi + j\sin\phi)\\ &= AB(\cos\theta\cos\phi + j\cos\theta\sin\phi +\\ &\quad j\sin\theta\cos\phi - \sin\theta\sin\phi).\end{aligned}$$

Using the double-angle formulae: $\cos(\theta + \phi) = \cos\theta\cos\phi - \sin\theta\sin\phi$
$\sin(\theta + \phi) = \cos\theta\sin\phi + \sin\theta\cos\phi$

$$A\angle\theta \times B\angle\phi = AB(\cos(\theta + \phi) + j\sin(\theta + \phi)).$$

Converted to the polar form this is $AB\angle\theta + \phi$.

Hence to multiply two quantities in the polar form, multiply the two moduli together and add their arguments.

To demonstrate the method the figures from worked example 1.10 will be used.

$(3 + j4)$ in the polar form $= 5\angle 53.13°$.
$(5 - j12)$ in the polar form $= 13\angle -67.38°$.

The solution for this product in rectangular form has been shown to be $(63 - j16)$. In polar form this is $65\angle -14.25°$.

$$\begin{aligned}\text{Working in polar form:}\quad & 5\angle 53.13° \times 13\angle -67.38° \\ &= 5 \times 13\angle 53.13° + (-67.38°) \\ &= 65\angle -14.25°\end{aligned}$$

which is the required result.

WORKED EXAMPLE 1.11

Multiply $(3 + j2)$ by $(-2 - j4)$ using both rectangular and polar forms.

(a) Rectangular

$$\begin{aligned}(3 + j2) \times (-2 - j4) &= -6 - j12 - j4 - j^2 8 \\ &= -6 + 8 - j16 \\ &= (2 - j16)\ (\text{in polar form } 16.1\angle -82.9°)\end{aligned}$$

(b) Polar

$(3 + j2) = 3.6\angle 33.7°$ $\quad (-2 - j4) = 4.47\angle -116.6°$.
Product $= 3.6 \times 4.47\angle 33.7° + (-116.6°) = 16.1\angle -82.9°$

which agrees with the result obtained using rectangular co-ordinates.

SELF-ASSESSMENT EXAMPLE 1.12

Express the answers to the following products in both rectangular and polar forms.

(a) $(10 - j15) \times 6\angle 225°$, $\quad$ (b) $(-5 + j6) \times (3 - j10)$.

1.7 Division of complex numbers

1.7.1 Using rectangular co-ordinates

Consider performing the calculation $(x + jy)/(a - jb)$.
The division cannot be carried out while the expression is in this form. It is possible to divide by real numbers alone or by j terms alone but not simultaneously by a combination of both.

To turn the expression into a suitable form it is necessary to multiply both the numerator and denominator by a quantity which will turn the denominator into a pure number. This process is called rationalisation. The quantity involved is called the complex conjugate of the denominator.

The complex conjugate of the denominator is the denominator itself with the sign of the j term changed.

The original quotient becomes $\dfrac{(x + jy)}{(a - jb)} \times \dfrac{(a + jb)}{(a + jb)}$.

Note that the operation has not changed the value of the expression since the two $(a + jb)$ terms could be cancelled.

Multiply out the numerator: $ax + jbx + jay - by$
$= (ax - by) + j(bx + ay)$.

Multiply out the denominator: $a^2 + jab - jab + b^2$
$= a^2 + b^2$.

Thus multiplying the denominator by the complex conjugate has eliminated the j term. Therefore

$$\frac{(x + jy)}{(a - jb)} = \frac{(ax - by) + j(bx + ay)}{a^2 + b^2}$$
$$= \frac{(ax - by)}{a^2 + b^2} + \frac{j(bx + ay)}{a^2 + b^2}.$$

WORKED EXAMPLE 1.13

Divide (3 + j2) by (2 + j4).

$\dfrac{(3 + j2)}{(2 + j4)}$ becomes $\dfrac{(3 + j2)\,(2 - j4)}{(2 + j4)\,(2 - j4)}$ after rationalising.

Multiplying out both numerator and denominator

$$\frac{6 - j12 + j4 + 8}{2^2 + 4^2} = \frac{(14 - j8)}{20} = \frac{14}{20} - j\frac{8}{20}.$$

Thus $\dfrac{(3 + j2)}{(2 + j4)} = \dfrac{14}{20} - j\dfrac{8}{20} = 0.7 - j0.4$

WORKED EXAMPLE 1.14

Evaluate $(-7 - j9)/(5 - j12)$.

Rationalise the expression by multiplying numerator and denominator by (5 + j12). This gives

$$\frac{(-7 - j9)\,(5 + j12)}{5^2 + 12^2} = \frac{-35 - j84 - j45 + 108}{169}$$
$$= \frac{(73 - j129)}{169}$$
$$= (0.43 - j0.76).$$

1.7.2 Using the polar form

Consider dividing $A\angle\theta$ by $B\angle\phi$.
Converting to rectangular form gives

$$\frac{A\angle\theta}{B\angle\phi} = \frac{A(\cos\theta + \mathrm{j}\sin\theta)}{B(\cos\phi + \mathrm{j}\sin\phi)}.$$

Rationalise the expression by multiplying numerator and denominator by $(\cos\phi - \mathrm{j}\sin\phi)$. This gives

$$\frac{A(\cos\theta + \mathrm{j}\sin\theta)(\cos\phi - \mathrm{j}\sin\phi)}{B(\cos^2\phi + \sin^2\phi)}.$$

But $\cos^2\phi + \sin^2\phi = 1$. Hence the expression becomes

$$\frac{A}{B}(\cos\theta\cos\phi - \mathrm{j}\cos\theta\cos\phi + \mathrm{j}\sin\theta\cos\phi + \sin\theta\sin\phi)$$

$$= \frac{A}{B}((\cos\theta\cos\phi + \sin\theta\sin\phi) + \mathrm{j}(\sin\theta\cos\phi - \cos\theta\sin\phi)).$$

Using the double-angle formulae: $\cos(\theta - \phi) = \cos\theta\cos\phi + \sin\theta\sin\phi$
$\sin(\theta - \phi) = \sin\theta\cos\phi - \cos\theta\sin\phi$,

$$\frac{A\angle\theta}{B\angle\phi} = \frac{A}{B}(\cos(\theta - \phi) + \mathrm{j}\sin(\theta - \phi))$$

which in the polar form is $(A/B)\angle\theta - \phi$.

Division in the polar form is accomplished by dividing the moduli and subtracting the argument of the denominator from that of the numerator, having due regard for the signs.

Using the quantities from worked example 1.14:

$$(-7 - \mathrm{j}9) = 11.4\angle-127.87^\circ \qquad (5 - \mathrm{j}12) = 13\angle-67.38^\circ.$$

$$\text{Therefore } \frac{(-7 - \mathrm{j}9)}{(5 - \mathrm{j}12)} = \frac{11.4\angle-127.87^\circ}{13\angle-67.38^\circ} = \frac{11.4}{13}\angle-127.87^\circ - (-67.38^\circ)$$

$$= 0.877\angle-60.49^\circ.$$

Converting this to rectangular form yields $(0.43 - \mathrm{j}0.76)$, which is the result previously obtained.

WORKED EXAMPLE 1.15

Divide $50\angle-126^\circ$ by $7.5\angle36^\circ$.

$$\frac{50\angle-126^\circ}{7.5\angle36^\circ} = \frac{50}{7.5}\angle-126^\circ - 36^\circ = 6.67\angle-162^\circ.$$

SELF-ASSESSMENT EXAMPLE 1.16

Divide $(125 + \mathrm{j}70)$ by $14\angle65^\circ$. Express your answer in both rectangular and polar forms.

From the analyses performed so far, it should be observed that:

(a) to *add* or *subtract* complex numbers they have to be in the rectangular $(a + \mathrm{j}b)$, form

(b) when *multiplying* or *dividing* complex numbers it is far simpler to work in the polar form. Although $(a + jb)(x + jy)/(p + jq)$ can be evaluated, it is better expressed as $(c\angle\theta \times d\angle\phi)/e\angle\alpha$ when the solution is $[(c \times d)/e] \angle\theta + \phi - \alpha$.

1.8 The significance of j in a.c. circuits

1.8.1 Resistance

When an alternating voltage is applied to a resistance, the current flowing is in phase with the voltage (see Fig. 1.12 in which, for this case, the load is considered to be purely resistive).

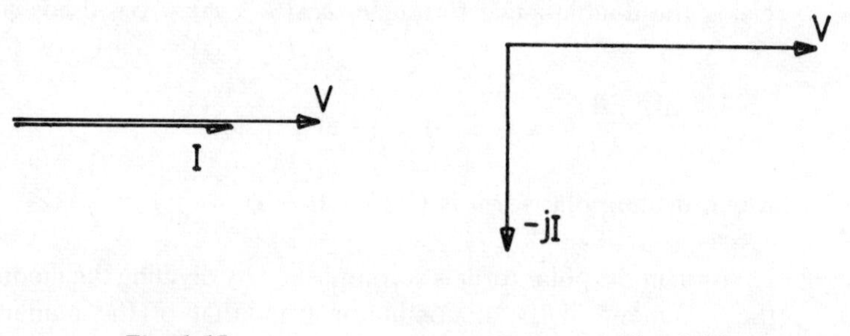

Fig. 1.15 *Fig. 1.16*

As shown in Fig 1.15, both voltage and current phasors are in the $+x$ direction. The voltage can be expressed as $(V + j0)$ and the current as $(I + j0)$

$$R = \frac{(V + j0)}{(I + j0)} \quad \text{or simply} \quad \frac{|V|}{|I|} = R\ \Omega.$$

There are no j terms involved.

1.8.2 Inductance

When an alternating voltage is applied to an inductance with negligible resistance the current flowing lags the voltage by 90° (take Fig. 1.12 to have such an inductance as its load). The phasors are shown in Fig. 1.16.

$$X_L = 2\pi fL = \frac{|V|}{|I|}.$$

$$\text{In rectangular form, impedance} = \frac{(V + j0)}{(0 - jI)} = \frac{|V|}{-j|I|} = \frac{X_L}{-j}.$$

Multiplying the numerator and the denominator by j, the impedance becomes

$$\frac{jX_L}{-j^2} = jX_L\ \Omega.$$

In rectangular form the impedance of an inductive reactance with negligible resistance may be represented as jX_L Ω. In polar form this is $X_L\angle 90°$ Ω.

WORKED EXAMPLE 1.17

Determine the value of current flowing in an inductor with inductance 0.1 H and negligible resistance when it is connected to 200 V, 50 Hz mains.

$$X_L = 2\pi fL = 2\pi \times 50 \times 0.1 = 31.41\ \Omega.$$

In rectangular form, impedance = j31.41 Ω.

$$I = \frac{V}{Z} = \frac{200}{j31.41}.$$

Multiplying numerator and denominator by −j gives

$$I = \frac{-j200}{-j^2 31.41} = \frac{-j200}{31.41} = -j6.37\ \text{A}.$$

or working in polar form

$$I = \frac{200\angle 0°}{31.41\angle 90°} = \frac{200}{31.41}\angle 0° - 90°$$

$$= 6.37\angle -90°\ \text{A}.$$

Both methods place the current in the correct phasor position.

Further to section 1.5, although peak or rms values for voltage and current can equally well be used it is more normal to use rms values, which is the case in this example.

1.8.3 Capacitance

In a capacitor the current leads the applied voltage by 90° (in Fig. 1.12 the load is now considered to be a capacitor). As shown in Fig. 1.17, the voltage is again $(V + j0)$. The current phasor is vertically upwards and is therefore jI amperes.

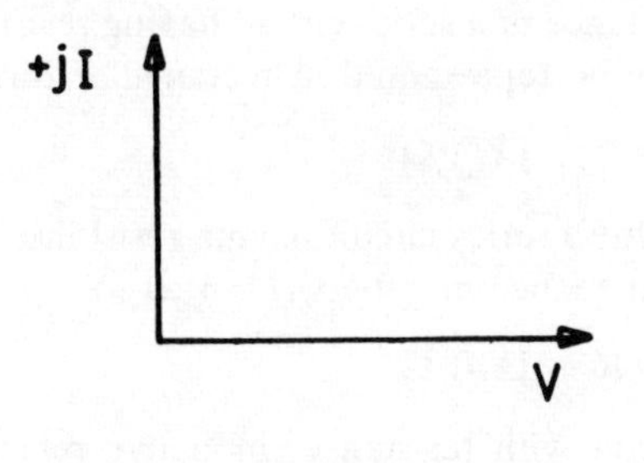

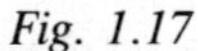

Fig. 1.17

$$X_C = \frac{1}{2\pi fC} = \frac{|V|}{|I|}\ \Omega.$$

$$\text{In rectangular form, impedance} = \frac{(V + \text{j}0)}{(0 + \text{j}I)} = \frac{|V|}{+\text{j}|I|} = \frac{X_C}{\text{j}}.$$

Multiplying numerator and denominator by $-\text{j}$ gives

$$\text{Impedance} = \frac{-\text{j}X_C}{-\text{j}^2} = -\text{j}X_C\ \Omega.$$

The impedance of a capacitive reactance may be represented as $-\text{j}X_C$ in rectangular form. In polar form this is $X_C\angle -90°$.

WORKED EXAMPLE 1.18

Determine the value of current flowing in a 40 μF capacitor connected to 200 V, 50 Hz mains.

$$X_C = \frac{1}{2\pi \times 50 \times 40 \times 10^{-6}} = 79.58\ \Omega.$$

In rectangular form, impedance $= -\text{j}79.58\ \Omega$.

$$I = \frac{V}{Z} = \frac{200}{-\text{j}79.58}.$$

Multiplying numerator and denominator by j gives

$$I = \frac{\text{j}200}{-\text{j}^2 79.58} = \text{j}2.07\ \text{A}.$$

or in polar form

$$I = \frac{200\angle 0°}{79.58\angle -90°} = \frac{200}{79.58}\angle 0° - (-90°) = 2.51\angle 90°\ \text{A}.$$

1.9 Series circuits

The impedance of a series circuit having resistance $R\ \Omega$ and inductive reactance $X_L\ \Omega$ may be represented in rectangular form as

$$Z = (R + \text{j}X_L)\ \Omega.$$

Similarly for a series circuit having resistance $R\ \Omega$ and capacitive reactance $X_C\ \Omega$, the impedance may be written as

$$Z = (R - \text{j}X_C)\ \Omega.$$

For a circuit with resistance, inductive reactance and capacitive reactance in series, the impedance in rectangular form will be

$$Z = (R + \text{j}X_L - \text{j}X_C) = (R + \text{j}(X_L - X_C))\ \Omega.$$

WORKED EXAMPLE 1.19

A circuit consists of a 3 Ω resistor in series with an inductor having 0.02 H inductance and negligible resistance. It is connected to a 100 V, 50 Hz supply. Calculate (a) the value of the circuit current, (b) the phase angle between current and voltage.

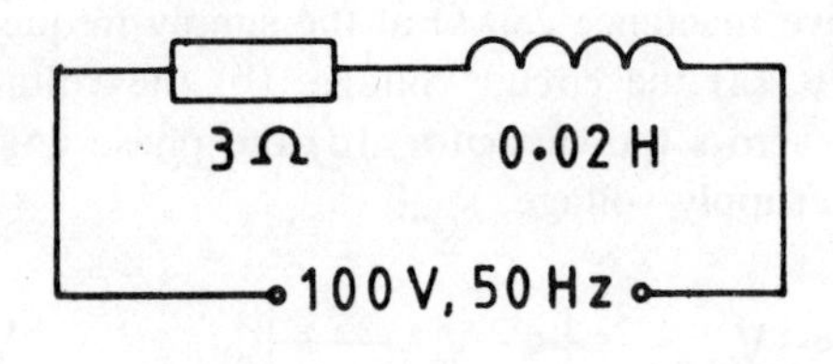

Fig. 1.18

Inductive reactance = $2\pi \times 50 \times 0.02 = 6.28\ \Omega$.

Using the correct j notation this is written as j6.28 Ω.

Total impedance of the circuit = (3 + j6.28) Ω.

$$I = \frac{V}{Z} = \frac{100}{(3 + j6.28)} = \frac{100(3 - j6.28)}{(3^2 + 6.28^2)} = \frac{300 - j628}{48.44}$$
$$= (6.19 - j12.96)\ \text{A}.$$

Converting to polar form

$$I = 14.37\angle{-64.46^\circ}\ \text{A}$$

or working directly in polar form

$$I = \frac{100}{(3 + j6.28)} = \frac{100}{6.96\angle 64.46^\circ} = 14.37\angle{-64.46^\circ}\ \text{A}.$$

Modulus of the current = 14.37 A (answer (a)) (this is the value of current indicated on an a.c. ammeter included in the circuit).

The current lags the voltage by 64.46° (answer (b)).

WORKED EXAMPLE 1.20

The current in a circuit is (4.5 + j12) A, when the applied voltage is (100 + j150) V. Calculate the value of the circuit impedance.

$$I = \frac{V}{Z}, \quad \text{re-arranging gives } Z = \frac{V}{I}.$$

$$Z = \frac{(100 + j150)}{(4.5 + j12)} = \frac{180.27\angle 56.3^\circ}{12.82\angle 69.4^\circ}$$
$$= 14.06\angle{-13.1^\circ}\ \Omega$$
$$= (13.7 - j3.2)\ \Omega.$$

The circuit consists of a resistance of value 13.7 Ω in series with a capacitor with capacitive reactance 3.2 Ω at the particular frequency being used.

WORKED EXAMPLE 1.21

A circuit consists of a resistance of value 120 Ω in series with a capacitor with capacitive reactance 250 Ω at the supply frequency. The current is $0.9\angle 0°$ A. Calculate (a) the circuit voltage, (b) the voltage across the resistor, (c) the voltage across the capacitor, (d) the phase angle between the circuit current and the supply voltage.

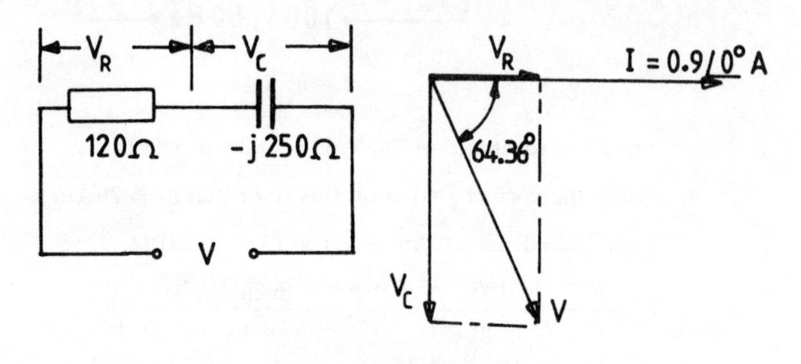

Fig. 1.19

(a) $V = IZ = 0.9(120 - j250) = (108 - j225)$ V $= 249.6\angle -64.36°$ V.
(b) V_R is in phase with the circuit current and is equal to 108 V.
(c) V_C lags the circuit current and is equal to 225 V.
(d) The phase angle between the circuit current and the voltage is 64.36°.

SELF-ASSESSMENT EXAMPLE 1.22

A current of 2.5 A flows in a circuit consisting of a resistance of value 50 Ω in series with an inductor with inductance 0.2 H when connected to 50 Hz mains. Determine (a) the modulus of the supply voltage, (b) the phase angle between the circuit current and the supply voltage, (c) the value of the circuit current and its phase angle with respect to the supply voltage if the inductor is replaced by a capacitor with capacitance 80 μF, the supply voltage meanwhile remaining unchanged.

WORKED EXAMPLE 1.23

Calculate the value of the current flowing in the circuit shown in Fig. 1.20.

$$\begin{aligned} Z &= 10 + j25 - j15 \\ &= 10 + j10 \\ &= 14.14\angle 45°. \end{aligned}$$

$$I = \frac{V}{Z} = \frac{25}{14.14\angle 45°} = 1.77\angle -45° \text{ A}.$$

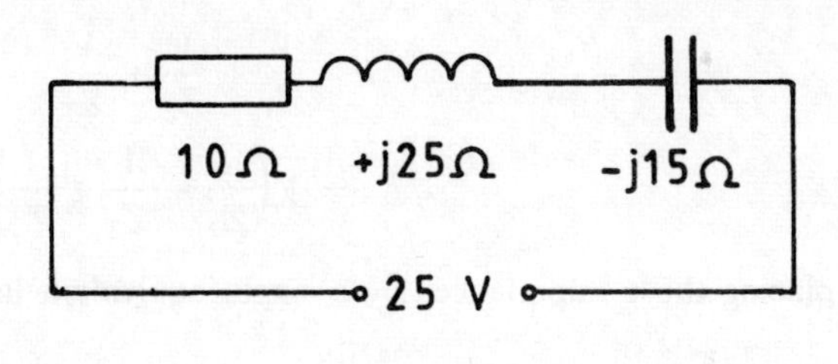

Fig. 1.20

SELF-ASSESSMENT EXAMPLE 1.24

Calculate the value of capacitance necessary to cause 1 A to flow in the circuit shown in Fig. 1.21. (*Hint*: What must be the value of the total circuit impedance Z?)

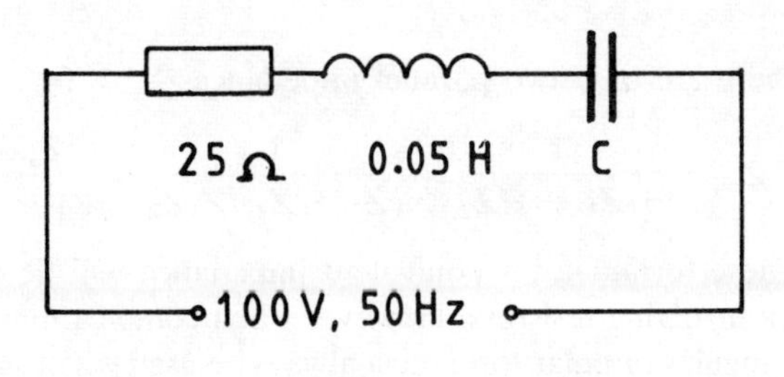

Fig. 1.21

1.10 Parallel circuits

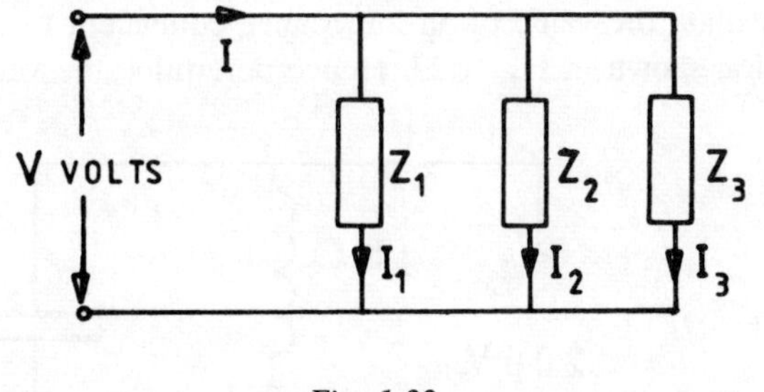

Fig. 1.22

Figure 1.22 shows three impedances Z_1, Z_2 and Z_3 connected in parallel across a supply of V volts. The potential difference across each impedance is the same, namely the supply voltage V.

$$I_1 = \frac{V}{Z_1} \qquad I_2 = \frac{V}{Z_2} \qquad I_3 = \frac{V}{Z_3}$$

The total supply current, $I = I_1 + I_2 + I_3$

$$I = \frac{V}{Z_1} + \frac{V}{Z_2} + \frac{V}{Z_3}$$

$$= V\left(\frac{1}{Z_1} + \frac{1}{Z_2} + \frac{1}{Z_3}\right).$$

Replacing these impedances by a single equivalent impedance Z_{eq} gives

$$I = \frac{V}{Z_{eq}}.$$

$$\text{Hence } \frac{V}{Z_{eq}} = V\left(\frac{1}{Z_1} + \frac{1}{Z_2} + \frac{1}{Z_3}\right).$$

Cancelling V on both sides of the equation and rearranging gives

$$Z_{eq} = \frac{1}{\dfrac{1}{Z_1} + \dfrac{1}{Z_2} + \dfrac{1}{Z_3}}.$$

If there are only two parallel impedances

$$Z_{eq} = \frac{1}{1/Z_1 + 1/Z_2} = \frac{1}{(Z_2 + Z_1)/Z_1Z_2} = \frac{Z_1Z_2}{Z_1 + Z_2}.$$

These formulae for equivalent impedance will be recognised from previous work involving resistors. However when complex impedances are involved, the rectangular or polar form must always be used when substituting in the formulae to take account of the angles involved. Worked examples 1.25 and 1.27 will make this clear.

WORKED EXAMPLE 1.25

Calculate the value of an impedance equivalent to that of the parallel combination shown in Fig. 1.23. Hence determine the value of the circuit current.

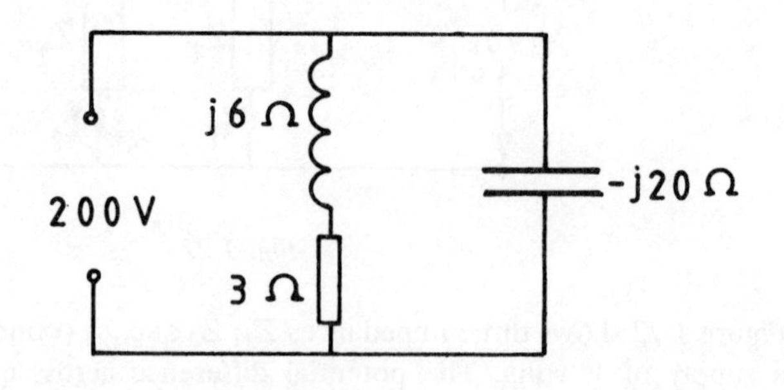

Fig. 1.23

$$Z_{eq} = \frac{Z_1Z_2}{Z_1 + Z_2} = \frac{(3 + j6)\,(-j20)}{3 + j6 - j20} = \frac{6.71\angle 63.4° \times 20\angle -90°}{(3 - j14)}$$

$$Z_{eq} = \frac{134.2\angle -26.6°}{14.32\angle -77.9°}$$

$$= 9.37\angle 51.3° \ \Omega.$$

$$I = \frac{V}{Z} = \frac{200\angle 0°}{9.37\angle 51.3°}$$

$$= 21.34\angle -51.3° \text{ A}.$$

SELF-ASSESSMENT EXAMPLE 1.26

Determine the value of a single impedance equivalent to the parallel combination shown in Fig. 1.24.

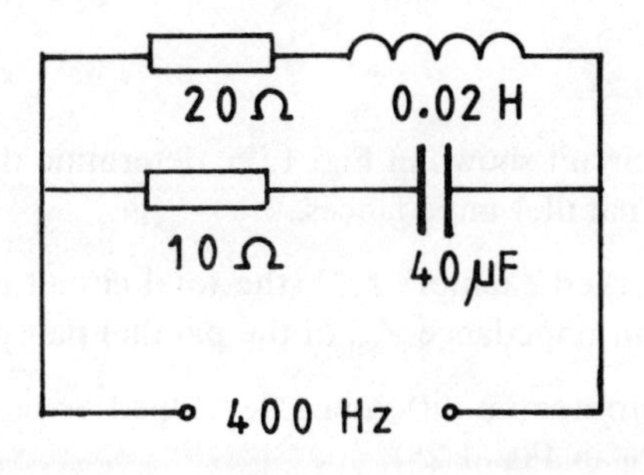

Fig. 1.24

1.11 Series parallel connection

WORKED EXAMPLE 1.27

Determine the value of the circuit current for the series parallel combination of impedances shown in Fig. 1.25.

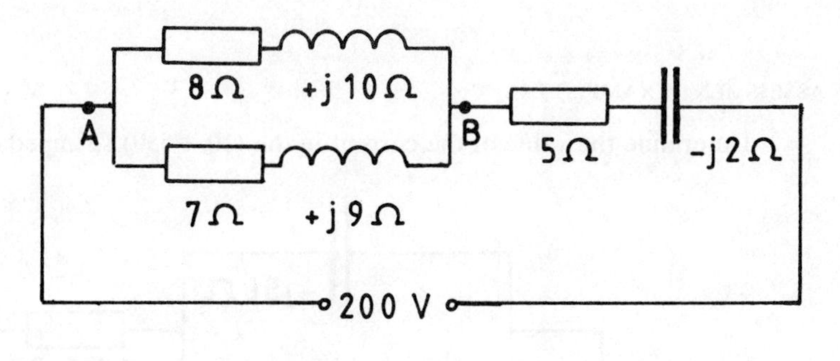

Fig. 1.25

It is necessary firstly to reduce the parallel pair of impedances to a single equivalent impedance. This may be added to (5 − j2) using the rectangular form.

$$Z_{eq} = \frac{(8 + j10)\ (7 + j9)}{(8 + j10 + 7 + j9)} = \frac{12.81\angle 51.3° \times 11.4\angle 52.12°}{(15 + j19)}$$

$$= \frac{164.2\angle 103.42°}{24.2\angle 51.7°}$$

$$= 6.03\angle 51.7°\ \Omega = (3.72 + j4.75)\ \Omega.$$

This is added to $(5 - j2)\ \Omega$

$$Z_{total} = (3.72 + j4.75) + (5 - j2) = (8.72 + j2.75)\ \Omega$$

$$= 9.15\angle 17.4°\ \Omega$$

$$I = \frac{V}{Z} = \frac{200}{9.15\angle 17.4°} = 21.85\angle{-17.4°}\ \text{A}.$$

WORKED EXAMPLE 1.28

For the circuit shown in Fig. 1.25, determine the value of the current in each of the two parallel impedances.

From worked example 1.27, the total circuit current $= 21.85\angle{-17.4°}$ A. The equivalent impedance Z_{eq} of the parallel pair of impedances $= 6.03\angle 51.76°\ \Omega$.

The potential difference developed across $Z_{eq} = IZ_{eq}$ (V_{AB} in Fig. 1.25)

$$V_{AB} = 21.85\angle{-17.4°} \times 6.03\angle 51.76°$$

$$= 131.8\angle 34.36°\ \text{V}$$

$$\text{Current in } (8 + j10)\ \Omega = \frac{131.8\angle 34.36°}{(8 + j10)} = \frac{131.8\angle 36.36°}{12.81\angle 51.3°}$$

$$= 10.28\angle{-17.04°}\ \text{A}.$$

$$\text{Current in } (7 + j9)\ \Omega = \frac{131.8\angle 34.36°}{(7 + j9)} = \frac{131.8\angle 34.36°}{11.4\angle 52.12°}$$

$$= 11.56\angle{-17.76°}\ \text{A}.$$

SELF-ASSESSMENT EXAMPLE 1.29

Determine the value of the current in the $(10 + j30)\ \Omega$ impedance in Fig. 1.26.

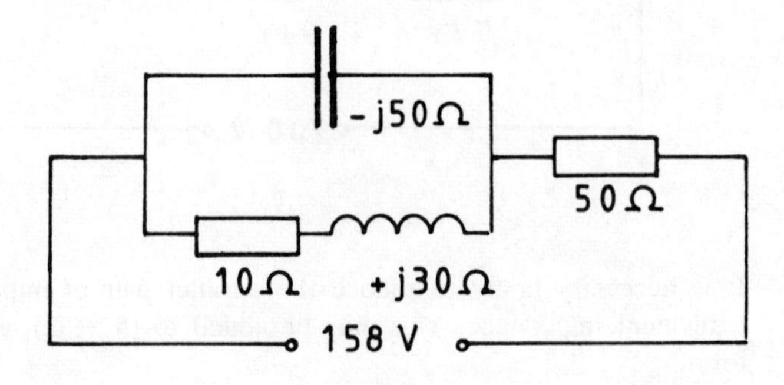

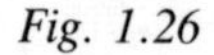

Fig. 1.26

1.12 Admittance, conductance and susceptance

For the circuit shown in Fig. 1.27, the impedance $Z = (R + jX_L)\ \Omega$. The admittance Y is defined as $1/Z$. The unit of Y is the siemen (S).

$$Y = \frac{1}{(R + jX_L)}\ \mathrm{S}$$

Rationalising

$$Y = \frac{(R - jX_L)}{R^2 + X_L^2} = \frac{(R - jX_L)}{Z^2}$$

$$= \frac{R}{Z^2} - \frac{jX_L}{Z^2}$$

$$= G - jB$$

where $G = \dfrac{R}{Z^2}$, the conductance of the circuit

$B = \dfrac{X_L}{Z^2}$, the inductive susceptance of the circuit.

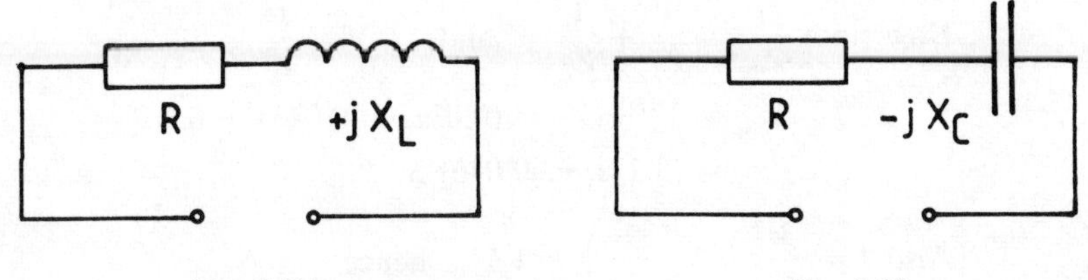

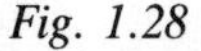

Fig. 1.27

Fig. 1.28

Similarly for the resistive capacitive circuit in Fig. 1.28;

$$Y = \frac{1}{(R - jX_C)}\ \mathrm{S}$$

$$= \frac{(R + jX_C)}{R^2 + X_C^2} = \frac{(R + jX_C)}{Z^2}$$

$$= \frac{R}{Z^2} + \frac{jX_C}{Z^2}$$

$$= G + jB$$

where $G = \dfrac{R}{Z^2}$, the circuit conductance, as in the inductive case

$B = \dfrac{X_C}{Z^2}$, the capacitive susceptance of the circuit.

Notice that the sign of the j term is reversed when considering admittance as opposed to impedance.

Admittance is particularly useful where circuits have several parallel arms as the next two worked examples will show.

WORKED EXAMPLE 1.30

For the circuit shown in Fig. 1.29, determine (a) the total admittance, (b) the total circuit current when the supply voltage is 25 V.

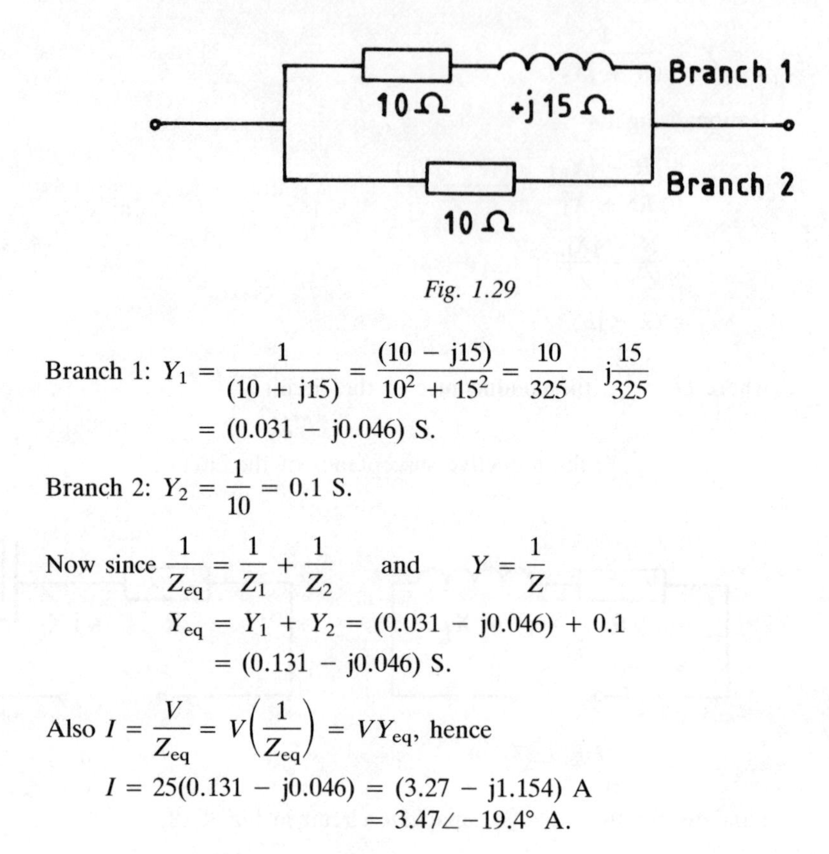

Fig. 1.29

$$\text{Branch 1: } Y_1 = \frac{1}{(10 + j15)} = \frac{(10 - j15)}{10^2 + 15^2} = \frac{10}{325} - j\frac{15}{325}$$

$$= (0.031 - j0.046) \text{ S.}$$

$$\text{Branch 2: } Y_2 = \frac{1}{10} = 0.1 \text{ S.}$$

$$\text{Now since } \frac{1}{Z_{eq}} = \frac{1}{Z_1} + \frac{1}{Z_2} \quad \text{and} \quad Y = \frac{1}{Z}$$

$$Y_{eq} = Y_1 + Y_2 = (0.031 - j0.046) + 0.1$$

$$= (0.131 - j0.046) \text{ S.}$$

$$\text{Also } I = \frac{V}{Z_{eq}} = V\left(\frac{1}{Z_{eq}}\right) = VY_{eq}, \text{ hence}$$

$$I = 25(0.131 - j0.046) = (3.27 - j1.154) \text{ A}$$

$$= 3.47\angle -19.4^\circ \text{ A.}$$

WORKED EXAMPLE 1.31

For the circuit shown in Fig. 1.30, determine (a) the total conductance, (b) the total susceptance, (c) the total admittance in polar form, (d) the current drawn from the supply, and (e) the circuit impedance.

$$\text{Branch 1: } Y_1 = \frac{1}{(30 - j62.8)} = \frac{(30 + j62.8)}{30^2 + 62.8^2} = 0.0062 + j0.0129$$

$$\text{Branch 2: } Y_2 = \frac{1}{(25 + j125.6)} = \frac{(25 - j125.6)}{25^2 + 125.6^2} = 0.0015 - j0.0076$$

$$\text{Branch 3: } Y_3 = \frac{1}{40 + j157 - j159} = \frac{(40 + j2)}{40^2 + 2^2} = 0.0249 + j0.0012$$

$$\text{Branch 4: } Y_4 = \frac{1}{100} = 0.01$$

$$\text{Adding } Y_{total} = 0.0426 + j0.0065$$

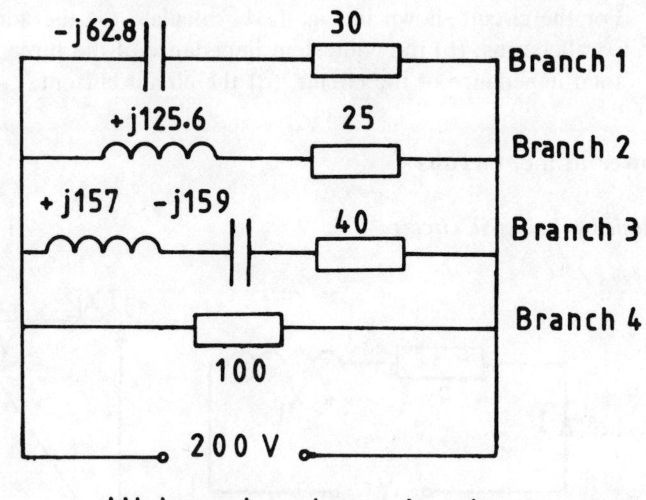

Fig. 1.30

Answers

(a) $G = 0.0426$ S.
(b) $B = 0.0065$ S (+j, so that the circuit overall is capacitive).
(c) $Y = (0.0426 + j0.0065) = 0.043\angle 8.4°$ S in the polar form.
(d) $I = VY = 200 \times 0.043\angle 8.4° = 8.6\angle 8.4°$ A.
(e) $Z = \frac{1}{Y}$

$$= \frac{1}{0.043\angle 8.4°} = 23.26\angle -8.4° \ \Omega$$

$$= (23 - j3.4)\ \Omega.$$

SELF-ASSESSMENT EXAMPLE 1.32

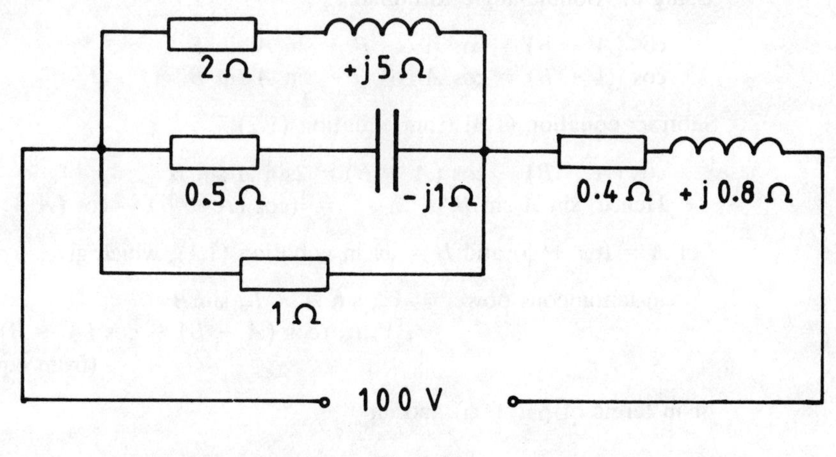

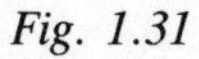

Fig. 1.31

For the circuit shown in Fig. 1.31, calculate (a) the admittance of the three parallel arms, (b) the equivalent impedance of the three parallel arms, (c) the total impedance of the circuit, (d) the circuit current.

1.13 Power in a.c. circuits

1.13.1 Resistive inductive circuit

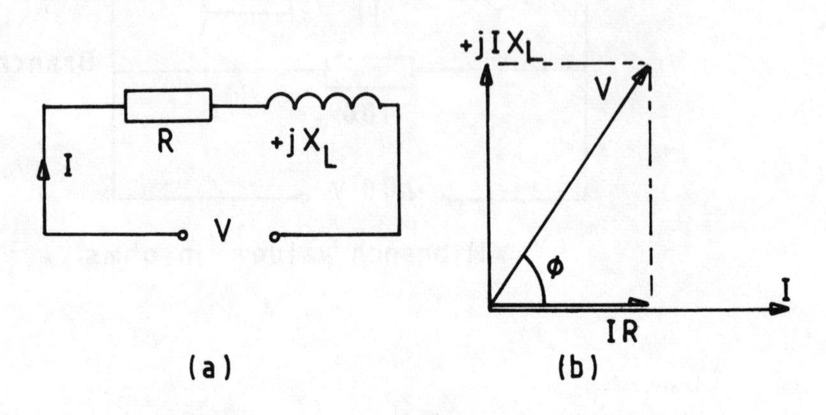

Fig. 1.32

In the RL circuit shown in Fig. 1.32(a), the current lags the voltage by an angle ϕ as shown in the phasor diagram Fig. 1.32(b).

$$v = V_m \sin(\omega t + \phi) \text{ V}$$
$$i = I_m \sin \omega t \text{ A.}$$

The instantaneous power p in a circuit is the product of instantaneous voltage and current.

$$p = vi$$
$$= V_m \sin(\omega t + \phi) \times I_m \sin \omega t \text{ W.} \qquad (1.1)$$

Using the double-angle formulae:

$$\cos(A - B) = \cos A \cos B + \sin A \sin B \qquad (1.2)$$
$$\cos(A + B) = \cos A \cos B - \sin A \sin B \qquad (1.3)$$

Subtract equation (1.3) from equation (1.2)

$$\cos(A - B) - \cos(A + B) = 2\sin A \sin B$$
$$\text{Hence, } \sin A \sin B = \tfrac{1}{2}(\cos(A - B) - \cos(A + B)) \qquad (1.4)$$

Let $A = (\omega t + \phi)$ and $B = \omega t$ in equation (1.1), which gives

$$\text{instantaneous power} = V_m \sin A \times I_m \sin B$$
$$= \tfrac{1}{2}V_m I_m (\cos(A - B) - \cos(A + B))$$

(from equation (1.4))

or in terms of $(\omega t + \phi)$ and ωt

$$\text{instantaneous power} = \tfrac{1}{2}V_m I_m\,(\cos(\omega t + \phi - \omega t) - \cos(\omega t + \phi + \omega t))$$
$$= \tfrac{1}{2}V_m I_m\,(\cos\phi - \cos(2\omega t + \phi))\ \text{W}.$$

Now ϕ is constant for a particular circuit so that $\frac{1}{2}V_m I_m \cos\phi$ is constant throughout the cycle. The mean value of $\frac{1}{2}V_m I_m \cos(2\omega t + \phi)$ is zero over any number of complete cycles since this produces equal areas above and below the time axis. Both these quantities are shown in Fig. 1.33(a).

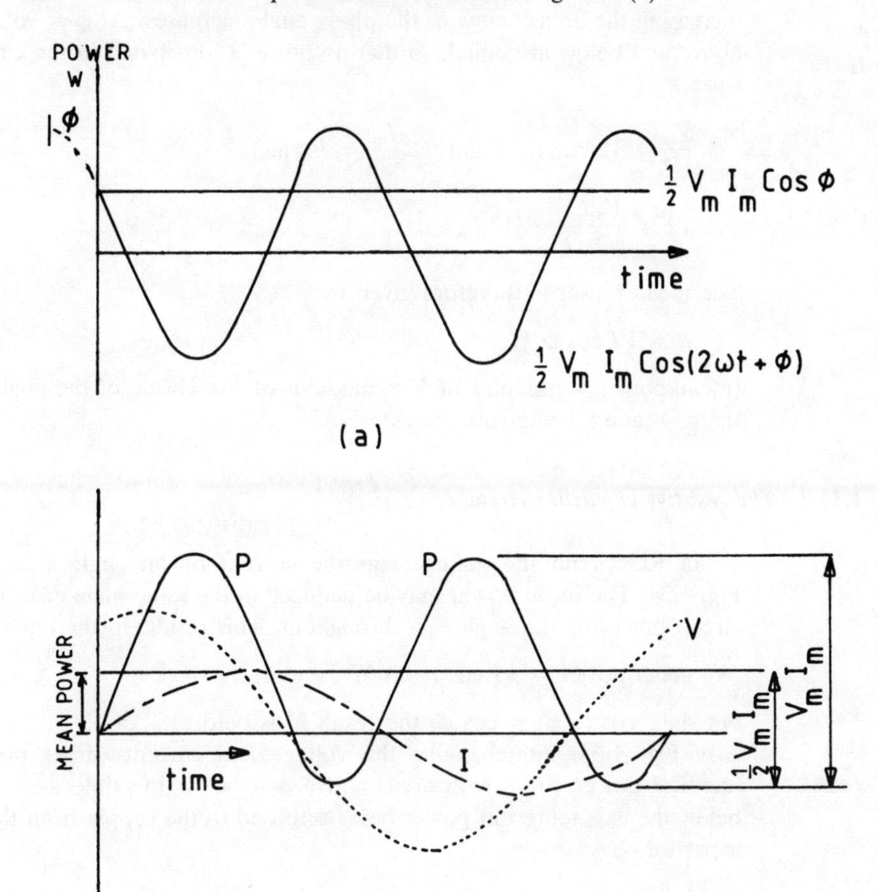

Fig. 1.33

The power curve P in Fig. 1.33(b) is derived by adding $-\frac{1}{2}V_m I_m \cos(2\omega t + \phi)$ to the constant term $\frac{1}{2}V_m I_m \cos\phi$. The voltage and current waves have also been added.

The power to the circuit can be seen to be a pulsating quantity. It pulsates at twice the supply frequency. The areas of the power curve above the time axis represent power supplied to the circuit, while the small lobes beneath the axis represent power being returned to the supply from the circuit inductance as the magnetic field collapses. The difference between the areas above and below the time axis represents the heat loss due to the circuit resistance.

The mean or average value of the difference between the power supplied and the power returned is $\frac{1}{2}V_m I_m \cos\phi$ watts.

Figure 1.33 has been drawn for $\phi = 60°$. The shapes are typical however for all lagging phase angles, the power curve remaining a sinusoid but moving downward with respect to the time axis so decreasing the areas above and increasing the areas below as the phase angle increases. At $\phi = 90°$, the areas above and below are equal, so that no power is dissipated in the circuit. Now since

$$\frac{V_m}{\sqrt{2}} = V \text{ (rms)} \quad \text{and} \quad \frac{I_m}{\sqrt{2}} = I \text{ (rms)}$$

$$VI = \frac{V_m I_m}{\sqrt{2}\sqrt{2}} = \tfrac{1}{2}V_m I_m.$$

The mean power is therefore given by

$$P = VI \cos\phi \text{ W} \tag{1.5}$$

(mean power = modulus of V × modulus of I × cosine of the angle between them, V and I being rms values).

1.13.2 *The resistive capacitive circuit*

In an RC circuit the voltage lags the current by an angle ϕ as shown in Fig. 1.34. The mean power may be deduced in the same manner as for the RL circuit but using the angle $-\phi$ throughout. This results in the equation

$$\text{mean power} = VI\cos(-\phi) \text{ W}$$

but since $\cos(-\phi) = \cos\phi$, the result is as before.

In Fig. 1.33, interchanging the voltage and current curves produces an identical power curve with areas above and below the time axis. The areas below the axis represent power being returned to the supply from the charged capacitor.

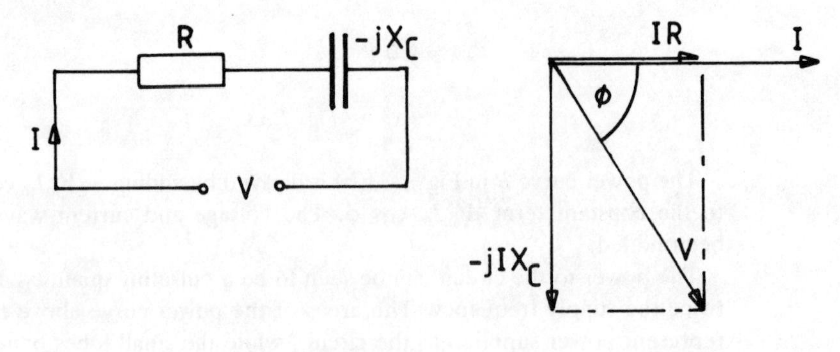

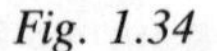

Fig. 1.34

1.13.3 *Resistance only*

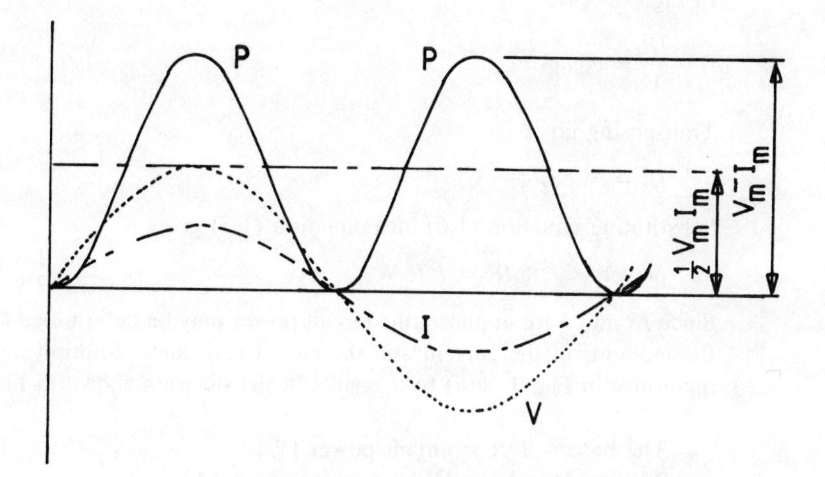

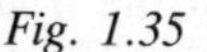
Fig. 1.35

Using the formula

$P = VI \cos \phi$ W

For resistance only, $\phi = 0$, hence

$P = VI$ W.

The curves of voltage, current and power are shown in Fig. 1.35.

1.14 Watts, volt-amperes and volt-amperes reactive (W, VA and VA_R)

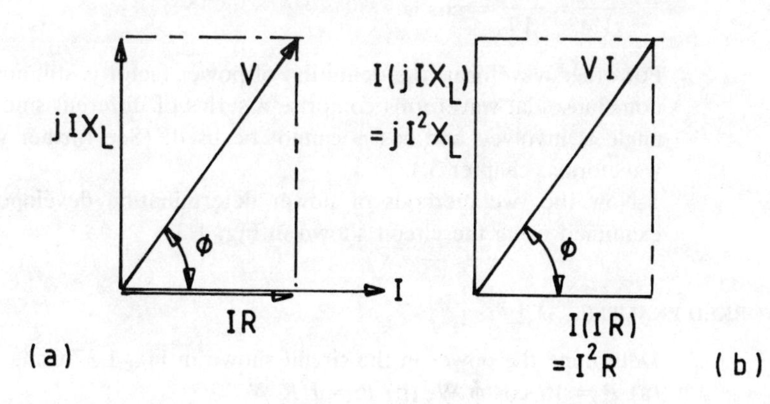

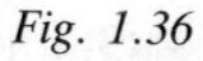
Fig. 1.36

Consider the RL circuit and its phasor diagram in Fig. 1.32. The phasor diagram is reproduced in Fig. 1.36(a). It has been shown that

$$\text{power} = VI \cos \phi \text{ W}. \qquad \text{(equation 1.5)}$$

In Fig. 1.36(a),

$$\frac{IR}{V} = \cos \phi.$$

Transposing gives

$$IR = V \cos \phi \qquad (1.6)$$

Substituting equation (1.6) into equation (1.5) gives

$$\text{power} = I \times IR = I^2R \text{ W}.$$

Since IR and I are in phase, the circuit power may be determined knowing only the modulus of the current and the circuit resistance. Multiplying each of the quantities in Fig. 1.36(a) by I results in the diagram shown in Fig. 1.36(b).

The base $= I^2R =$ circuit power (W).
The hypotenuse $= VI =$ volt-amperes (VA).
The vertical side of the triangle becomes jI^2X_L.
$I^2X_L/VI = \sin \phi$, hence $I^2X_L = VI \sin \phi =$ volt-amperes reactive (VA_R).

These are present because the circuit has reactance. In a circuit possessing inductive reactance these are known as lagging VA_R (often spoken of as 'lagging vars'). In a capacitive circuit the analysis is similar, the triangle being inverted giving rise to leading VA_R.

$$\frac{\text{power (watts)}}{\text{voltage} \times \text{current}} = \frac{W}{VA} \text{ is defined as the circuit power factor.} \qquad (1.7)$$

Phasors are used to represent sinusoidal quantities so that in Fig. 1.36

$$\frac{W}{VA} = \frac{I^2R}{VI} = \cos \phi.$$

For other waveforms the definition of power factor is still applicable but since non-sinusoidal waveforms comprise a series of different sine waves, no single angle is involved and $\cos \phi$ cannot be used. (See further work on complex waveforms, chapter 3.)

Now the two methods of power determination developed so far will be examined using the circuit shown in Fig. 1.37.

WORKED EXAMPLE 1.33

Determine the power in the circuit shown in Fig. 1.37 using
(a) $P = VI \cos \phi$ W, (b) $P = I^2R$ W.

The impedance, $Z = (3 + j4) = 5\angle 53.13°\ \Omega$.

$$I = \frac{V}{Z} = \frac{20\angle 60°}{5\angle 53.13°} = 4\angle 6.87° \text{ A}.$$

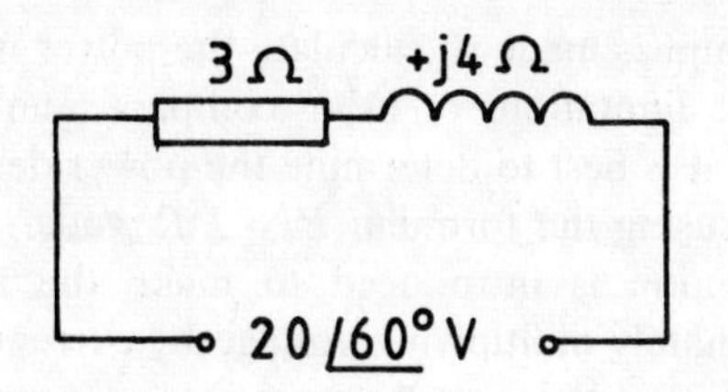

Fig. 1.37

The modulus of the current is 4 A. The modulus of the voltage is 20 V. The angle between them is 53.13°, which is the impedance angle. (Note that in Fig. 1.38 the impedance angle is the angle between the voltage and current phasors.)

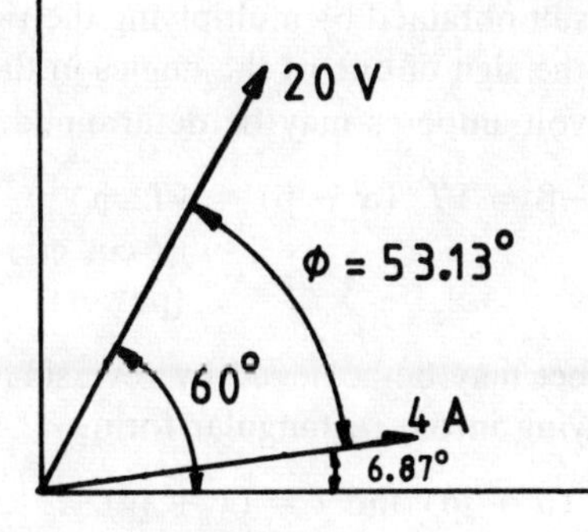

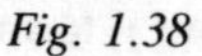

Fig. 1.38

(a) Power = $VI \cos \phi = 20 \times 4 \times \cos 53.13° = 48$ W.
(b) Power = $I^2R = 4^2 \times 3 = 48$ W ((modulus of current)^2 × resistance).

1.15 Determination of power using complex numbers

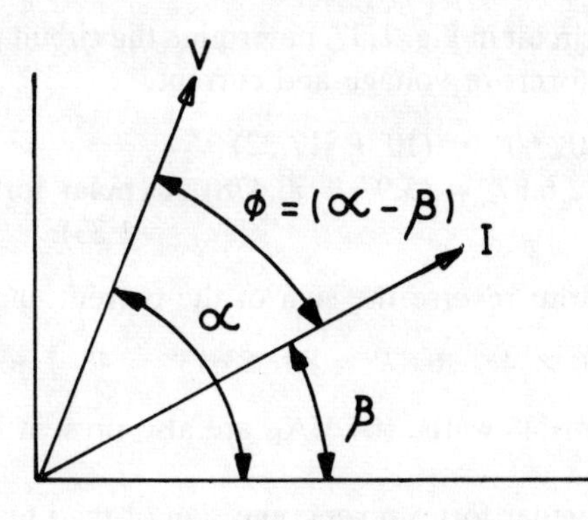

Fig. 1.39

If an attempt is made to calculate the power in an a.c. circuit as a *VI* product, one of the limitations of using complex numbers will be discovered. Almost invariably it is best to determine the power developed in an alternating current circuit by using the formula: $P = I^2R$ watts.

This section is introduced to make the reader aware of the pitfalls of indiscriminately multiplying voltage by current believing that the outcome will be the value of the circuit power in every case.

In Fig. 1.39,

$$\text{voltage} = V\angle\alpha°, \text{ and current} = I\angle\beta°, \text{ in polar form.}$$

$$\text{The product, voltage} \times \text{current} = V\angle\alpha\, I\angle\beta = VI\angle(\alpha + \beta)$$
$$= VI(\cos(\alpha + \beta) + \text{j} \sin(\alpha + \beta)).$$

However, power $= VI \cos(\alpha - \beta)$, and reactive volt-amperes $= VI \sin(\alpha - \beta)$,

so that the result obtained by multiplying the two complex quantities is invalid. By reversing the sign of one of the angles in the polar forms above, the power and reactive volt-amperes may be determined. Thus by changing $\angle\beta$ to $\angle-\beta$

$$V\angle\alpha\, I\angle-\beta = VI\angle(\alpha - \beta) = VI\angle\phi.$$
$$= \underset{\text{(power)}}{VI\cos\phi} + \underset{(\text{VA}_\text{R})}{\text{j}VI\sin\phi}.$$

The same effect may be achieved by reversing one of the signs of the j terms when multiplying in the rectangular form.

Let $V = (a + \text{j}b)$ and $I = (x + \text{j}y)$.

Reversing the sign of the *y* term the product becomes

$$(a + \text{j}b)(x - \text{j}y) = ax - \text{j}ay + \text{j}bx + by$$
$$= \underset{\text{(power)}}{(ax + by)} + \underset{(\text{VA}_\text{R})}{\text{j}(bx - ay)}.$$

WORKED EXAMPLE 1.34

Using the circuit in Fig. 1.37, determine the circuit power using both rectangular and polar forms of voltage and current.

$$V = 20\angle60° = (10 + \text{j}17.32)\text{ V}$$
$$I = 4\angle6.87° = (3.97 + \text{j}0.478)\text{ A (polar forms from worked example 1.33).}$$

In polar form: reverse the sign of the current angle.

$$20\angle60° \times 4\angle-6.87° = 80\angle53.13° = 48 + \text{j}64.$$

The power is 48 watts. (64 VA_R are also present but this result was not called for.)

In rectangular form: reverse the sign of the j term in the current expression.

$$(10 + \text{j}17.32)(3.97 - \text{j}0.478) = 39.72 - \text{j}4.78 + \text{j}68.78 - \text{j}^2 8.28$$
$$= 48 + \text{j}64.$$

Again the power is 48 W.

Generally it is to be recommended that power be calculated as I^2R or $VI \cos \phi$. Worked example 1.34 is carried out so that the reader will not, in ignorance of the shortcomings of the method, simply multiply voltage and current in the complex form with the expectancy that the result will be power.

WORKED EXAMPLE 1.35

Calculate the value of power developed in each arm of the circuit shown in Fig. 1.40.

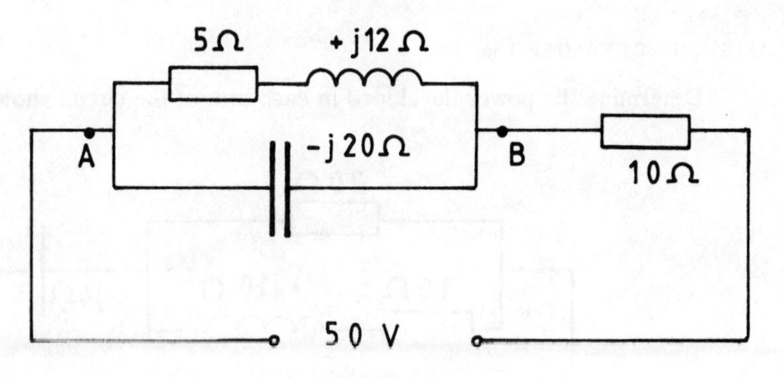

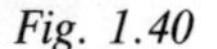

Fig. 1.40

Equivalent impedance of the two parallel arms,

$$\begin{aligned} Z_{eq} &= \frac{(5 + j12)(-j20)}{5 + j12 - j20} \\ &= \frac{13\angle 67.4^\circ \times 20\angle -90^\circ}{9.43\angle -58^\circ} \\ &= 27.57\angle 35.4^\circ \\ &= (22.47 + j15.97)\ \Omega. \end{aligned}$$

$$\begin{aligned} \text{Total impedance of the circuit} &= 22.47 + j15.97 + 10 \\ &= (32.47 + j15.97) = 36.2\angle 26.2^\circ\ \Omega. \end{aligned}$$

$$\text{Circuit current} = \frac{V}{Z} = \frac{50}{36.2\angle 26.2^\circ} = 1.38\angle -26.2^\circ \text{ A.}$$

Power developed in the 10 Ω resistor $= I^2R = 1.38^2 \times 10 = 19.04$ W.

Potential difference developed across the 10 Ω resistor

$$\begin{aligned} &= IR \\ &= 1.38\angle -26.2^\circ \times 10 \\ &= 13.8\angle -26.2^\circ \\ &= (12.38 - j6.1) \text{ V.} \end{aligned}$$

Potential difference across the two parallel arms (V_{AB} in Fig. 1.40)
= supply voltage − voltage drop across the 10 Ω resistor

$$= 50 - (12.38 - j6.1)$$
$$= (37.62 + j6.1) = 38.1\angle 9.21° \text{ V}.$$

$$\text{Current in } (5 + j12)\ \Omega \text{ impedance} = \frac{V_{AB}}{(5 + j12)} = \frac{38.1\angle 9.21°}{13\angle 67.4°}$$
$$= 2.93\angle -58.19° \text{ A}.$$

$$\text{Power developed in } (5 + j12)\ \Omega \text{ impedance} = I^2R = 2.93^2 \times 5$$
$$= 42.9 \text{ W}.$$

No power is developed in the capacitive branch. Power is only developed in resistance.

SELF-ASSESSMENT EXAMPLE 1.36

Determine the power developed in each arm of the circuit shown in Fig. 1.41.

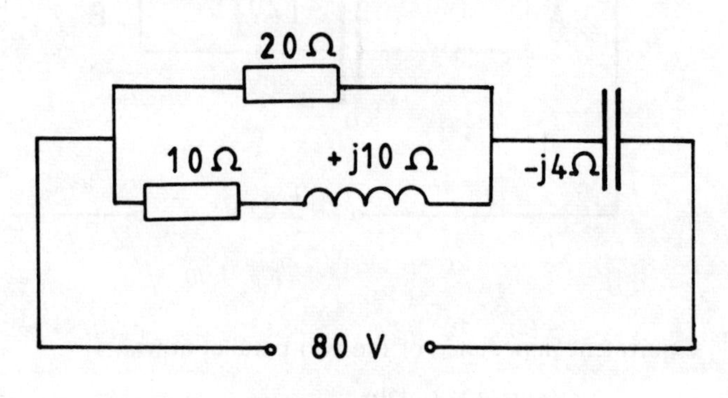

Fig. 1.41

FURTHER PROBLEMS

1.37 Perform the following calculations involving complex numbers expressing your answers in both polar and rectangular form.

(a) $\dfrac{(4 + j6)}{(10 - j8)}$, (b) $\dfrac{25\angle 60°}{10\angle -45°}$, (c) $\dfrac{(100 + j60)}{50\angle 45°}$,

(d) $65\angle 85° \times 40\angle -10°$, (e) $(5 + j20)(10 - j6)$,

(f) $(12 - j14) \times 50\angle 25°$.

1.38 Determine the value of the circuit current in Figs 1.42 to 1.45.

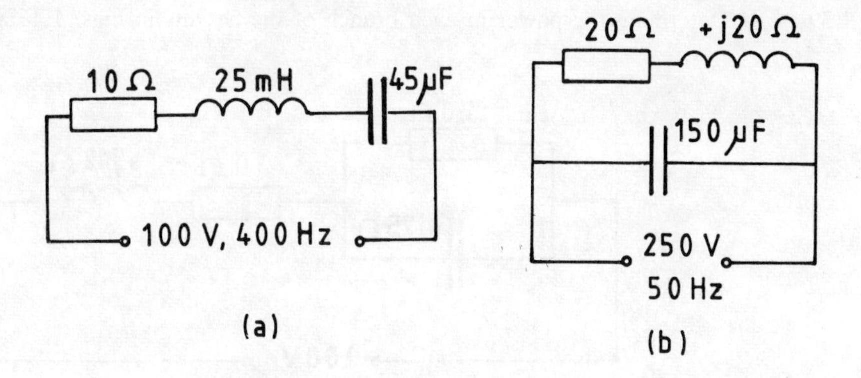

Fig. 1.42

Fig. 1.43

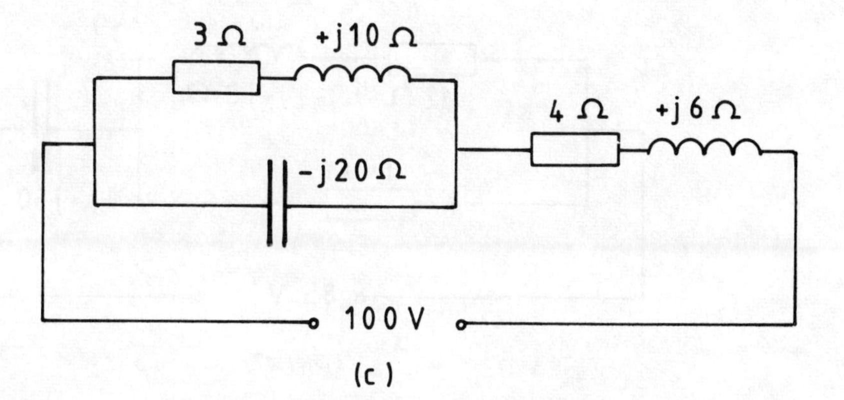

Fig. 1.44

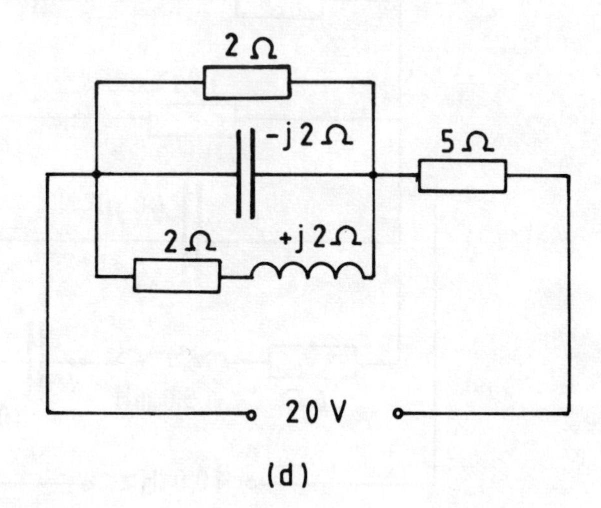

Fig. 1.45

1.39 Determine the power in each branch of the circuits in Figs. 1.46 and 1.47.

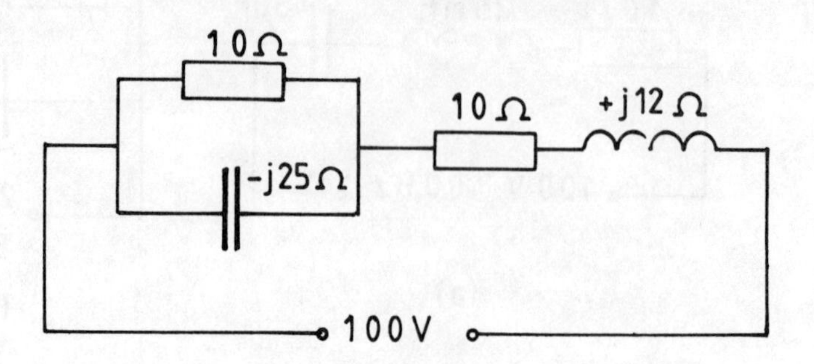

Fig. 1.46

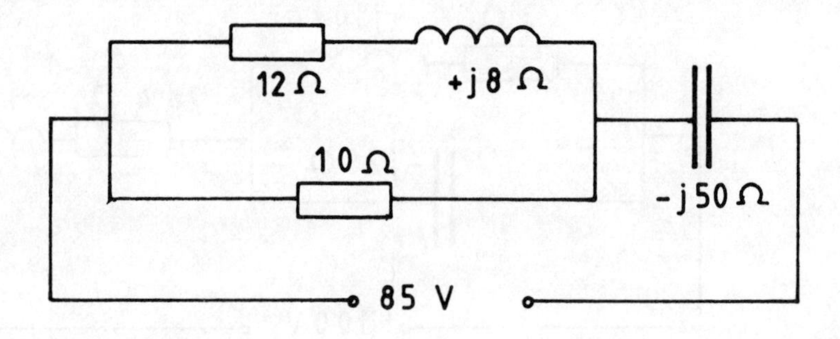

Fig. 1.47

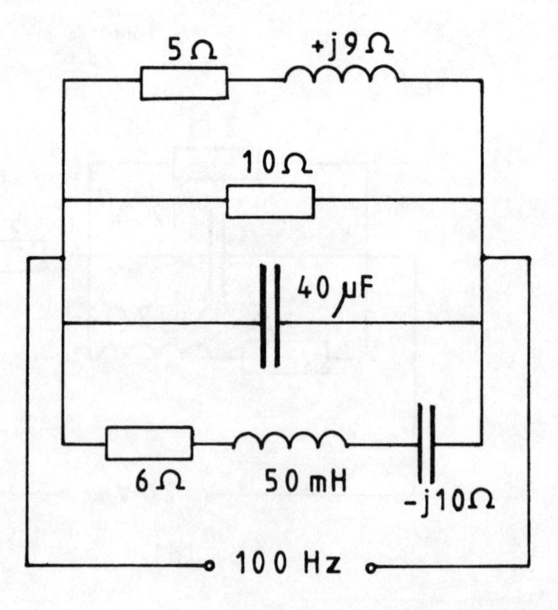

Fig. 1.48

1.40 Calculate the admittance of the circuit shown in Fig. 1.48. Use this value to determine what two components connected in series could be used to replace the given combination of impedances. (Answer in ohms and farads or ohms and henrys.)

1.41 For the circuit shown in Fig. 1.49, calculate (a) the circuit current, (b) the total circuit power, (c) the power developed in the 20 Ω resistor, (d) the reactive volt-amperes developed in the (4 + j6) Ω impedance, and (e) the overall circuit power factor.

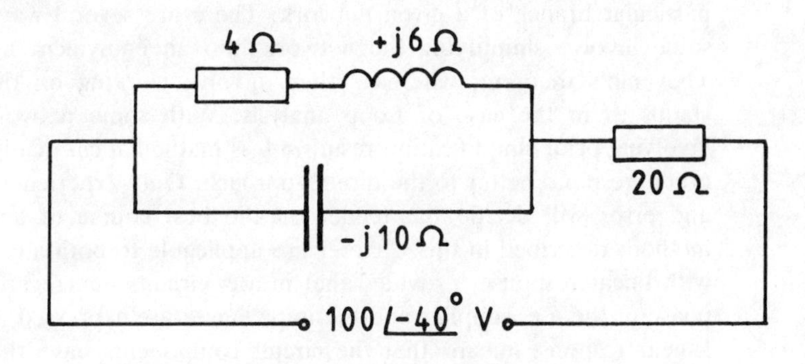

Fig. 1.49

Answers

1.4 (2.5 − j4.33)
1.7 5∠−53.1°
1.8 (a) 10.2∠11.3° (b) 7.2∠−123.7° (c) 9.85∠156° (d) 13∠−67.38°
1.9 (a) (2.18 + j24.9) (b) (−3.06 − j2.57) (c) (−5.8 + j1.55) (d) (9.84 − j1.74) (e) (3.62 − j13.5)
1.12 (a) 108.2∠168.7°; (−106.1 + j21.2) (b) 81.54∠56.5°; (45 + j68)
1.16 10.23∠−35.75°; (8.3 − j5.98)
1.22 (a) 200.7 V (b) 51.5° (c) I = 3.14 A, leading V by 38.5°
1.24 Z = 100 Ω; $|25 + j15.71 - jX_C| = 100$; X_C = 112.53 Ω; C = 28.29 μF
1.26 15.17∠−29.86° Ω; (13.16 − j7.55) Ω
1.29 I_{total} = 1∠−18.4° A; current in (10 + j30) Ω = 2.24∠−44.97° A
1.32 (a) (1.47 + j0.63) S (b) Z_{eq} = (0.57 − j0.25) Ω (c) (0.97 + j0.55) Ω (d) 89.7∠−29.8° A; (77.8 − j44.6) A
1.36 400 W; 400 W; zero
1.37 (a) 0.56∠94.95°; (−0.048 + j0.558) (b) 2.5∠105°; (−0.65 + j2.4) (c) 2.33∠−14°; (2.26 − j0.56) (d) 2600∠75°; (673 + j2511) (e) 240.3∠45°; (170 + j170) (f) 922∠−24.4°; (839.6 − j380.8)
1.38 (a) 1.82∠−79.5° A; (0.33 − j1.79) A (b) 9.2∠48.54° A; (6.09 + j6.89) A (c) 3.68∠−56.54° A; (2.02 − j3.07) A (d) 3.22∠3.69° A; (3.2 + j0.207) A
1.39 (a) In (10 + j12) Ω, 238.4 W; in 10 Ω, 205.7 W (b) In (12 + j8) Ω, 6.7 W; in 10 Ω, 11.5 W
1.40 Y = (0.159 − j0.104) S; Z_{eq} = (4.41 + j2.87) Ω; Z = 4.41 ohm resistor in series with a 4.57 mH inductor
1.41 (a) 3.07∠−44.4° A (b) 306.2 W (c) 188.5 W (d) 176.4 VA_R (e) cos 4.4° = 0.997

Chapter 2
Network Theorems

We are often faced with the necessity to determine the value of current in a particular branch of a given network. There are several ways of proceeding: some involve simplifying the network into an equivalent as in the case of Thevenin's theorem, whereas others involve working on the network as it stands as in the case of Loop analysis. With some networks the methods involving prior simplification result in less mathematical manipulation in total; others respond better to the direct approach. Only experience and a little trial and error will decide the reader on the best course of action. All of the methods described in this Chapter are applicable to both a.c. and d.c. circuits with linear response, provided that in a.c. circuits due regard is taken of the phase of the a.c. supplies and the impedances are expressed in complex form. Linear response means that the circuit components have the same value of impedance irrespective of the magnitude and direction of the current in them. A pure resistor is an example of a linear device. An iron-cored coil is a non-linear device in that at a certain value of current it becomes magnetically saturated and its impedance falls.

In the succeeding sections we will consider the following:

- the superposition theorem
- Thevenin's theorem
- Norton's theorem
- the star/delta and the delta/star transformations
- loop analysis
- node analysis.

One further theorem is the maximum power transfer theorem. This concerns the conditions which have to be satisfied in order that the maximum amount of power shall be transferred from a source of e.m.f., possibly through a transfer network, into a load.

2.1 The superposition theorem

In a network supplied by more than one source of e.m.f., each e.m.f. produces the same effect as if it were acting alone, all other e.m.f.s meanwhile being replaced by their respective internal impedances.

This means that in a network supplied by several sources of e.m.f. the current in each branch of the network due to one of the e.m.f.s with all the other e.m.f.s removed can be calculated, leaving only their internal impedances.

This will be repeated for each of the e.m.f.s in turn. Finally the individual currents in each branch are added algebraically to give the actual branch current. The word 'superposition' comes from superimpose which is adding one current to another in the same conductor. Worked example 2.1 should make the method clear.

WORKED EXAMPLE 2.1

In Fig. 2.1 two sources of e.m.f. E_1 and E_2 are connected in parallel to supply current to a 3 Ω load resistor. Source E_1 has an e.m.f. of 18 V and an internal resistance of 1 Ω. Source E_2 has an e.m.f. of 27 V and an internal resistance of 2 Ω and is connected in series with a 4 Ω resistor. The two e.m.f.s are in phase. Determine the value of current in each branch of the circuit and the potential difference developed across the 3 Ω load resistor.

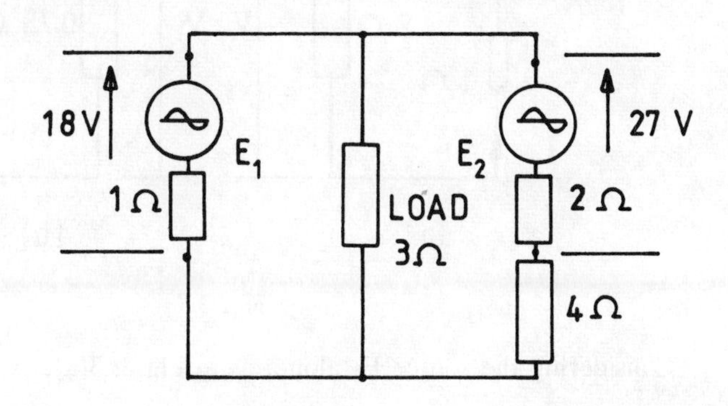

Fig. 2.1

Since each source of e.m.f. acts as though it were alone, the circuit can be considered firstly as in Fig. 2.2(a) and then as in Fig. 2.2(b).

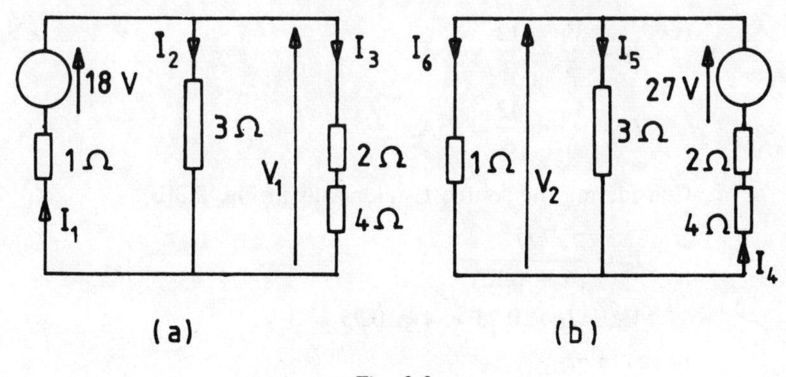

Fig. 2.2

Using Kirchhoff's first law:

In Fig. 2.2(a) $I_1 = I_2 + I_3$ and in Fig. 2.2(b) $I_4 = I_5 + I_6$.

The circuit in Fig. 2.2(a) may be further simplified by determining the equivalent resistance of 3 Ω and (4 + 2) Ω in parallel.

$$R_{eq} = \frac{3 \times 6}{3 + 6} = 2\ \Omega.$$

Likewise for the circuit in Fig. 2.2(b), 3 Ω in parallel with 1 Ω

$$R_{eq} = \frac{3 \times 1}{3 + 1} = 0.75\ \Omega.$$

The circuits then appear as in Fig. 2.3(a) and (b).

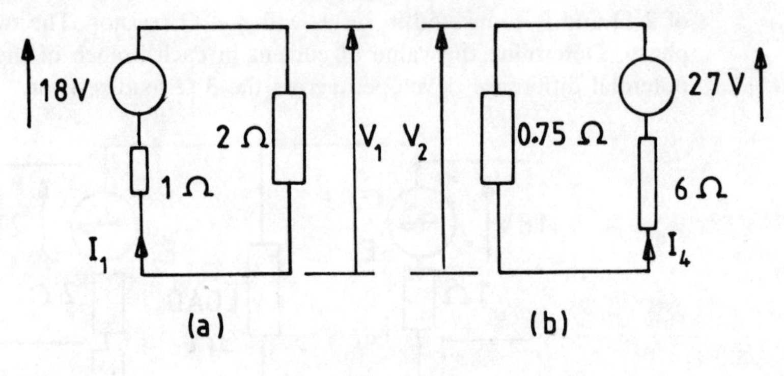

Fig. 2.3

Considering the source E_1 alone, as in Fig. 2.3(a):

$$I_1 = \frac{18}{(1 + 2)} = 6\ \text{A}.$$

$$V_1 = I_1 \times 2 = 6 \times 2 = 12\ \text{V}.$$

In Fig. 2.2(a):

$$I_2 = \frac{V_1}{3} = \frac{12}{3} = 4\ \text{A}.$$

$$I_3 = \frac{V_1}{6} = \frac{12}{6} = 2\ \text{A}.$$

Considering the source E_2 alone, as in Fig. 2.3(b):

$$I_4 = \frac{27}{(6 + 0.75)} = 4\ \text{A}.$$

$$V_2 = I_4 \times 0.75 = 4 \times 0.75 = 3\ \text{V}.$$

In Fig. 2.2(b):

$$I_5 = \frac{V_2}{3} = \frac{3}{3} = 1\ \text{A}.$$

$$I_6 = \frac{V_2}{1} = \frac{3}{1} = 3\ \text{A}.$$

Recombining the two circuits to form the original once more, there are six currents to consider (Fig. 2.4).

Adding the currents in each limb with due regard to direction gives:

Current in source E_1 = 6 A (up) + 3 A (down) = 3 A (up).
Current in the load resistor = 4 A (down) + 1 A (down)
= 5 A (down).

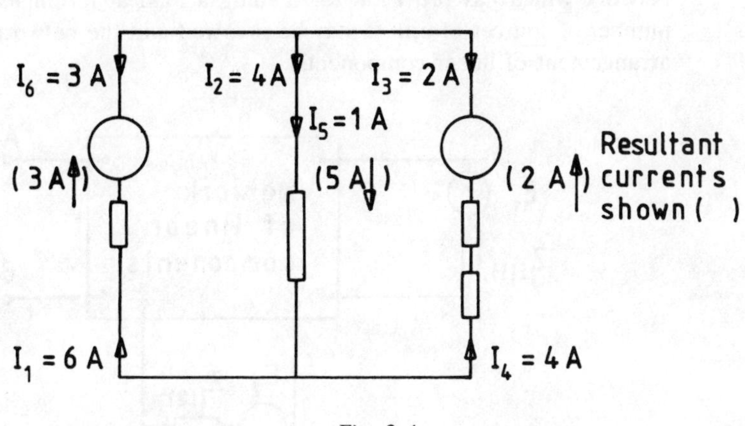

Fig. 2.4

Current in source E_2 = 2 A (down) + 4 A (up) = 2 A (up).

These results are shown in parentheses in Fig. 2.4. Checking, it can be seen that the two source currents sum to give the current in the load resistor.

Finally, the potential difference developed across the 3 Ω load resistor is

$IR = 5 \times 3 = 15$ V.

SELF-ASSESSMENT EXAMPLE 2.2

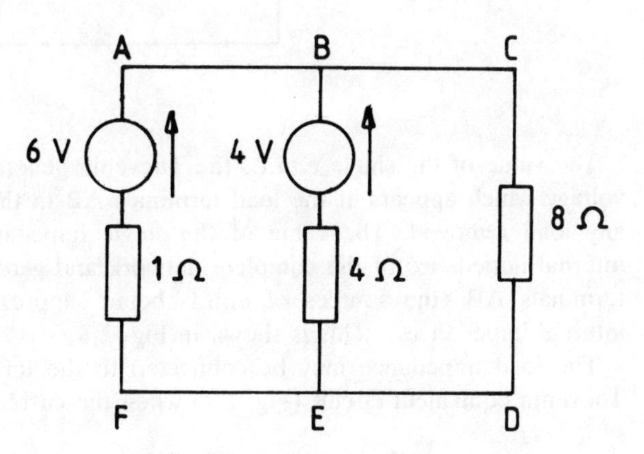

Fig. 2.5

For the circuit shown in Fig. 2.5, determine the current in each branch and the power developed in the 8 Ω resistor.

2.2 Thevenin's theorem

Consider the circuit shown in Fig. 2.6. Two sources of e.m.f. feed into a network which has provision for feeding a load at terminals A and B. Any number of sources of e.m.f. may be involved and the network consists of any arrangement of linear components.

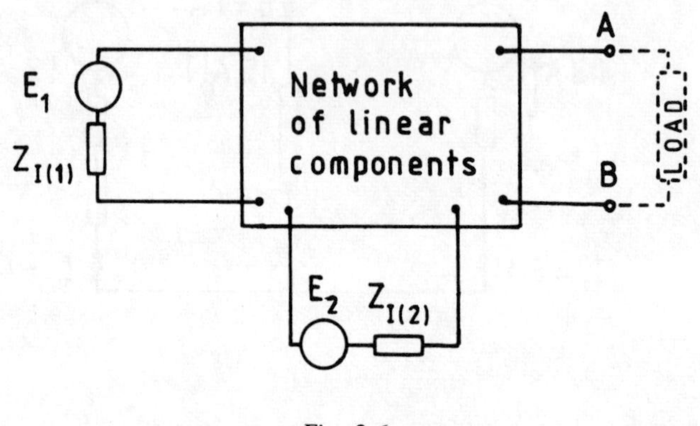

Fig. 2.6

Thevenin's theorem states that the whole circuit up to load terminals AB may be replaced by a single source of e.m.f. in series with a single impedance, as shown in Fig. 2.7.

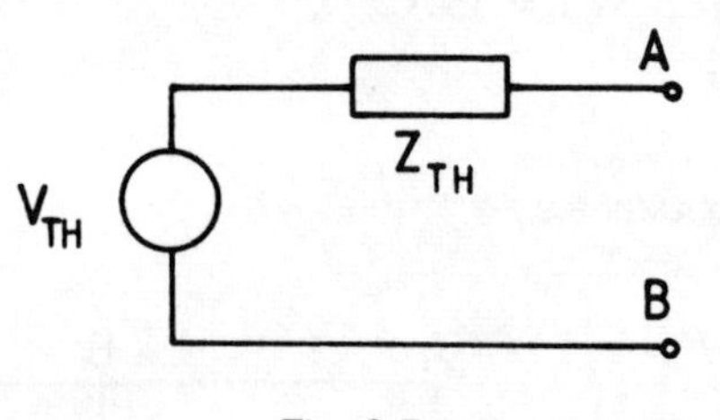

Fig. 2.7

The value of the single e.m.f., the Thevenin generator voltage V_{TH}, is the voltage which appears at the load terminals AB in the original network with any load removed. The value of the single impedance Z_{TH} is that of the internal impedance of the complete network and generators viewed from the terminals AB (the sources of e.m.f. being suppressed, leaving only their internal impedances). This is shown in Fig. 2.8.

The load impedance may be connected to the terminals A and B of the Thevenin equivalent circuit (Fig. 2.7) when the current will be given by

$$I_{load} = \frac{V_{TH}}{Z_{TH} + Z_{load}} \text{ A.}$$

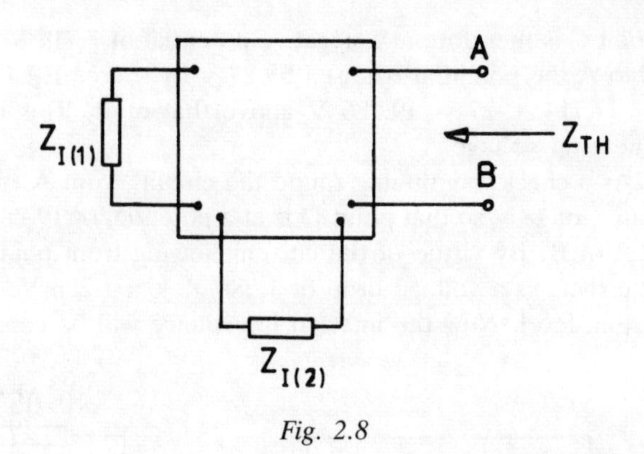

Fig. 2.8

It is therefore only necessary to calculate or measure V_{AB} and the internal impedance in order to be able to determine the current which would flow in any load connected to AB.

WORKED EXAMPLE 2.3

Determine the value of current in the 3 Ω resistor in Fig. 2.1 using Thevenin's theorem. (*Note*: the answer is already known from worked example 2.1 using the superposition theorem.)

The first step is to remove the load resistor (Fig. 2.9). There are 18 V attempting to drive current in a clockwise direction round the circuit and 27 V acting in the opposite direction.

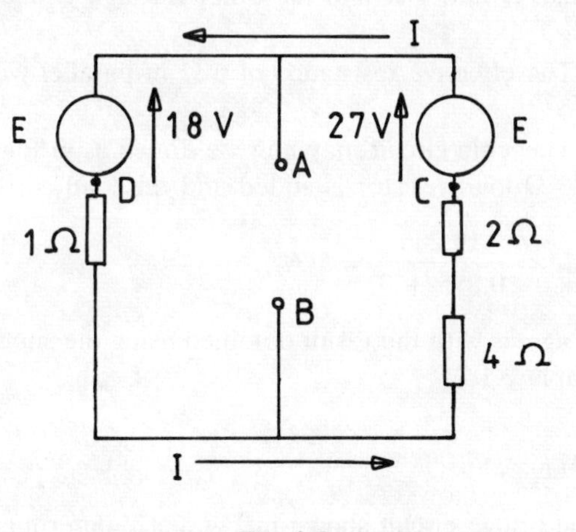

Fig. 2.9

The nett e.m.f. = 27 − 18 = 9 V (anti-clockwise).
The total resistance of the circuit = 1 + 2 + 4 = 7 Ω.
Hence I = 9/7 = 1.286 A (anti-clockwise: see Fig. 2.9).

Consider point B and the bottom line in the diagram to be at reference potential. From B to C there will be a voltage drop of 1.286 × 6 = 7.714 V.

Point C is therefore at a negative potential of 7.714 V with respect to B. From C to A the potential is raised by 27 V by source E_2. Hence the potential of A is $-7.714 + 27 = 19.286$ V above that of B. This is then the value of the Thevenin voltage.

As a check, continuing round the circuit, from A to D there is an opposing e.m.f. of 18 V so that point D is at a potential of $19.286 - 18 = 1.286$ V above that of B. By virtue of the current flowing from point D back to the bottom line there is a voltage drop of $1.286 \times 1 = 1.286$ V, which brings us back to datum level. Now the internal impedance will be considered.

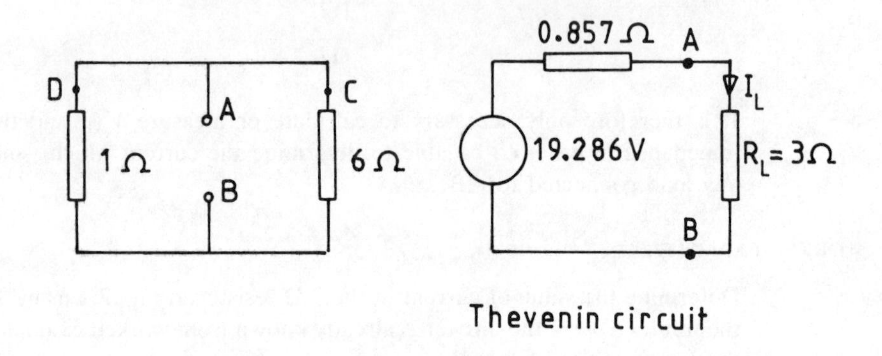

Fig. 2.10 *Fig. 2.11*

In Fig. 2.10 the e.m.f.s have been suppressed, leaving only their internal impedances. From point A to point B there are two parallel routes, one through D and 1 Ω and the other through C and 6 Ω.

The effective resistance of 6 Ω in parallel with 1 Ω $= \dfrac{6 \times 1}{6 + 1} = 0.875$ Ω.

The Thevenin circuit may now be drawn as in Fig. 2.11 up to points A and B. The 3 Ω load resistor is added and the load current becomes

$$I_L = \frac{19.286}{0.875 + 3} = 5 \text{ A}.$$

This agrees with the result obtained using the superposition theorem in worked example 2.1.

WORKED EXAMPLE 2.4

In the bridge circuit shown in Fig. 2.12, determine the value of the current in the 40 Ω resistor using Thevenin's theorem. (The voltage source has negligible internal resistance.)

The first step in the solution is to remove the resistor in which the current is to be found. The circuit is then as shown in Fig. 2.13.

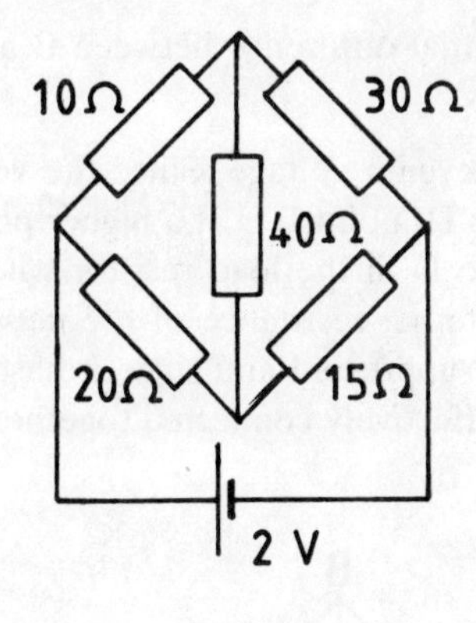

Fig. 2.12

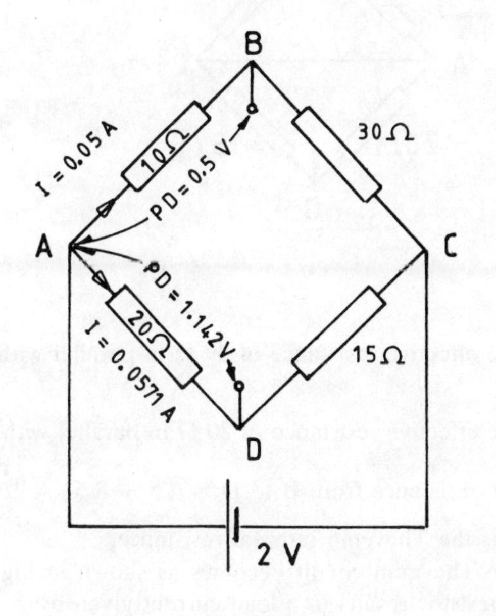

Fig. 2.13

To find the Thevenin voltage:

The current from the battery divides into two parts at point A.

The current in the top arms $= \dfrac{2}{10 + 30} = 0.05$ A.

Hence the potential difference between A and B $= IR = 0.05 \times 10 = 0.5$ V.

The current in the bottom arms $= \dfrac{2}{20 + 15} = 0.0571$ A.

Hence the potential difference between A and D $= IR = 0.0571 \times 20 = 1.142$ V.

The potential difference between B and D = $V_{AD} - V_{AB} = 1.142 - 0.5$ $= 0.642$ V.

This is the Thevenin voltage. Since the voltage drop from A to B is less than that from A to D, point B is at a higher potential than point D and current will flow from B to D in the load resistor when it is reconnected.

Next the internal resistance of the network must be found. The voltage of the battery is suppressed and since it has negligible internal resistance, points A and C are effectively connected together. The circuit then becomes as shown in Fig. 2.14.

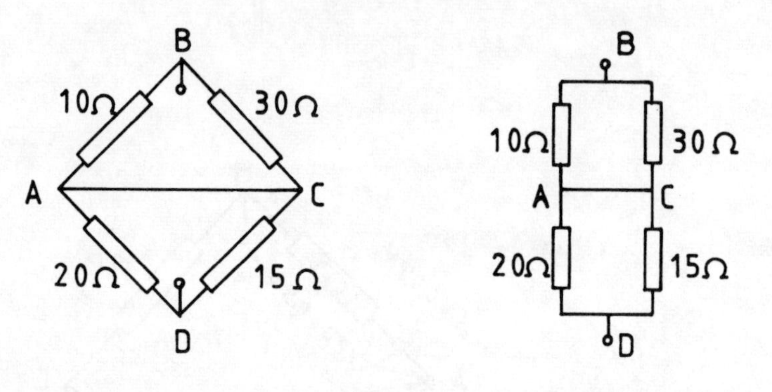

Fig. 2.14

The effective resistance of 30 Ω in parallel with 10 Ω $= \dfrac{30 \times 10}{30 + 10} = 7.5$ Ω.

The effective resistance of 20 Ω in parallel with 15 Ω $= \dfrac{20 \times 15}{20 + 15} = 8.57$ Ω.

The resistance from B to D = 7.5 + 8.57 = 16.07 Ω.

This is the Thevenin internal resistance.

The Thevenin circuit becomes as shown in Fig. 2.15 and reconnecting the load resistor results in a load current given by

$$I_L = \frac{0.642}{16.07 + 40} = 0.0115 \text{ A.}$$

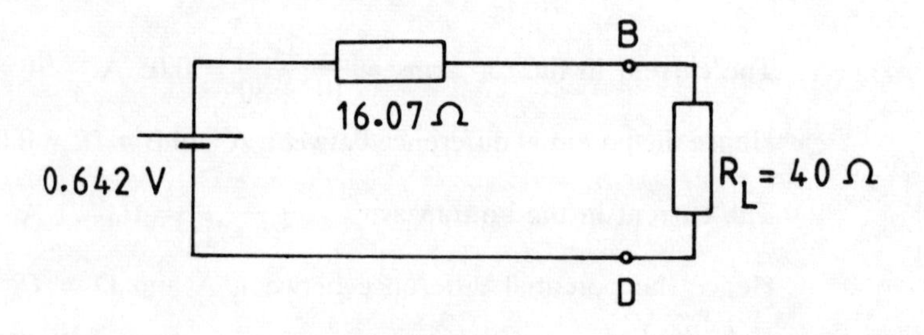

Fig. 2.15

SELF-ASSESSMENT EXAMPLE 2.5

Calculate the value of the current in the 8 Ω resistor in Fig. 2.16 using Thevenin's theorem.

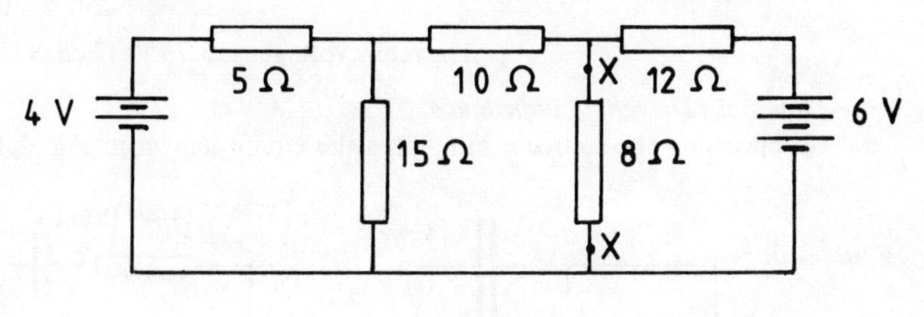

Fig. 2.16

(One method is to reduce the circuit to the left of XX to a Thevenin equivalent when, adding it to that to the right of XX, the circuit becomes as in worked example 2.1.)

In the preceding examples only resistive circuits have been considered. Thevenin's theorem is now applied to a circuit containing reactance.

WORKED EXAMPLE 2.6

Obtain the Thevenin equivalent circuit for the network shown in Fig. 2.17 and hence determine the value of current which would flow in an impedance of (60 + j300) Ω connected across AB.

What power will be dissipated in this impedance?

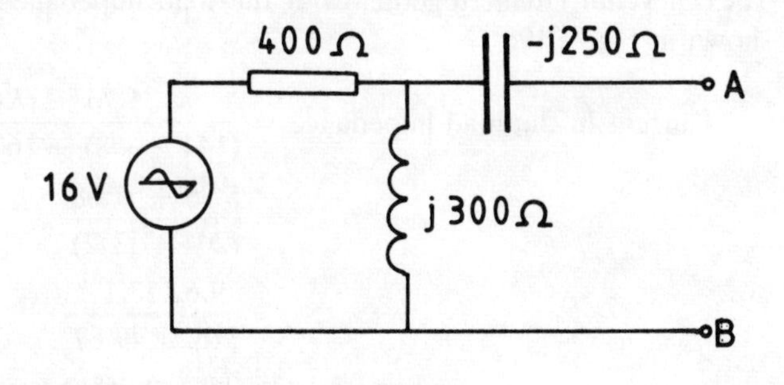

Fig. 2.17

Thevenin voltage:
Since there is no current in the capacitor with AB open-circuited, the potential difference across terminals AB is the same as that across the pure inductor, j300 Ω.

$$\text{Potential difference across the inductor} = \frac{16}{400 + j300} \times j300 \text{ V}$$

$$= \frac{j4800(400 - j300)}{400^2 + 300^2}$$

$$= \frac{1.44 \times 10^6 + j1.92 \times 10^6}{250 \times 10^3}$$

$$\text{Thevenin voltage} = 5.76 + j7.68 \text{ V.}$$

Internal (Thevenin) impedance:
Suppressing the source e.m.f. gives the circuit shown in Fig. 2.18.

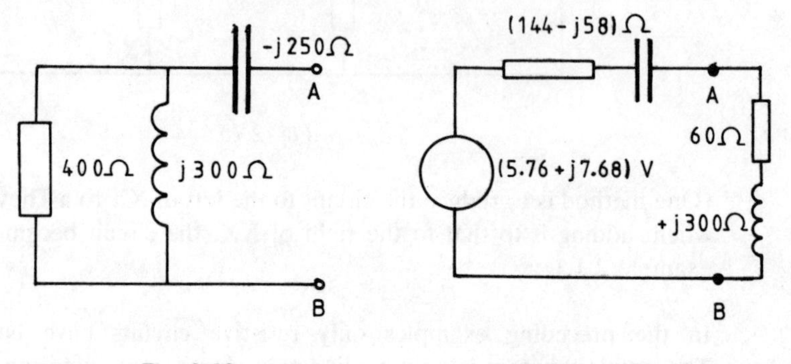

Fig. 2.18 *Fig. 2.19*

$$Z_{\text{internal}} = -j250 + \text{equivalent impedance of } 400\ \Omega \text{ and } j300\ \Omega \text{ in parallel}$$

$$= -j250 + \frac{400(j300)}{400 + j300} = -j250 + \frac{j120\,000(400 - j300)}{400^2 + 300^2}$$

$$= -j250 + 144 + j192$$

$$\text{Thevenin impedance} = (144 - j58)\ \Omega.$$

The Thevenin circuit together with the load impedance of $(60 + j300)\ \Omega$ is shown in Fig. 2.19.

$$\text{Current in the load impedance} = \frac{5.76 + j7.68}{(144 - j58) + (60 + j300)} \text{ A}$$

$$= \frac{5.76 + j7.68}{(204 + j242)}$$

$$= \frac{9.6\angle 53.13°}{316.5\angle 49.87°}$$

$$= 0.03\angle 3.26° \text{ A.}$$

The current in the load has a modulus of 0.03 A and it leads the supply voltage by 3.26°.

$$\text{Power in load impedance} = I^2R = (0.03)^2 \times 60 = 0.054 \text{ W.}$$

SELF-ASSESSMENT EXAMPLE 2.7

By the use of Thevenin's theorem determine the value of current in a 46.74 Ω resistor connected to terminals AB in Fig. 2.20. The two sources of e.m.f. have outputs which are in phase.

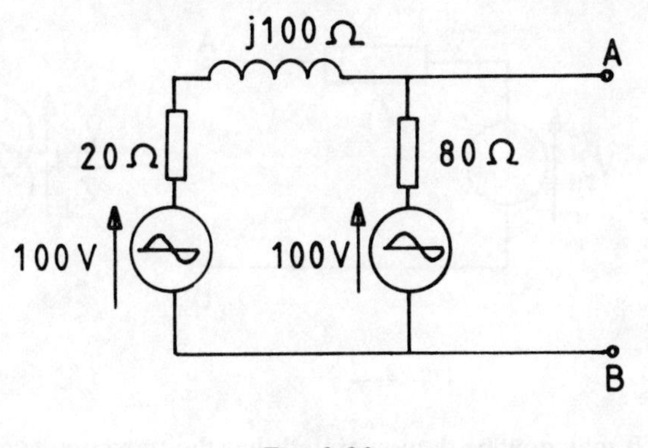

Fig. 2.20

2.3 Norton's theorem

Referring back to the network and sources of e.m.f. shown in Fig. 2.6, Norton's theorem states that the whole circuit up to the load terminals AB may be replaced by a single constant current source shunted by a single impedance as shown in Fig. 2.21. The value of current delivered by the current generator is that which would be delivered into a short-circuit across the terminals AB in the original network. The value of the single impedance is that of the internal impedance of the network (the sources of e.m.f. being suppressed leaving only their internal impedances), which is precisely the same as for Thevenin's theorem.

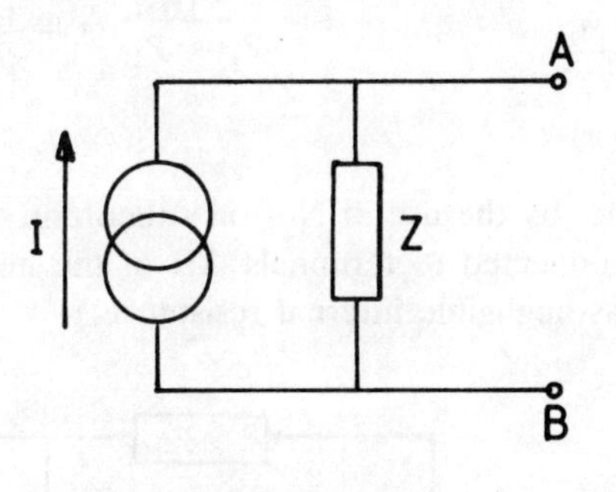

Fig. 2.21

Thevenin circuits may be simply transformed into Norton circuits in the following manner. Starting with a Thevenin circuit as shown in Fig. 2.22, short-circuiting terminals AB results in a current given by

$$I = \frac{V_{TH}}{Z_I}.$$

This is the Norton current. Since Z_I is the same for both theorems, the Norton circuit may be drawn as in Fig. 2.23.

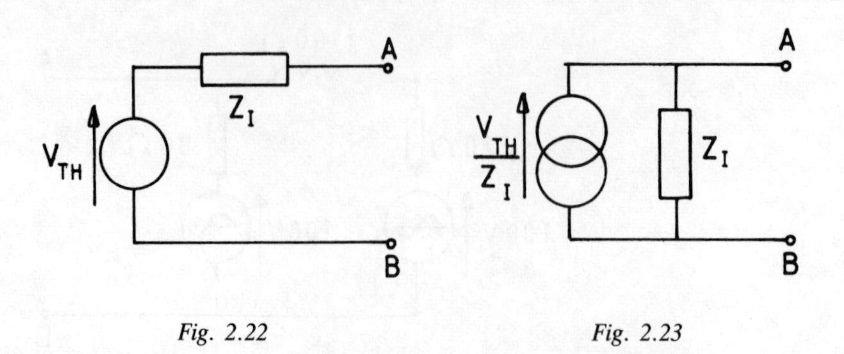

Fig. 2.22 *Fig. 2.23*

It may now be demonstrated that the two circuits are equivalent. Connecting a load impedance Z_L to the Thevenin circuit in Fig. 2.22 will result in a current in the load impedance of

$$I = \frac{V_{TH}}{Z_I + Z_L} \text{ A.}$$

Now considering the Norton circuit in Fig. 2.23, connecting the same impedance across the terminals AB puts Z_L in parallel with Z_I:

$$\text{The equivalent impedance } Z_{eq} = \frac{Z_I Z_L}{Z_I + Z_L}.$$

$$\text{Potential difference } V_{AB} = IZ_{eq} = \frac{V_{TH}}{Z_I}\left[\frac{Z_I Z_L}{Z_I + Z_L}\right].$$

$$\text{Current in } Z_L = \frac{V_{AB}}{Z_L} = \frac{V_{TH}}{Z_I}\left[\frac{Z_I Z_L}{Z_I + Z_L}\right]\frac{1}{Z_L}$$

$$= \frac{V_{TH}}{Z_I + Z_L} \text{ A as before.}$$

WORKED EXAMPLE 2.8

Determine, by the use of Norton's theorem, the value of current in the 10 Ω resistor connected to terminals AB of the network shown in Fig. 2.24. (The source has negligible internal resistance.)

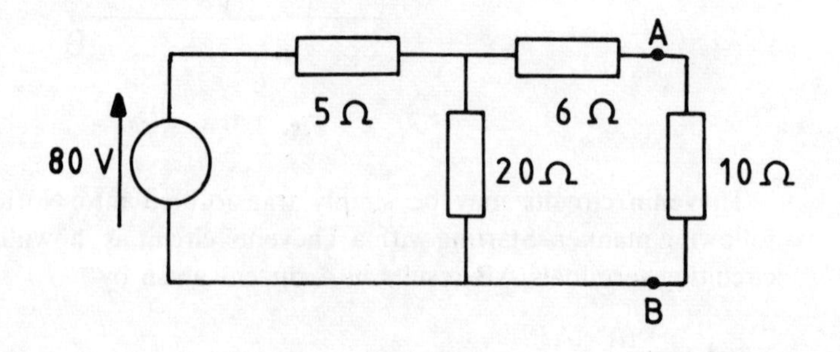

Fig. 2.24

Shorting terminals A and B effectively connects the 6 Ω resistor in parallel with the 20 Ω resistor and eliminates the 10 Ω resistor from the calculations.

$$\text{The equivalent resistance of 20 Ω and 6 Ω in parallel} = \frac{20 \times 6}{20 + 6} = 4.6\ \Omega.$$

Total resistance of the circuit = 4.6 + 5 = 9.6 Ω.

$$\text{Current delivered by the source} = \frac{V}{R_{\text{total}}} = \frac{80}{9.6} = 8.33\ \text{A}.$$

At junction X in Fig. 2.25, the current will divide between the 20 Ω and 6 Ω resistors.

$$I_{SC} = \frac{V_{XY}}{6} = \frac{8.33 \times 4.6}{6} = 6.39\ \text{A}.$$

This is the Norton current.

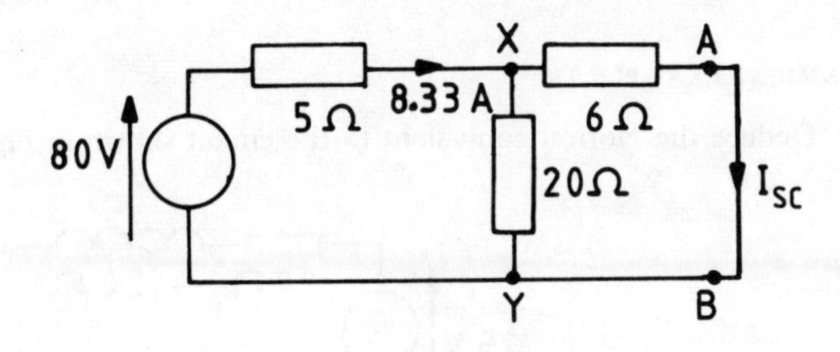

Fig. 2.25

The internal resistance is now needed looking back from the terminals AB with the source of e.m.f. removed and replaced by its internal resistance which is negligible.

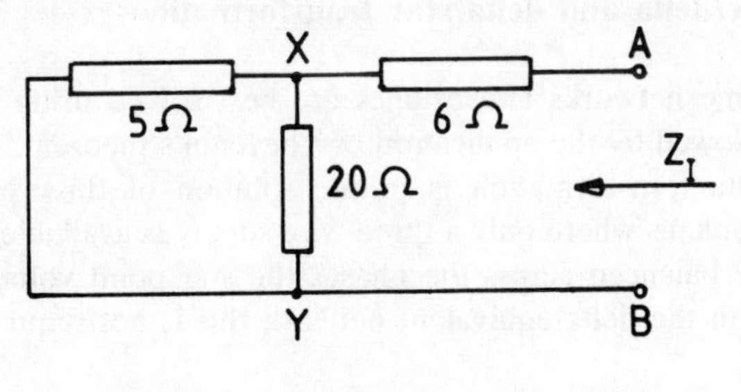

Fig. 2.26

In Fig. 2.26, looking left from terminals XY, 5 Ω is in parallel with 20 Ω.

$$Z_{eq} = \frac{20 \times 5}{20 + 5} = 4\ \Omega \qquad Z_I = 6 + 4 = 10\ \Omega.$$

The Norton circuit is as shown in Fig. 2.27.

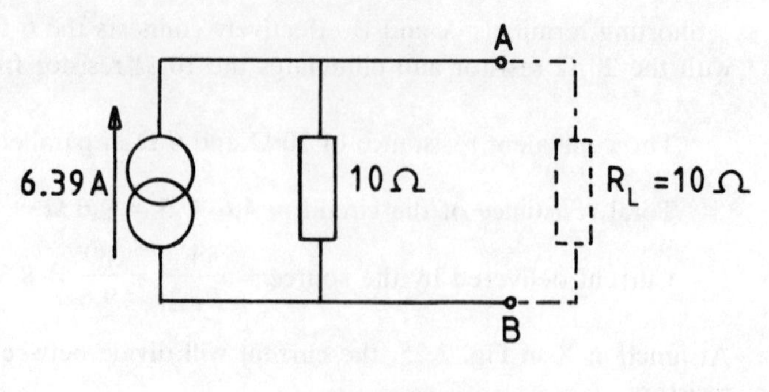

Fig. 2.27

With the load resistor of 10 Ω connected to AB, one half of the supply current flows in Z_I and the other half in Z_L.

$$\text{Current in } Z_L = 0.5 \times 6.39 = 3.195 \text{ A.}$$

SELF-ASSESSMENT EXAMPLE 2.9

Deduce the Norton equivalent to the circuit shown in Fig. 2.28.

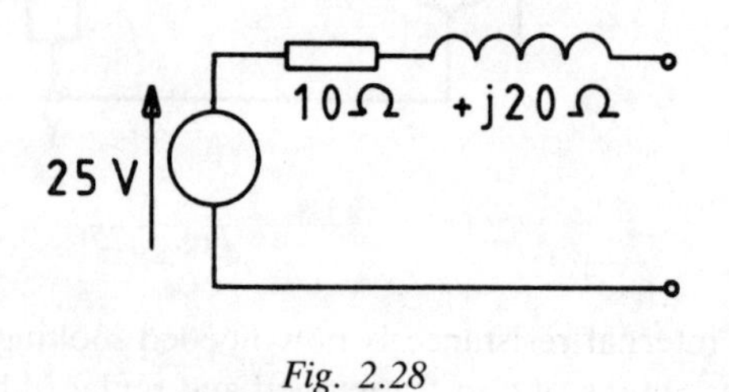

Fig. 2.28

2.4 The star/delta and delta/star transformations

Some networks and bridges are best solved using these transforms, possibly followed by the application of Thevenin's theorem. An application of the star/delta transformation is in the solution of three-phase, star-connected load problems where only a three-wire supply is available. In this case, if the load is not balanced across the phases the star point voltage is difficult to calculate. With the delta-equivalent network this is not required.

2.4.1 *The star/delta transformation*

Any star-connected network may be replaced with a delta (π)-connected network such that the impedance between any pair of terminals is unchanged. The equivalent network will consume the same power, operate at the same power factor and in all respects perform as did the original star-connected network.

Consider the three, star-connected impedances as shown in Fig. 2.29. The star may be replaced by a delta network as shown in Fig. 2.30 in which:

$$Z_A = Z_1Z_2\left(\frac{1}{Z_1} + \frac{1}{Z_2} + \frac{1}{Z_3}\right), \quad Z_B = Z_2Z_3\left(\frac{1}{Z_1} + \frac{1}{Z_2} + \frac{1}{Z_3}\right), \quad \text{and}$$

$$Z_C = Z_1Z_3\left(\frac{1}{Z_1} + \frac{1}{Z_2} + \frac{1}{Z_3}\right).$$

To calculate the value of Z_A, first multiply together Z_1 and Z_2. These are the two impedances in the original star connected to the same terminals as Z_A, i.e. terminals 1 and 2. For Z_B, the terminals are 2 and 3 so that the product Z_2Z_3 is required. Finally for Z_C, this is connected to terminals 1 and 3 so that the product Z_1Z_3 is used.

In each case the product is multiplied by the sum of the reciprocals (sum of the admittances) of the impedances in the original star.

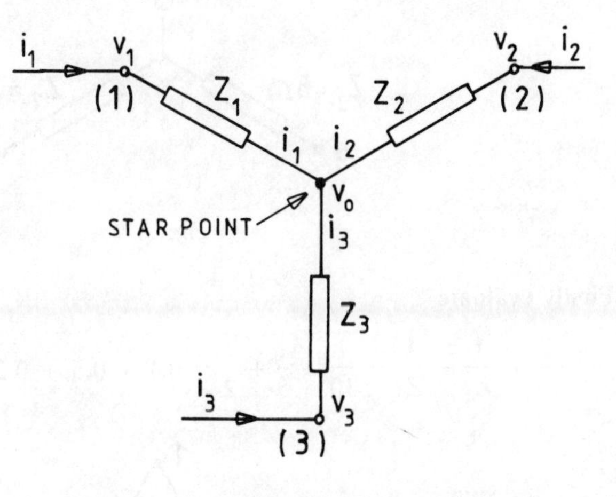

Fig. 2.29

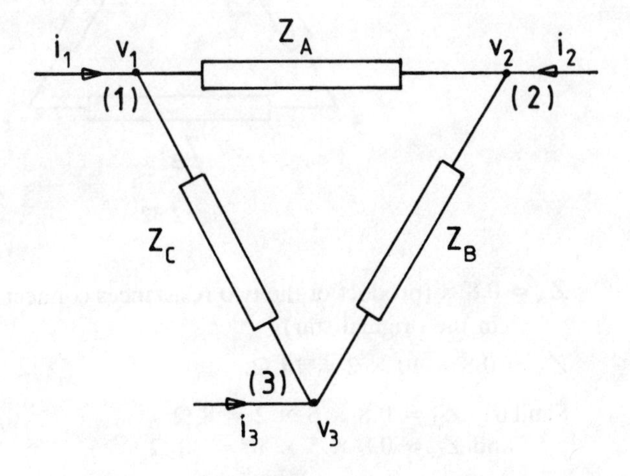

Fig. 2.30

This transformation may be extended for use with any number of impedances connected to a common point.

For a full proof of the transformation see appendix 1.

WORKED EXAMPLE 2.10

Deduce the equivalent delta to the star given in Fig. 2.31.

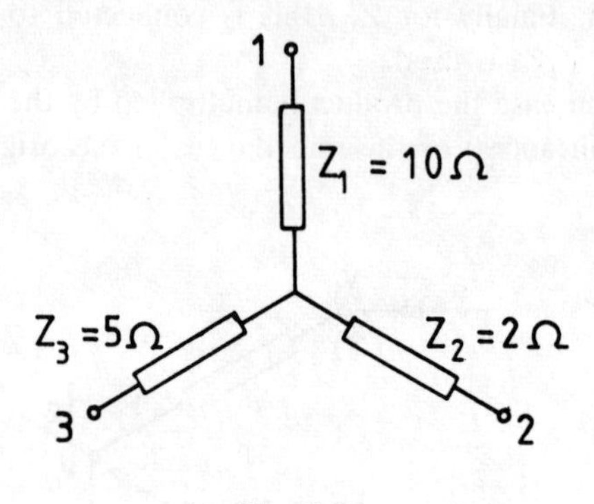

Fig. 2.31

Firstly evaluate

$$\frac{1}{Z_1} + \frac{1}{Z_2} + \frac{1}{Z_3} = \frac{1}{10} + \frac{1}{5} + \frac{1}{2} = 0.1 + 0.5 + 0.2 = 0.8$$

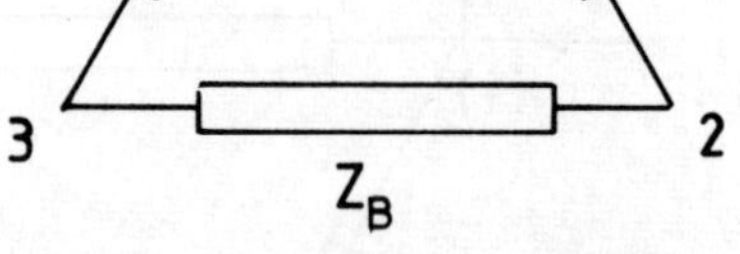

Fig. 2.32

$Z_A = 0.8 \times$ (product of the two resistances connected to terminals 1 and 2 in the original star)

$Z_A = 0.8 \times 10 \times 2 = 16\ \Omega.$

Similarly $Z_B = 0.8 \times 5 \times 2 = 8\ \Omega$

and $Z_C = 0.8 \times 5 \times 10 = 40\ \Omega.$

WORKED EXAMPLE 2.11

Determine the impedance Z_A in the equivalent delta to the star given in Fig. 2.33(a).

Firstly determine the value of the sum of the reciprocals of the impedances in the original star.

$$\frac{1}{20} + \frac{1}{(10 + j10)} + \frac{1}{-j25} \quad \left(\text{multiply } \frac{1}{-j25} \text{ by } \frac{j}{j}\right)$$

$$= 0.05 + \frac{(10 - j10)}{100 + 100} + j0.04$$

$$= 0.05 + 0.05 - j0.05 + j0.04$$

$$= 0.1 - j0.01$$

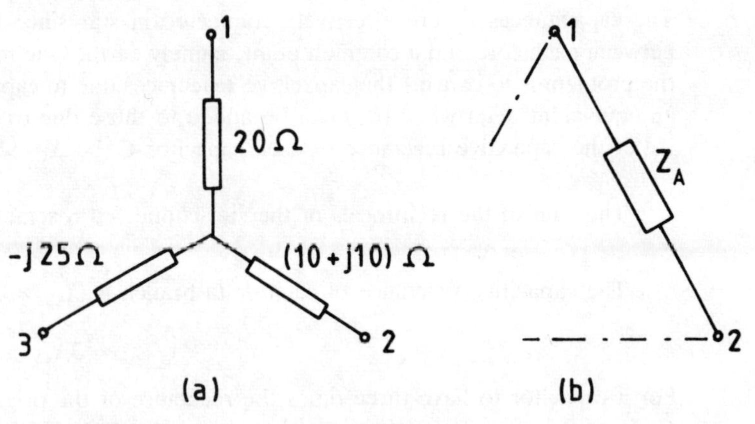

Fig. 2.33

$$\begin{aligned} Z_A &= 20 \times (10 + j10) \times (0.1 - j0.01) \\ &= 20 \times 14.14\angle 45° \times 0.1005\angle -5.71° \\ &= 28.42\angle 39.29° \\ &= (22 + j18)\ \Omega. \end{aligned}$$

SELF-ASSESSMENT EXAMPLE 2.12

Convert the star given in Fig. 2.34 into an exact equivalent delta.

WORKED EXAMPLE 2.13

A three-phase, three-core cable has capacitance C_c between cores and C_s between each core and earth (Fig. 2.35). Each capacitance C_c = 0.1 μF and each capacitance C_s = 0.2 μF. The voltage between cores = 11 kV at 50 Hz. Determine the value of charging current in each line.

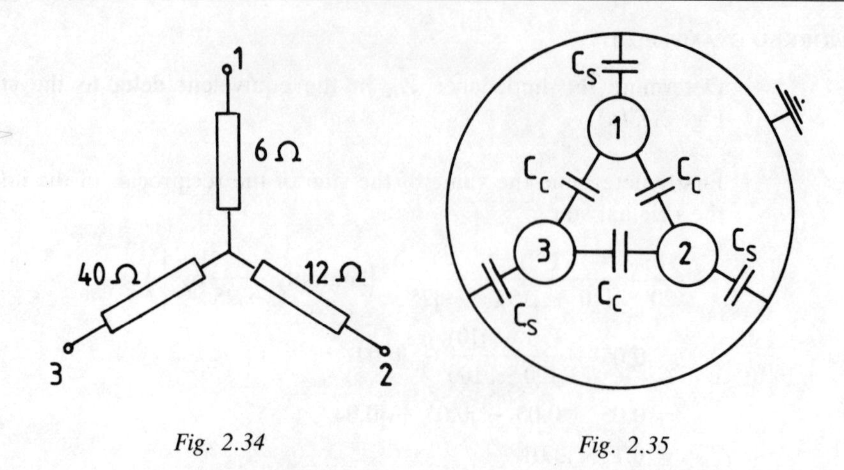

Fig. 2.34 *Fig. 2.35*

The capacitances C_s are effectively connected in star since they are formed between each core and a common point, namely earth. One method of solving the problem is to convert the capacitive reactances due to capacitances C_s into an equivalent delta when they can be added to those due to capacitances C_c.

Let the capacitive reactance of each capacitor C_s be X_{C_s} Ω.

The sum of the reciprocals of the star-connected reactances $= 3\left[\frac{1}{X_{C_s}}\right]$.

$$\text{The capacitive reactance of each delta branch} = X_{C_s} \times X_{C_s} \times 3\left[\frac{1}{X_{C_s}}\right]$$

$$= 3X_{C_s}.$$

For a capacitor to have three times the reactance of the original capacitance, C_s, it must have a capacitance equal to one third of C_s. Hence each branch of the delta equivalent has the value of $\frac{1}{3}C_s$. In Fig. 2.36 this is shown together with C_c to make up the total capacitance between one pair of cores.

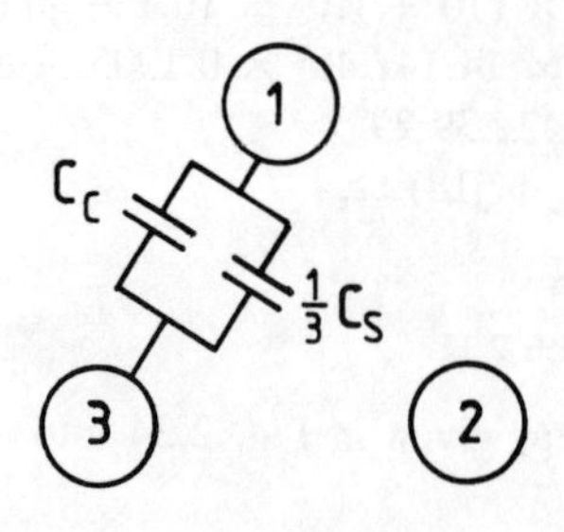

Fig. 2.36

$$\text{The total capacitance between cores} = C_c + \frac{1}{3}C_s = \left(0.1 + \frac{0.2}{3}\right) \mu\text{F}$$

$$= 0.1667 \ \mu\text{F}.$$

$$\text{Capacitive reactance, } X_c = \frac{1}{\omega C} = \frac{1}{2\pi \times 50 \times 0.1667 \times 10^{-6}} = 19\,099\ \Omega.$$

Since the cable is symmetrical, each pair of cores will yield a similar result.

The line current in conductor 1 will be the sum of the currents due to the capacitances between conductor 1 and conductor 2 and between conductor 1 and conductor 3. From previous studies on balanced, three-phase, delta-connected systems it should be remembered that the line current $I_L = \sqrt{3} \times$ (phase current). The currents I_{12} and I_{13} are phase currents in such a system so that the current in each line is $\sqrt{3}\ I_{12}$ amperes.

$$I_{12} = \frac{V_{\text{line}}}{X_C} = \frac{11\,000}{19\,099} = 0.576\ \text{A}.$$

Line charging current $= \sqrt{3} \times 0.576 = 0.997$ A.

2.4.2 *The delta/star transformation*

Any delta-connected network may be replaced by a star-connected network which is exactly electrically equivalent.

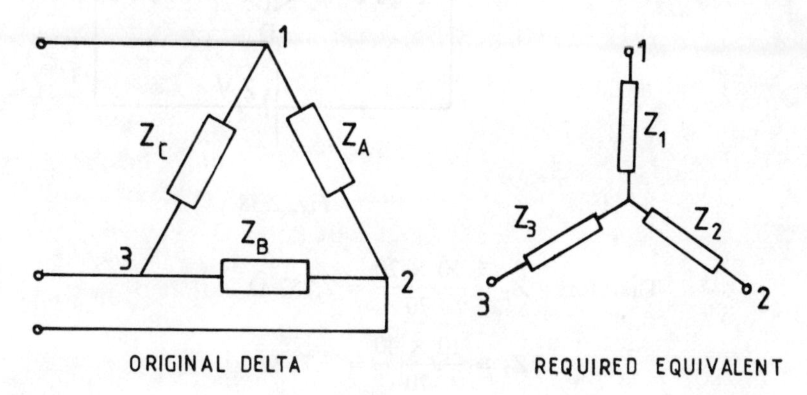

Fig. 2.37

In Fig. 2.37

$$Z_1 = \frac{Z_A Z_C}{Z_A + Z_B + Z_C} \quad Z_2 = \frac{Z_A Z_B}{Z_A + Z_B + Z_C} \quad \text{and } Z_3 = \frac{Z_B Z_C}{Z_A + Z_B + Z_C}.$$

The branches of the equivalent star are found by multiplying together the two impedances in the original delta connected to the relevant terminal and then dividing by the sum of the delta-connected impedances. (Compare this with the result of the star/delta transform and take particular note that $1/(Z_A + Z_B + Z_C)$ is NOT the same as the sum of the individual reciprocals.)

For a proof of the transformation see appendix 2.

WORKED EXAMPLE 2.14

Determine the value of current in the 40 Ω resistor in the Wheatstone bridge arrangement shown in Fig. 2.38 using the delta/star transformation. (Note: this current has already been deduced using Thevenin's theorem in worked example 2.4 so it will be interesting to compare the complexity of the solutions and indeed to see whether the same result is achieved.)

First the values of the impedances must be calculated which form an equivalent star to the delta ABD in Fig. 2.38. Let Z_1 be connected to A, Z_2 to B, and Z_3 to D.

The sum of the three delta impedances = 10 + 20 + 40 = 70 Ω.

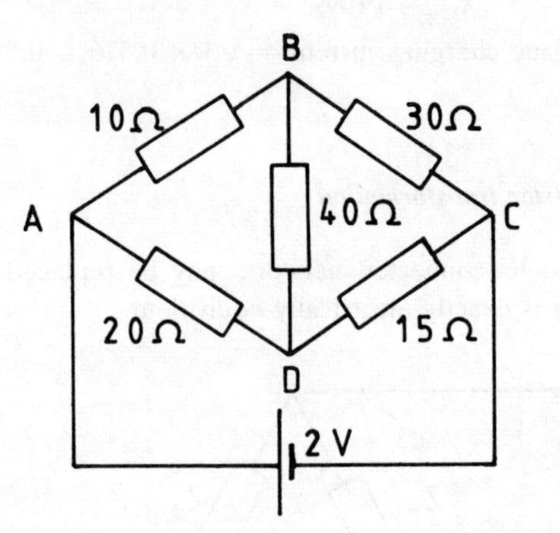

Fig. 2.38

$$\text{Therefore} \quad Z_1 = \frac{10 \times 20}{70} = 2.857\ \Omega$$

$$Z_2 = \frac{10 \times 40}{70} = 5.71\ \Omega$$

$$Z_3 = \frac{20 \times 40}{70} = 11.43\ \Omega.$$

The circuit is redrawn as in Fig. 2.39.

From S to C there are two parallel routes: (5.71 + 30) Ω in parallel with (11.43 + 15) Ω.

$$Z_{eq} = \frac{(5.71 + 30)\ (11.43 + 15)}{5.71 + 30 + 11.43 + 15} = 15.2\ \Omega.$$

Total resistance A to C = 15.2 + 2.857 = 18.057 Ω.

$$I = \frac{2}{18.057} = 0.111\ \text{A}.$$

Hence p.d. SC = 0.111 × 15.2 = 1.687 V.

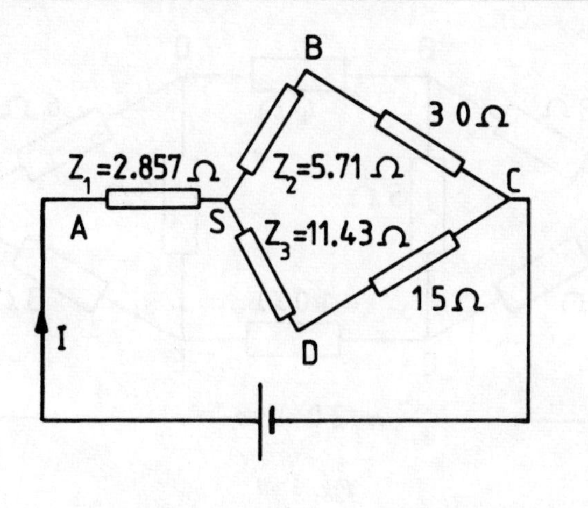

Fig. 2.39

The p.d. between B and D must be found since this p.d. exists across the same two points and the 40 Ω resistor in the original network.

$$\text{Current in top branch SBC} = \frac{1.687}{35.71} = 0.0472 \text{ A.}$$

$$\text{p.d. SB} = 0.0472 \times 5.71 = 0.27 \text{ V.}$$

$$\text{Current in bottom branch SDC} = \frac{1.687}{26.43} = 0.064 \text{ A.}$$

$$\text{p.d. SD} = 0.064 \times 11.43 = 0.73 \text{ V.}$$

$$\text{Therefore p.d. BD} = 0.73 - 0.27 = 0.46 \text{ V.}$$

Since this p.d. also exists in the original network,

$$\text{current in the 40 Ω resistor} = \frac{0.46}{40} = 0.0115 \text{ A.}$$

This is as found in worked example 2.4.

Most problems can be solved using a selection of network theorems. The ease of solution often depends on the correct choice of theorem.

SELF-ASSESSMENT PROBLEM 2.15

Convert the two deltas ABC and DEF in Fig. 2.40 into equivalent stars. Hence deduce the value of current flowing in the 10 Ω resistor connected between terminals C and E.

2.5 Loop analysis

In the circuit shown in Fig. 2.41 the battery raises charge to a positive potential of V volts with respect to the datum point. Two methods of indicating battery

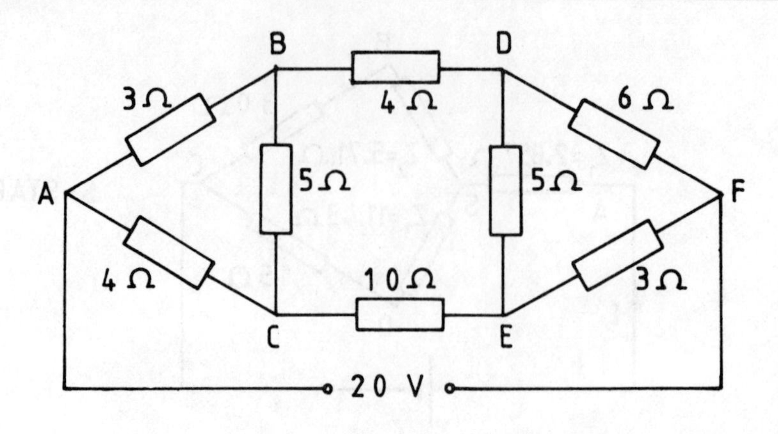

Fig. 2.40

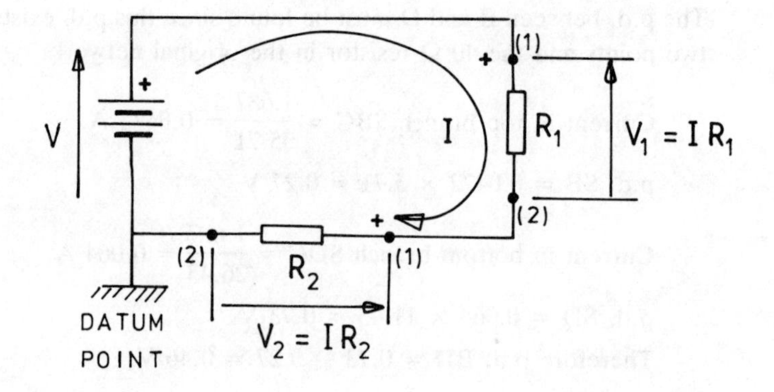

Fig. 2.41

polarity are used. One is the + sign at the positive terminal, the other is the arrow, the head of which indicates the high-potential side. Current is driven by the battery through R_1 and the potential falls by V_1 volts.

$$V_1 = IR_1.$$

End (1) of resistor R_1 is more positive than end (2) and this fact is indicated by using the + sign and an arrow with its head pointing to the high-potential end. Similarly there is a fall of potential V_2 volts across R_2.

$$V_2 = IR_2.$$

Again the + sign and the arrow are used to indicate the high-potential end of this resistor. Observe that when considering circuit resistances (and later complex impedances) the arrow indicating potential points against the direction of the current.

The battery voltage is matched by the two falls in potential.

$$\begin{aligned} V &= V_1 + V_2 \\ &= IR_1 + IR_2. \end{aligned}$$

Alternatively, it may be argued that the sum of the voltages in the closed loop is zero since charge starts at the datum level, is raised in potential by the battery, and then falls in potential through the resistors, arriving back at the datum level once more. Regarding potential arrows pointing round the loop in a clockwise direction as being positive and those in an anti-clockwise direction as negative we may write

$$V - V_1 - V_2 = 0$$

which when transposed gives the result $V = V_1 + V_2$ as before.

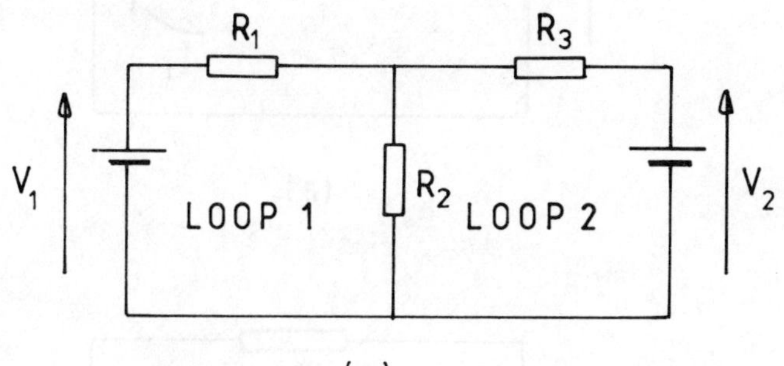

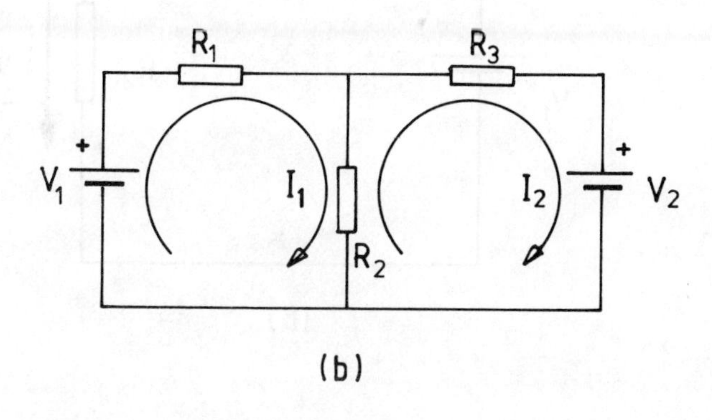

Fig. 2.42

Now consider a circuit with two loops as shown in Fig. 2.42(a). We cannot be certain of correct directions in either loop since these will depend on the relative magnitudes of the two voltages V_1 and V_2. Completely at random two loop currents are drawn as shown in Fig. 2.42(b).

In Fig. 2.43(a) we consider Loop 1 and current I_1, alone. The falls of potential are indicated using the + signs and arrows in exactly the same manner as in Fig. 2.41. In Fig. 2.43(b) Loop 1 is again shown but this time the effect of I_2 is considered and it will be observed that the high-potential end of R_2 caused by this current is at the bottom. *The Loop 2 current has an effect in Loop 1 and this must be considered.*

In Loop 1 there are two potential arrows pointing in a clockwise direction, the battery voltage V_1 and I_2R_2, and two in an anti-clockwise direction, I_1R_1 and I_1R_2. These are shown on a clock face in Fig. 2.43(c).

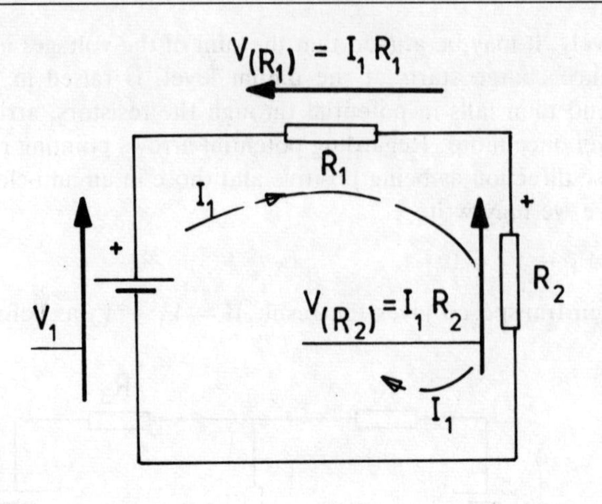
$V_{(R_1)} = I_1 R_1$
R_1
I_1
+
+
R_2
V_1
$V_{(R_2)} = I_1 R_2$
I_1

(a)

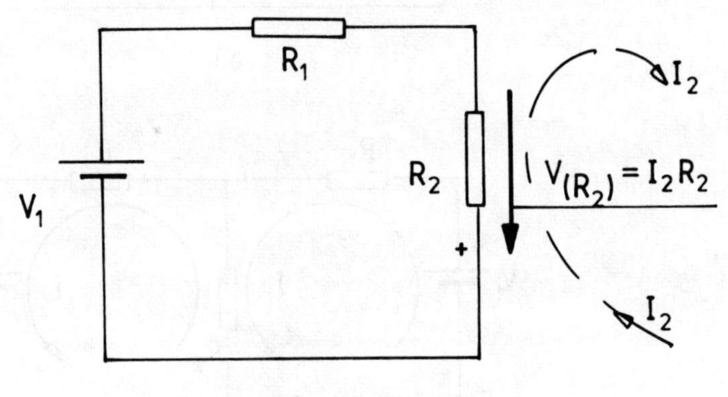
R_1
I_2
R_2
$V_{(R_2)} = I_2 R_2$
V_1
+
I_2

(b)

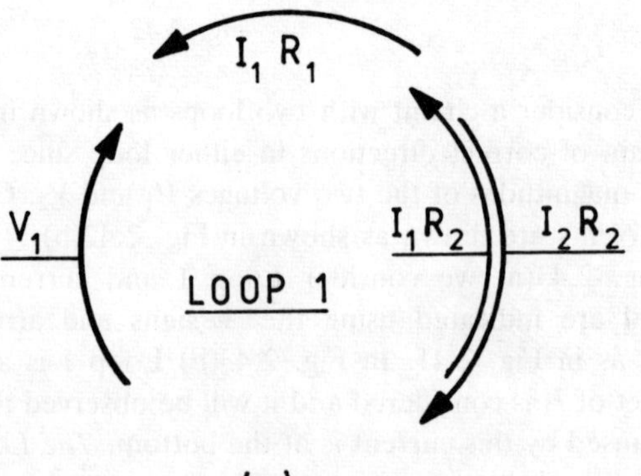
$I_1 R_1$
V_1
$I_1 R_2$
$I_2 R_2$
LOOP 1

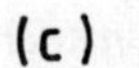

(c)

Fig. 2.43

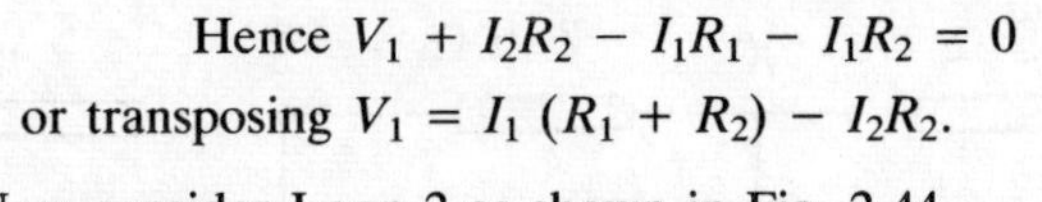

$$\text{Hence } V_1 + I_2R_2 - I_1R_1 - I_1R_2 = 0$$
$$\text{or transposing } V_1 = I_1\,(R_1 + R_2) - I_2R_2.$$

Now consider Loop 2 as shown in Fig. 2.44.

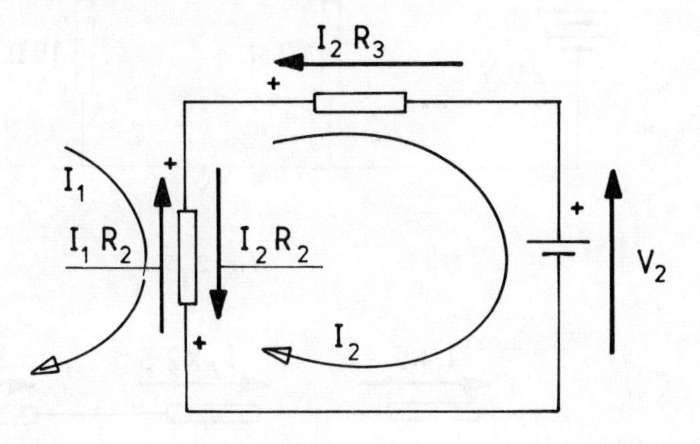

Fig. 2.44

For the direction of I_2 shown the two potential arrows point in an anti-clockwise direction as does the battery voltage V_2. The Loop 1 current has an effect in Loop 2 and its effect must be considered. The direction of I_1, shown outside the loop, gives rise to a clockwise voltage arrow. Considering all these effects, the equation is written

$$-V_2 - I_2R_3 - I_2R_2 + I_1R_2 = 0.$$

Having set up the two equations containing the two unknowns I_1 and I_2 it is necessary to solve these simultaneously by any method known. In any network it is necessary to set up as many equations as there are loop currents, so that in networks containing many loops the main difficulty is in solving this number of equations at once. We will now determine the values of currents in a three-loop network.

WORKED EXAMPLE 2.16

For the network shown in Fig. 2.45, determine (i) the magnitude of the current in each of the resistors and (ii) the potential differences between points X and Y and between Z and Y.

Mark three loop currents I_1, I_2 and I_3.

Indicate the high potential end of each of the resistors due to these currents and add the potential arrows. These quantities are added to Fig. 2.45 to form Fig. 2.46.

Again, considering clockwise potential arrows to be positive quantities and anti-clockwise ones to be negative the equations are drawn up.

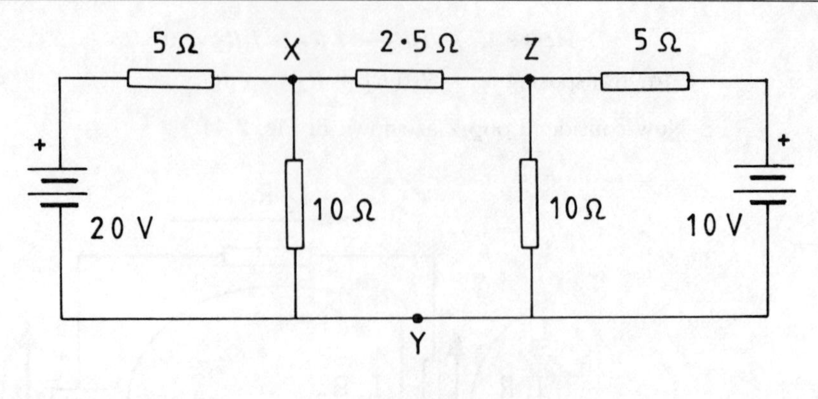

Fig. 2.45

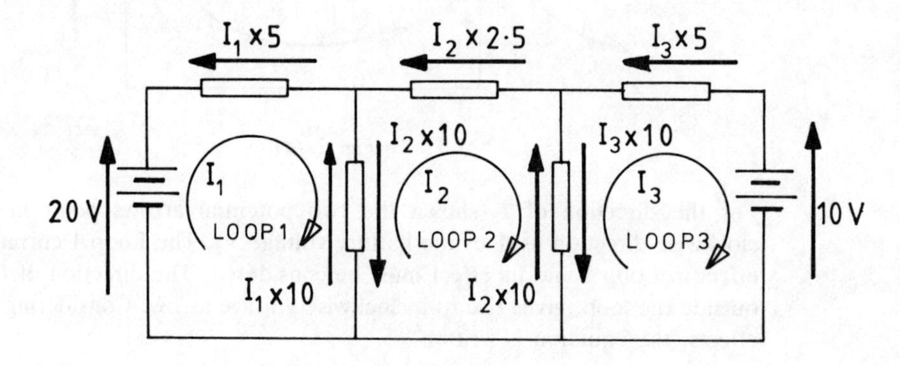

Fig. 2.46

LOOP 1

$$20 - 5I_1 - 10I_1 + 10I_2 = 0$$

$$20 - 15I_1 + 10I_2 = 0 \qquad \text{(i)}$$

LOOP 2

$$10I_1 - 10I_2 - 2.5I_2 - 10I_2 + 10I_3 = 0$$

$$10I_1 - 22.5I_2 + 10I_3 = 0 \qquad \text{(ii)}$$

LOOP 3

$$10I_2 - 10I_3 - 5I_3 - 10 = 0$$

$$10I_2 - 15I_3 - 10 = 0 \qquad \text{(iii)}$$

From equation (iii) $10I_2 - 10 = 15I_3$

$$I_3 = \frac{10I_2}{15} - \frac{10}{15}$$

Substitute in equation (ii)

$$10I_1 - 22.5I_2 + 10(10I_2/15 - 10/15) = 0$$

$$10I_1 - 15.833I_2 - 6.667 = 0 \qquad \text{(iv)}$$

Bring down equation (i) and rearrange: $-15I_1 + 10I_2 + 20 = 0$
To eliminate the I_1 terms multiply equation (iv) by 15/10 which gives

$$15I_1 - 23.75I_2 - 10 = 0 \qquad \text{(v)}$$

Equation (i) $\quad -15I_1 + 10I_2 + 20 = 0$
Add equations (v) and (i) $\quad 0 - 13.75I_2 - 10 = 0$

$$I_2 = 0.7272 \text{ A}$$

Substitute in equation (i) $\quad -15I_1 + 10 \times 0.7272 + 20 = 0$

$$I_1 = 1.8182 \text{ A}$$

Substitute in equation (iii) $\quad 10 \times 0.7272 - 15I_3 - 10 = 0$

$$I_3 = -0.1818 \text{ A}$$

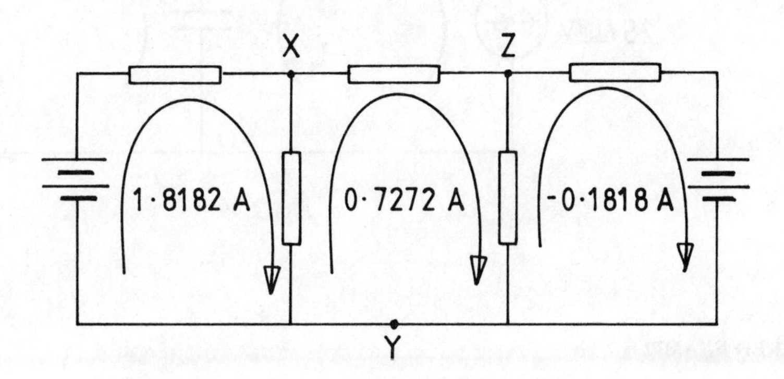

Fig. 2.47

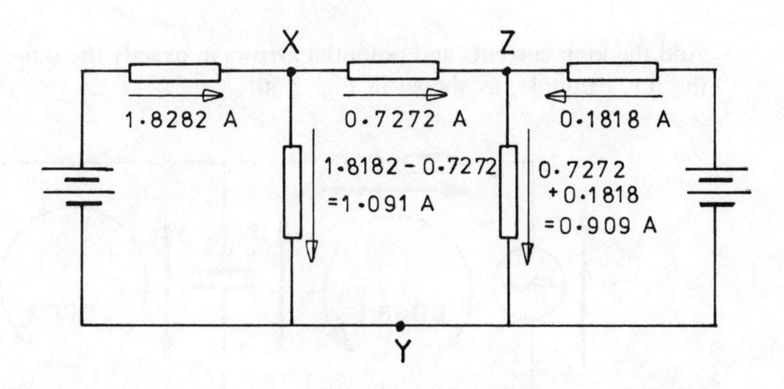

Fig. 2.48

Add the loop currents to Fig. 2.45 giving Fig. 2.47. A negative value for I_3 indicates that the original loop current direction was incorrect and needs to be reversed. The current values in each resistor are shown with its derivation in Fig. 2.48.

The potential difference XY = 1.091 × 10 = 10.91 V
The potential difference ZY = 0.909 × 10 = 9.09 V.

They should differ by the volt drop in the 2.5 Ω resistor.

This is 0.7272 × 2.5 = 1.82 V and 10.91 − 9.09 = 1.82 V. This is a useful check on the original current results. The reader might care to check that these potential differences may be obtained starting from the batteries. $V_{XY} = 20 - 1.8182 \times 5$ and similarly $V_{ZY} = 10 - 0.1818 \times 5$.

The same method may be applied to alternating current circuits with complex impedances.

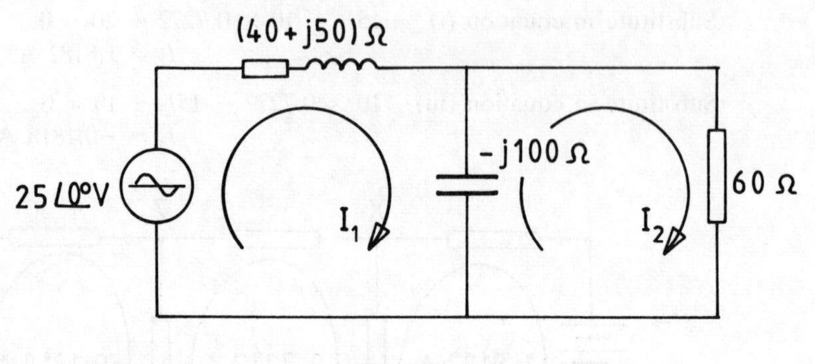

Fig. 2.49

WORKED EXAMPLE 2.17

For the circuit shown in Fig. 2.49, calculate the magnitude and phase of each branch current.

Add the loop currents and potential arrows in exactly the same manner as with the d.c. example, as shown in Fig. 2.50.

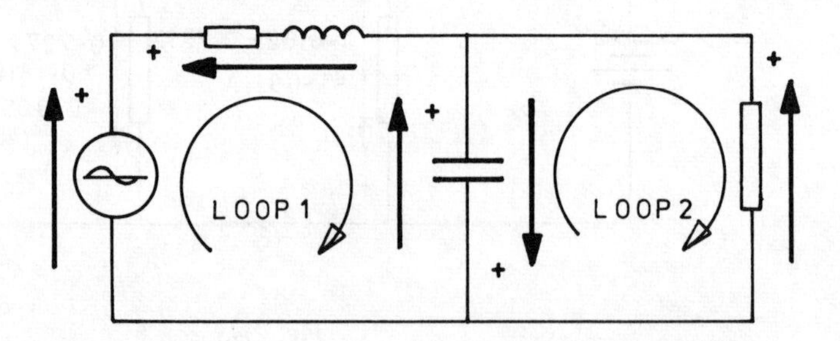

Fig. 2.50

Write the loop equations.

LOOP 1

$$25 - I_1(40 + j50) - (-j100I_1) + (-j100I_2) = 0$$

$$25 - 40I_1 - j50I_1 + j100I_1 - j100I_2 = 0$$

$$25 - I_1\,(40 - j50) - j100I_2 = 0 \qquad \text{(i)}$$

LOOP 2

$$-60I_2 - (-j100I_2) + (-j100I_1) = 0$$
$$-60I_2 + j100I_2 - j100I_1 = 0$$
$$-I_2\,(60 - j100) - j100I_1 = 0$$

$$I_2 = \frac{-j100I_1}{(60 - j100)} = \frac{100\angle-90^\circ\ I_1}{116.62\angle-59^\circ} = 0.8575\angle-31^\circ\ I_1 \qquad \text{(ii)}$$

Substitute the value for I_2 in equation (i)

$$25 - I_1\,(40 - j50) - j100 \times 0.8575\angle-31^\circ\ I_1 = 0$$
$$25 - 40I_1 + j50I_1 - 85.75\angle59^\circ\ I_1 = 0 \quad (j100 = 100\angle90^\circ)$$
$$25 - 40I_1 + j50I_1 - 44.16I_1 - j73.5I_1 = 0$$
$$25 - I_1\,(84.16 + j23.5) = 0$$

$$I_1 = \frac{25}{(84.16 + j23.5)} = \frac{25}{87.38\angle15.6^\circ} = 0.286\angle-15.6^\circ\ \text{A}$$

Now since $I_2 = 0.8575\angle-31^\circ\ I_1$

$$I_2 = 0.8575\angle-31^\circ \times 0.286\angle-15.6^\circ = 0.2452\angle-46.6^\circ\ \text{A}$$

The current in the centre limb $= (I_1 - I_2)$ (downwards)

$$= 0.286\angle-15.6^\circ - 0.2452\angle-46.6^\circ$$
$$= (0.2755 - j0.0769) - (0.1685 - j0.1782)$$
$$= 0.107 + j0.1013$$
$$= 0.1473\angle43.43^\circ\ \text{A}$$

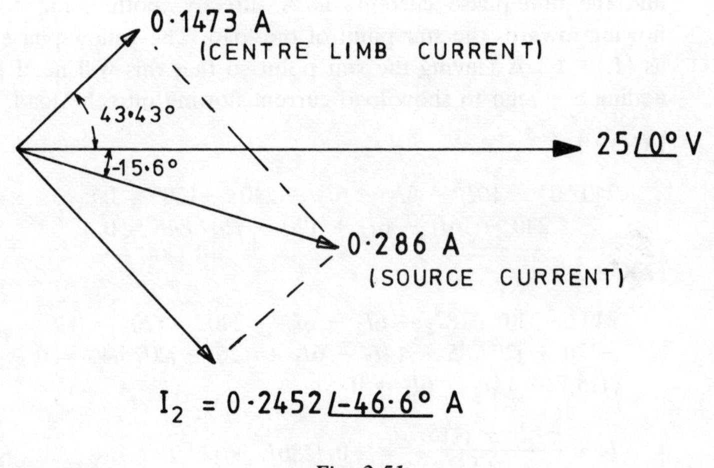

Fig. 2.51

Figure 2.51 shows the three currents with reference to the supply voltage.

WORKED EXAMPLE 2.18

A three-phase symmetrical system has voltages $V_{red} = 240\angle0^\circ$ V, $V_{yellow} = 240\angle-120^\circ$ V and $V_{blue} = 240\angle-240^\circ$ V. It supplies a star-connected asym-

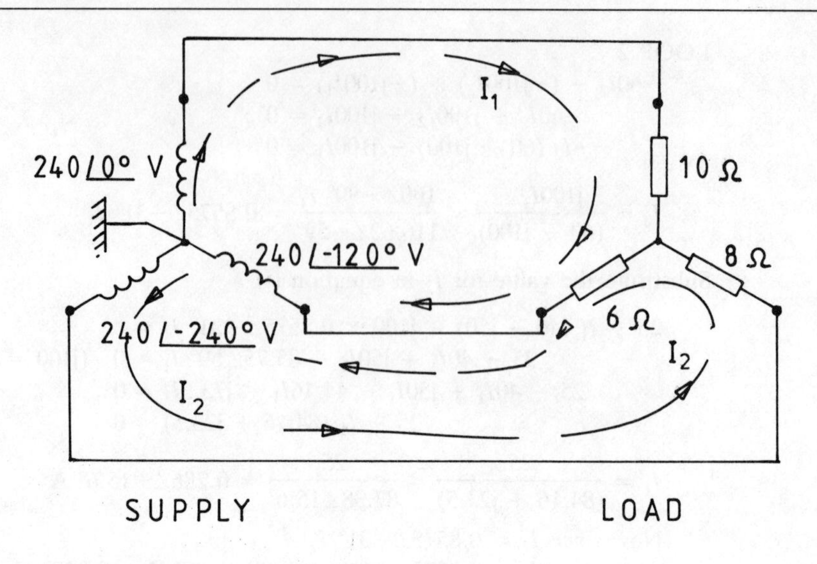

Fig. 2.52

metrical load as shown in Fig. 2.52. Determine the values of the currents in the three phases.

The two loop currents marked will suffice, giving the red-phase current, I_1 A and the blue-phase current, I_2 A directly, both being correctly shown as flowing towards the star point of the load. The yellow-phase current is shown as $(I_1 + I_2)$ A leaving the star point so that this will need to be reversed by adding a − sign to show load current flowing into the load.

LOOP 1

$$240\angle 0° - 10I_1 - 6I_1 - 6I_2 - 240\angle -120° = 0$$

$$240 - 16I_1 - 6I_2 + 120 + j207.846 = 0 \qquad \text{(i)}$$

LOOP 2

$$240\angle -240° - 8I_2 - 6I_2 - 6I_1 - 240\angle -120° = 0$$

$$-120 + j207.85 - 14I_2 - 6I_1 + 120 + j207.846 = 0$$

$$j415.7 - 14I_2 - 6I_1 = 0$$

$$I_2 = \frac{-6I_1 + j415.7}{14} = -0.4286I_1 + j29.7 \qquad \text{(ii)}$$

Substitute this value for I_2 in equation (i).

$$240 - 16I_1 + 2.5716I_1 - j178.2 + 120 + j207.846 = 0$$

$$360 + j29.646 - 13.43I_1 = 0$$

$$I_1 = \frac{360 + j29.646}{13.43} = 26.8 + j2.207 = 26.89\angle 4.71° \text{ A}$$

Substitute this value for I_1 in equation (ii).

$$
\begin{aligned}
I_2 &= -0.4286\ (26.8 + j2.207) + j29.7 \\
&= -11.486 - j0.946 + j29.7 \\
&= 30.96\angle 111.775^\circ \text{ A}
\end{aligned}
$$

$$
\begin{aligned}
I_3 &= -(I_1 + I_2) \\
&= -(26.8 + j2.207 - 11.486 + j28.754) \\
&= -(15.314 + j30.961) \\
&= -34.54\angle 63.68^\circ \\
&= 34.54\angle -116.33^\circ \text{ A}
\end{aligned}
$$

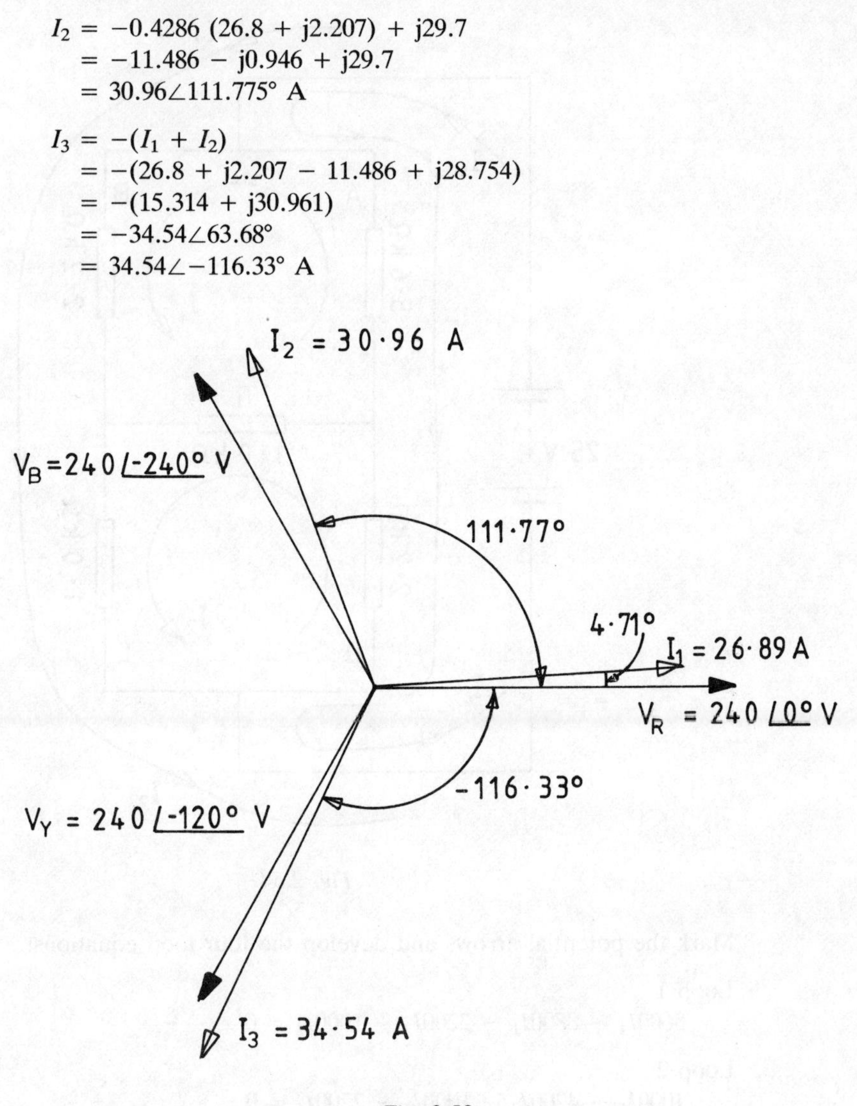

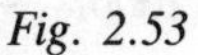

Fig. 2.53

SELF-ASSESSMENT PROBLEM 2.19

Figure 2.54 shows a Wheatstone bridge. Use loop analysis to determine the values of the currents in each section of the network.

There are six currents to determine; the battery current and the currents in five different resistors. This can be accomplished using four loops as shown. The battery current is $(I_3 + I_4)$ and the current in the 1 kΩ galvanometer arm G is $(I_1 - I_2)$.

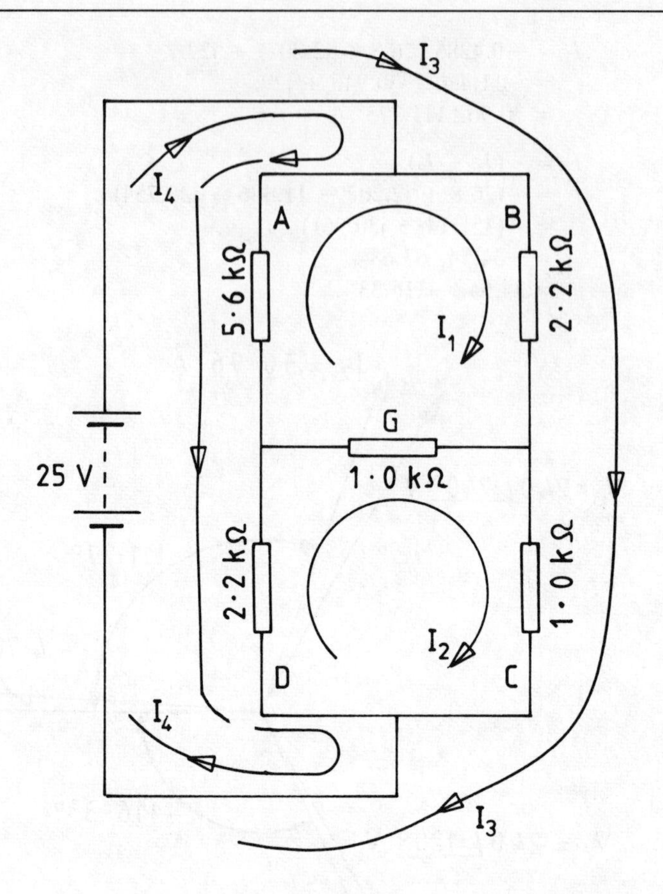

Fig. 2.54

Mark the potential arrows and develop the four loop equations:

Loop 1

$$5600I_4 - 8800I_1 - 2200I_3 + 1000I_2 = 0$$

Loop 2

$$1000I_1 - 4200I_2 - 1000I_3 + 2200I_4 = 0$$

Loop 3

$$25 - 3200I_3 - 2200I_1 - 1000I_2 = 0$$

Loop 4

$$25 - 7800I_4 + 5600I_1 + 2200I_2 = 0$$

2.6 Node analysis

To commence nodal analysis it is necessary to choose one point or *node* in the circuit the potential of which is taken as a reference. This point is called N_0. All other points or nodes in the circuit which are not at this potential, i.e. all points in the circuit which have a source of e.m.f. or an impedance between

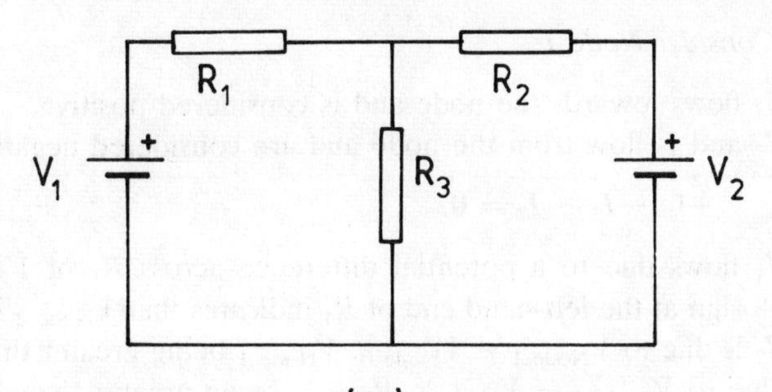

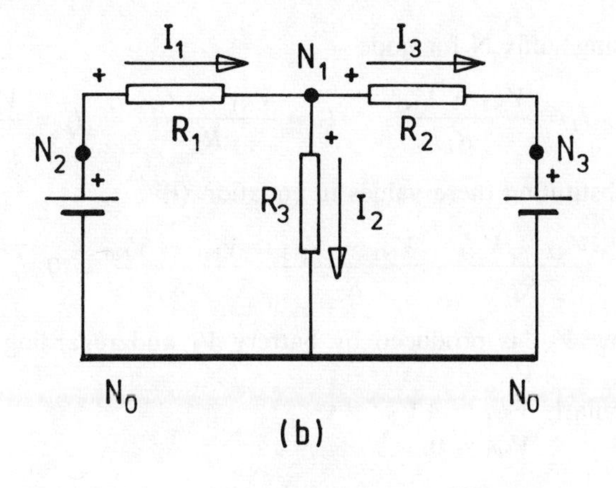

Fig. 2.55

them and the reference point are labelled N_1, N_2, etc. in sequence. In Fig. 2.55(a) a suitable choice for N_0 is the bottom line since both batteries have their negative terminals connected to it. There are three points in the circuit which have potential differences between them and the bottom line. N_0, N_1, N_2 and N_3 are indicated in Fig. 2.55(b).

The next step is to mark currents entering and leaving the nodes using suffix numbers in sequence. The directions chosen are completely at random. Should the directions prove to be incorrect the numerical solution will contain a negative sign indicating that the current direction must be reversed. The current from Node 2 is shown flowing towards Node 1 and this is described I_1. Current from Node 1 to Node 3 is described I_3 and that from Node 1 to Node 0 I_2. Current I_3 could well have been indicated in the opposite direction, the final result will depend on the relative sizes of the two battery e.m.f.s.

Currents flowing towards a node are considered to be positive and those away from a node, negative. The sum of the currents at any node must be zero; currents entering being balanced by currents leaving since a point on a circuit cannot store charge. Mark the high-potential ends of the resistors as in the loop-analysis method. The high-potential end is the one at which the current enters each resistor.

Consider Node 1

I_1 flows towards the node and is considered positive.
I_3 and I_2 flow from the node and are considered negative.

$$+I_1 - I_3 - I_2 = 0. \qquad \text{(i)}$$

I_1 flows due to a potential difference across R_1 of $V_{\text{Node 1}} - V_{\text{Node 2}}$. The + sign at the left-hand end of R_1 indicates that $V_{\text{Node 2}}$ is greater than $V_{\text{Node 1}}$. I_2 is due to $V_{\text{Node 1}} - V_{\text{Node 0}}$, $V_{\text{Node 1}}$ being greater than $V_{\text{Node 0}}$. Again I_3 is due to $V_{\text{Node 1}} - V_{\text{Node 3}}$, $V_{\text{Node 1}}$ being greater than $V_{\text{Node 3}}$ for the current direction chosen, the + being at the Node 1 end of R_2.

Using suffix N for node

$$I_1 = \frac{V_{N2} - V_{N1}}{R_1}, \quad I_3 = \frac{V_{N1} - V_{N3}}{R_2}, \quad I_2 = \frac{V_{N1} - V_{N0}}{R_3}.$$

Substituting these values in equation (i)

$$\frac{V_{N2} - V_{N1}}{R_1} - \frac{V_{N1} - V_{N3}}{R_2} - \frac{V_{N1} - V_{N0}}{R_3} = 0.$$

Now V_{N2} is produced by battery V_1 and regarding V_{N0} as zero or datum, $V_{N2} = +V_1$.
Similarly $V_{N3} = +V_2$
$V_{N0} = 0$.

$$\therefore \quad \frac{V_1 - V_{N1}}{R_1} - \frac{V_{N1} - V_2}{R_2} - \frac{V_{N1} - 0}{R_3} = 0$$

$$\frac{V_1}{R_1} - \frac{V_{N1}}{R_1} - \frac{V_{N1}}{R_2} + \frac{V_2}{R_2} - \frac{V_{N1}}{R_3} = 0$$

V_1 and V_2 will be known so that the only unknown is V_{N1}. The equation may be solved to determine its value and hence the values of the three currents. More complicated circuits will contain more than one unknown and these are determined by the use of simultaneous equations.

WORKED EXAMPLE 2.20

For the circuit shown in Fig. 2.56 determine the values of the currents in each resistor using node analysis.

Consider Node 1

$$+I_2 - I_3 - I_1 = 0 \quad \therefore \quad \frac{V_{N2} - V_{N1}}{5} - \frac{V_{N1} - V_{N0}}{50} - \frac{V_{N1} - V_{N3}}{20} = 0$$

Since $V_{N0} = 0$, separating the terms yields

$$\frac{V_{N2}}{5} - \frac{V_{N1}}{5} - \frac{V_{N1}}{50} - \frac{V_{N1}}{20} + \frac{V_{N3}}{20} = 0$$

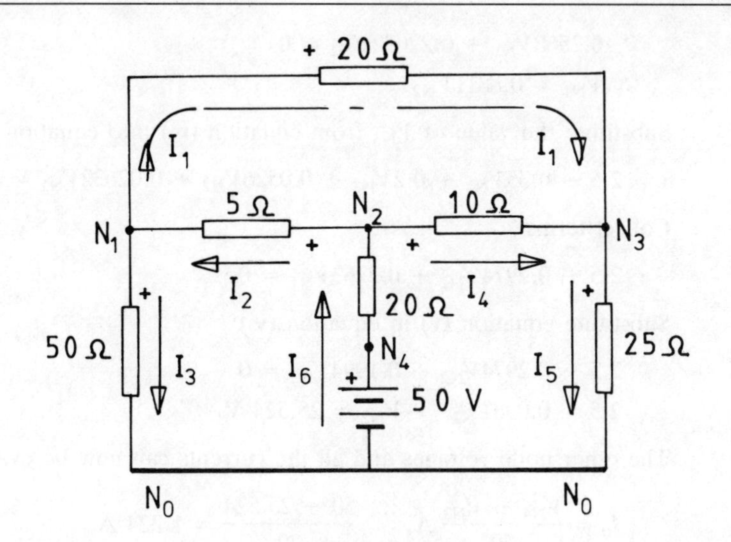

Fig. 2.56

$$\text{or } -V_{N1}\left(\frac{1}{5}+\frac{1}{50}+\frac{1}{20}\right)+\frac{V_{N2}}{5}+\frac{V_{N3}}{20}=0$$

$$-0.27V_{N1}+\frac{V_{N2}}{5}+\frac{V_{N3}}{20}=0 \qquad \text{(i)}$$

Consider Node 2

$$I_6 - I_2 - I_4 = 0 \quad \therefore \quad \frac{50 - V_{N2}}{20} - \frac{V_{N2} - V_{N1}}{5} - \frac{V_{N2} - V_{N3}}{10} = 0$$

$$2.5 - \frac{V_{N2}}{20} - \frac{V_{N2}}{5} + \frac{V_{N1}}{5} - \frac{V_{N2}}{10} + \frac{V_{N3}}{10} = 0$$

$$2.5 - 0.35V_{N2} + \frac{V_{N1}}{5} + \frac{V_{N3}}{10} = 0 \qquad \text{(ii)}$$

Consider Node 3

$$I_4 + I_1 - I_5 = 0 \quad \therefore \quad \frac{V_{N2} - V_{N3}}{10} + \frac{V_{N1} - V_{N3}}{20} - \frac{V_{N3} - V_{N0}}{25} = 0$$

$$\frac{V_{N2}}{10} - \frac{V_{N3}}{10} + \frac{V_{N1}}{20} - \frac{V_{N3}}{20} - \frac{V_{N3}}{25} = 0$$

$$-0.19V_{N3} + \frac{V_{N2}}{10} + \frac{V_{N1}}{20} = 0 \qquad \text{(iii)}$$

From equation (iii) $\quad V_{N3} = 0.526V_{N2} + 0.263V_{N1}$. (iv)

Substitute the value of V_{N3} from equation (iv) into equation (i).

$$-0.27V_{N1} + 0.2V_{N2} + 0.02632V_{N2} + 0.01316V_{N1} = 0$$

$$-0.2568V_{N1} + 0.22632V_{N2} = 0$$

$$\therefore V_{N1} = 0.8811V_{N2}. \qquad \text{(v)}$$

Substitute the value of V_{N3} from equation (iv) into equation (ii)

$$2.5 - 0.35V_{N2} + 0.2V_{N1} + 0.0526V_{N2} + 0.02632V_{N1} = 0.$$

Collect terms

$$2.5 - 0.2974V_{N2} + 0.2263V_{N1} = 0. \qquad \text{(vi)}$$

Substitute equation (v) in equation (vi)

$$2.5 - 0.2974V_{N2} + 0.1994V_{N2} = 0$$

$$2.5 = 0.098V_{N2} \qquad V_{N2} = 25.524 \text{ V}.$$

The other node voltages and all the currents can now be evaluated.

$$I_6 = \frac{V_{N4} - V_{N2}}{20} \text{ A}, \qquad \frac{50 - 25.524}{20} = 1.224 \text{ A}$$

$$V_{N1} = 0.8811V_{N2} \qquad \text{(equation (v))} \qquad \therefore V_{N1} = 22.49 \text{ V}$$

From equation (iv) $V_{N3} = 0.526V_{N2} + 0.263V_{N1} = 19.352$ V

$$I_4 = \frac{25.524 - 19.352}{10} = 0.617 \text{ A}$$

Similar calculations yield the values shown in Fig. 2.57.

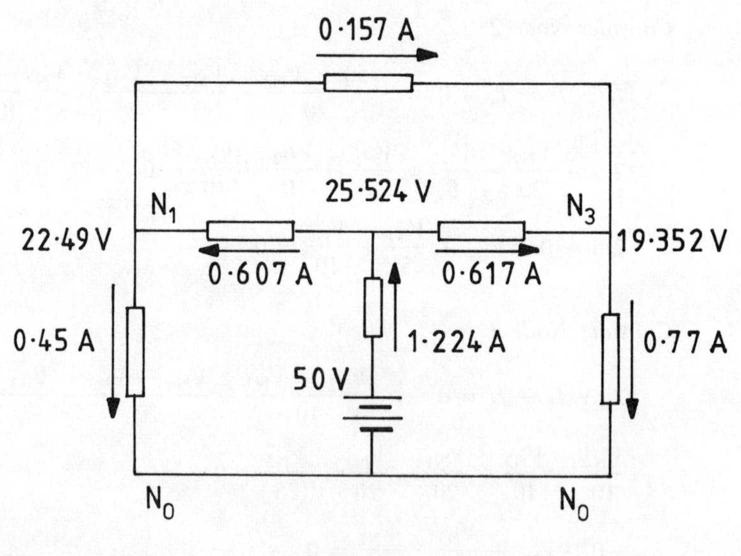

Fig. 2.57

SELF-ASSESSMENT PROBLEM 2.21

Using the nodes and currents marked in Fig. 2.58, determine the node voltages and the branch currents in the bridge network. (This is the same bridge as used

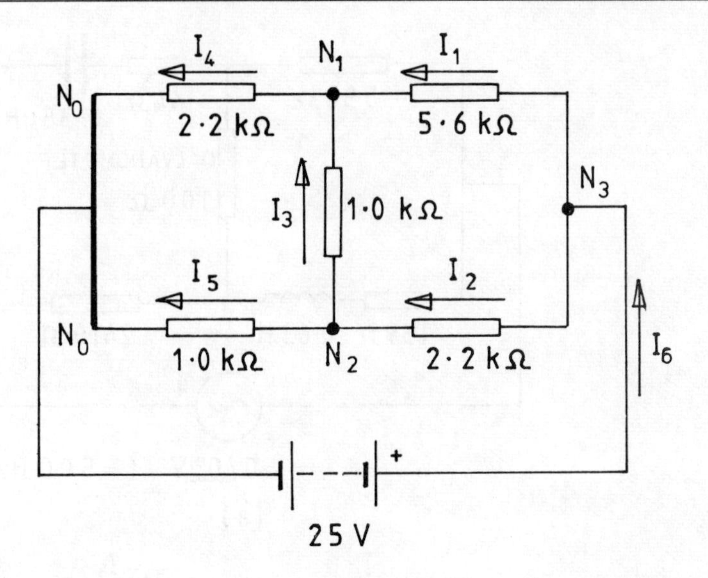

Fig. 2.58

in self-assessment problem 2.19, and the reader might like to compare the amount of work involved in the two methods. Many networks may be solved in a number of different ways, some involving little mathematical manipulation, whereas in others the route to the solution is much longer.)

Node analysis may be applied to circuits containing complex impedances which is now demonstrated using a Hay bridge at balance.

WORKED EXAMPLE 2.22

Figure 2.59(a) shows a Hay bridge used for the measurement of the resistance and inductance of a coil. Using the method of node analysis, establish that the bridge is balanced in the condition shown, i.e. that the current in the galvanometer is zero.

The capacitive reactance of the capacitor and the inductive reactance of the coil are determined at 500 Hz and the circuit redrawn as in Fig. 2.59(b). The nodes and currents are added.

Consider Node 1

$$I_1 + I_3 - I_4 = 0$$

$$0 = \frac{V_{N3} - V_{N1}}{(64 - j909.5)} + \frac{V_{N2} - V_{N1}}{100} - \frac{V_{N1} - V_{N0}}{750}$$

$$0 = \frac{50\angle 0^\circ}{(64 - j909.5)} - \frac{V_{N1}}{(64 - j909.5)} + \frac{V_{N2}}{100} - \frac{V_{N1}}{100} - \frac{V_{N1}}{750}$$

$$= \frac{50\angle 0^\circ}{911.75\angle -85.9^\circ} - V_{N1}\left(\frac{1}{911.75\angle -85.9^\circ} + \frac{1}{100} + \frac{1}{750}\right) + \frac{V_{N2}}{100}$$

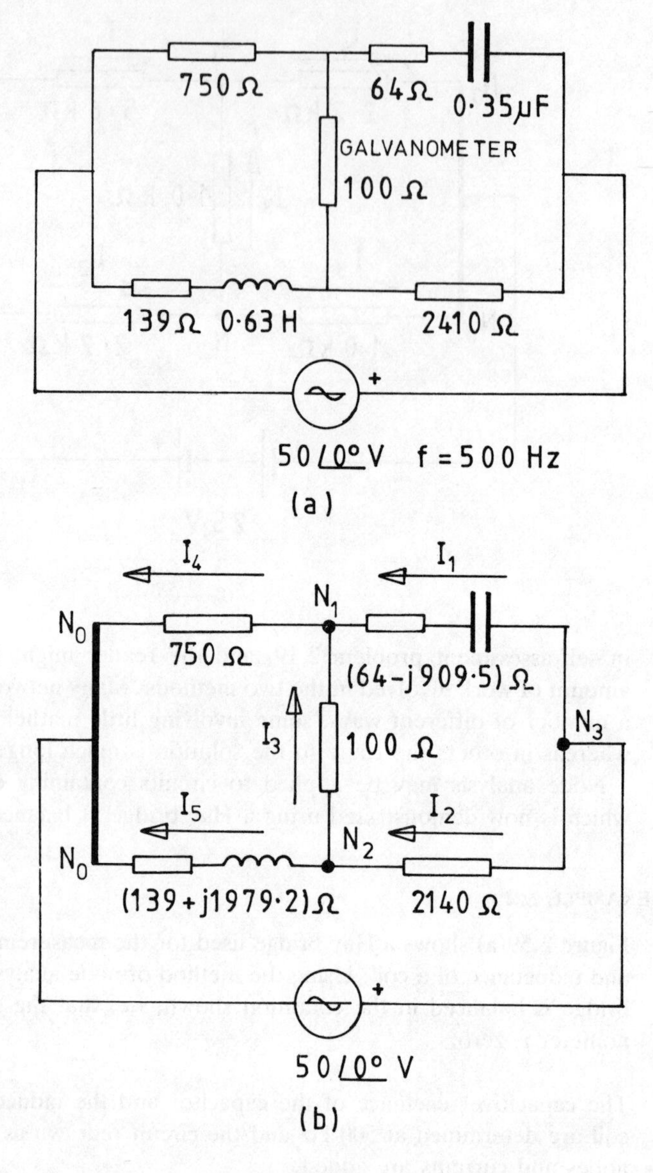

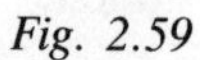

Fig. 2.59

$$= 0.0548\angle 85.9° - V_{N1}\,(1.0968 \times 10^{-3}\angle 85.9° + 0.01 + 1.33 \times 10^{-3}) + 0.01V_{N2}$$

$$= 0.0548\angle 85.9° - V_{N1}\,(0.0114 + j1.094 \times 10^{-3}) + 0.01V_{N2}$$

$$V_{N1} = \frac{0.01V_{N2} + 0.0548\angle 85.9°}{0.0115\angle 5.48°}$$

$$= 0.872\angle -5.476°\ V_{N2} + 4.784\angle 80.42°. \qquad \text{(i)}$$

Consider Node 2

$$I_2 - I_3 - I_5 = 0$$

$$0 = \frac{V_{N3} - V_{N2}}{2410} - \frac{V_{N2} - V_{N1}}{100} - \frac{V_{N2} - V_{N0}}{(139 + j1792.2)}$$

$$0 = \frac{50}{2410} - \frac{V_{N2}}{2410} - \frac{V_{N2}}{100} + \frac{V_{N1}}{100} - \frac{V_{N2}}{(139 + j1979.2)}. \qquad \text{(ii)}$$

Substitute the value for V_{N1} from equations (i) and (ii)

$$0 = 0.0207 - 414.94 \times 10^{-6} V_{N2} - 0.01 V_{N2} - \frac{V_{N2}}{(139 + j1979.2)}$$

$$+ \frac{(0.872\angle{-5.476°}\ V_{N2} + 4.784\angle 80.42°)}{100}$$

$$0 = 0.0207 - 0.0104 V_{N2} - 35.31 \times 10^{-6} V_{N2} + j502.77 \times 10^{-6} V_{N2}$$

$$+ 8.683 \times 10^{-3} V_{N2} - j832.4 \times 10^{-6} V_{N2} + 7.958 \times 10^{-3} + j0.04717$$

$$0 = 0.0287 + j0.04717 + V_{N2}(-1.767 \times 10^{-3} - j329.6 \times 10^{-6})$$

$$V_{N2} = \frac{0.0552\angle 58.685°}{1.798 \times 10^{-3}\angle 10.567°} = 30.72\angle 48.1° \text{ V}.$$

Using equation (i)

$$V_{N1} = 0.872\angle{-5.476°} \times 30.72\angle 48.1° + 4.784\angle 80.42°$$

$$= 19.7125 + j18.15 + 0.7958 + j4.717$$

$$= 30.72\angle 48.1° \text{ V}$$

$$I_3 = \frac{V_{N2} - V_{N1}}{100} = \frac{30.72\angle 48.1° - 30.72\angle 48.1°}{100} = 0.$$

It has been established that the bridge is balanced. The currents in each arm of the bridge can be evaluated if required.

$$I_2 = \frac{50 - 30.72\angle 48.1°}{2410} \text{ A}, \qquad I_1 = \frac{50 - 30.72\angle 48.1°}{(64 - j909.5)} \text{ A}.$$

2.7 The maximum power transfer theorem

When considering equipment supplied with power from the mains it is never necessary to consider the conditions for maximum power transfer. The power supply is treated as being infinitely great compared with the demand being made, and an impedance to suit the user rather than to suit the supply network is connected.

However when a small source of power is involved, such as the output from an audio-amplifier or a telephone system, it is often important that an impedance of the correct magnitude, and where possible with the correct phase angle, be used in order that the maximum use of the available power is made by the load.

The conditions for maximum power transfer into an impedance with fixed phase angle will now be considered, followed by a discussion of the effects of variable phase angle.

Consider the source shown in Fig. 2.60 with internal impedance $(R_I + jX_I)$ Ω supplying a load with impedance $(R_L + jX_L)$ Ω, which may be varied in magnitude while maintaining the phase angle constant. This would be the case if the load comprised a tapped coil, a variable wire-wound rheostat or a fixed impedance being fed through a variable-ratio transformer.

From the impedance triangle for the load

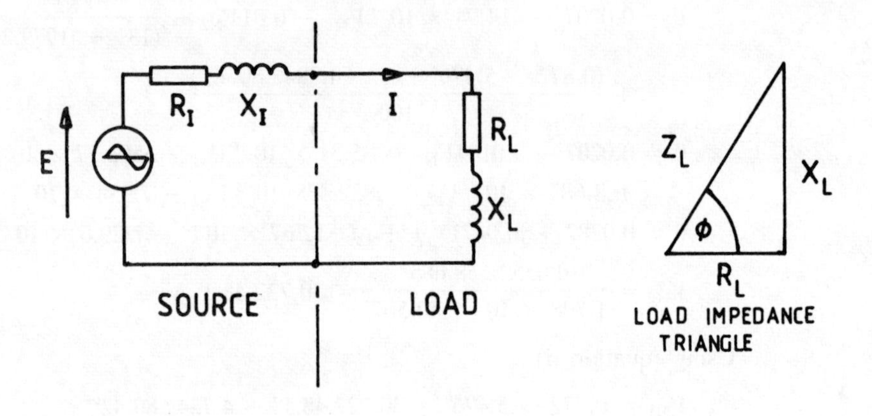

Fig. 2.60

$|Z_L| = \sqrt{(R_L^2 + X_L^2)}, \qquad R_L = Z_L \cos\phi, \qquad X_L = Z_L \sin\phi.$

The total circuit resistance $= R_I + R_L = R_I + Z_L \cos\phi$.

The total circuit reactance $= X_I + X_L = X_I + Z_L \sin\phi$.

Hence, total circuit impedance $= \sqrt{[(R_I + Z_L \cos\phi)^2 + (X_I + Z_L \sin\phi)^2]}$. (2.1)

Load power $= I^2R$ and $I = E/Z$

Therefore,

$$\text{load power} = \left[\frac{E}{\sqrt{[(R_I + Z_L\cos\phi)^2 + (X_I + Z_L\sin\phi)^2]}}\right]^2 \times Z_L\cos\phi.$$

Squaring the denominator eliminates the root sign and multiplying out the brackets yields

$$\text{power} = \frac{E^2 Z_L \cos\phi}{R_I^2 + 2R_I Z_L\cos\phi + Z_L^2\cos^2\phi + X_I^2 + 2X_I Z_L\sin\phi + Z_L^2\sin^2\phi}.$$

Divide numerator and denominator by Z_L and collect $\cos^2\phi$ and $\sin^2\phi$.

$$\text{Power} = \frac{E^2\cos\phi}{R_I^2/Z_L + 2R_I\cos\phi + X_I/Z_L + 2X_I\sin\phi + Z_L(\cos^2\phi + \sin^2\phi)}$$

But $\cos^2\phi + \sin^2\phi = 1$, hence

$$\text{power} = \frac{E^2 \cos \phi}{R_I^2/Z_L + 2R_I \cos \phi + X_I^2/Z_L + 2X_I \sin \phi + Z_L} \text{ W.}$$

For the power to be a maximum, the denominator (= y, say) must be made into a minimum. $|Z_L|$ is the only variable; E, $\cos \phi$, R_I and X_I all being constant. Differentiating the denominator with respect to Z_L gives

$$\frac{dy}{dZ_L} = -\frac{R_I^2}{Z_L^2} - \frac{X_I^2}{Z_L^2} + 1$$

d^2y/dZ_L^2 is positive which indicates a minimum value.

Equating dy/dZ_L to zero gives

$$-\frac{R_I^2}{Z_L^2} - \frac{X_I^2}{Z_L^2} + 1 = 0.$$

Transposing gives

$$Z_L^2 = R_I^2 + X_I^2$$

$$|Z_L| = \sqrt{(R_I^2 + X_I^2)}.$$

Therefore, maximum power transfer is achieved when the modulus of the load impedance is equal to the modulus of the source impedance.

WORKED EXAMPLE 2.23

For the circuit shown in Fig. 2.61 calculate:
(a) the value of load impedance to give maximum power transfer, (b) the value of load resistance, (c) the power developed in the load, (d) the power loss in the source.

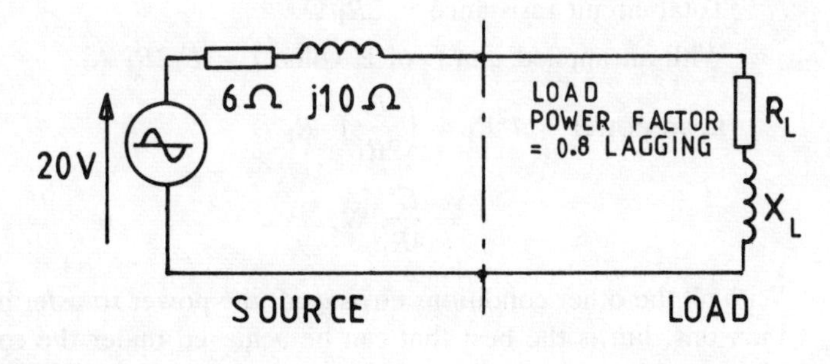

Fig. 2.61

(a) For maximum power transfer $|Z_L| = \sqrt{(R_I^2 + X_I^2)} = \sqrt{(6^2 + 10^2)}$
$= 11.66\ \Omega$.

For power factor 0.8, $\cos \phi = 0.8$, hence $\phi = 36.87°$. (Fig. 2.60.)

(b) $R_L = Z_L \cos \phi = 9.33\ \Omega$
$X_L = Z_L \sin \phi = 7\ \Omega$.

The total impedance of the circuit $= \sqrt{[(R_I + R_L)^2 + (X_I + X_L)^2]}$ (equation 2.1)

$= \sqrt{[(6 + 9.33)^2 + (10 + 7)^2]}$

$= 22.9\ \Omega.$

Therefore the circuit current $I = V/Z = 20/22.9 = 0.873$ A.

(c) Power in the load $= I^2R_L = 0.873^2 \times 9.33 = 7.12$ W.
(d) Power loss in the source $= I^2R_T = 0.873^2 \times 6 = 4.57$ W.

Three other possibilities will now be considered:
(a) Where the load is purely resistive.

Maximum power transfer will occur when $R_L = |Z_I|$.

(b) Where the load consists of a variable resistance in series with a fixed inductive reactance $X_F\ \Omega$, maximum power transfer will occur when

$R_L = \sqrt{[(R_I)^2 + (X_I + X_F)^2]}$.

The load resistance has to be made equal to the modulus of the impedance of the rest of the circuit. It could be envisaged that a resistive load is being fed from an inductive source through a transmission line or network which is also inductive. The load resistance would then be made equal to the modulus of the impedance of the rest of the system complete.

(c) Where a completely free choice may be made, the maximum possible power transferred into the load is obtained when the load impedance is the complex conjugate of the impedance of the source. For example, if the source has impedance $(R_I + jX_I)\ \Omega$, then maximum power transfer occurs when the load impedance is $(R_I - jX_I)\ \Omega$.

Total circuit resistance $= 2R_I\ \Omega$.
With an applied e.m.f. of E volts, $I = E/2R_I$ A.

$$\text{Load power} = I^2R_I = \left(\frac{E}{2R_I}\right)^2 R_I$$

$$= \frac{E^2}{4R_I}\ \text{W.} \qquad (2.2)$$

With all the other conditions envisaged, the power transfer into the load is less than this, but is the best that can be achieved under the constraints detailed. However, there is a price to pay since such a circuit will be selective, i.e. it exhibits resonance which implies that this maximum power transfer occurs at one particular frequency only. Where a reasonable bandwidth is required, such as in audio work, this method cannot be used. It could be employed in circuits required to perform a single function, for example, closing a relay.

WORKED EXAMPLE 2.24

(a) A 5 V, 50 Hz source has an internal impedance which is equivalent to a 600 Ω resistance in series with a 3.183 μF capacitor. Determine the load components required for maximum power transfer when the load is (i) variable in reactance and resistance, (ii) a variable resistance only. (b) Determine the load power developed in (a) (i) and (ii) above.

(a) 3.183 μF at 50 Hz gives capacitive reactance

$$X_C = \frac{1}{2\pi \times 50 \times 3.183 \times 10^{-6}}$$
$$= 1000\ \Omega\ (-j1000\ \Omega).$$

(i) Variable reactance allows tuning of the circuit.
A coil is required with inductive reactance 1000 Ω (+j1000 Ω)

$$L = \frac{1000}{2\pi \times 50} = 3.183\ \text{H}.$$

Maximum power transfer occurs when the load consists of a 600 Ω resistor in series with a coil with inductance 3.183 H.
(ii) For resistance only

$$R_L = Z_I$$
$$= \sqrt{(600^2 + 1000^2)}$$
$$= 1166\ \Omega.$$

(b) (i) Power $= \dfrac{E^2}{4R_I}$ (equation 2.2)

$$= \frac{5^2}{4 \times 600} = 0.0104\ \text{W}.$$

(ii) Circuit impedance $= \sqrt{[(1166 + 600)^2 + 1000^2]} = 2029\ \Omega.$

$$I = \frac{5}{2029}\ \text{A}.$$

$$\text{Load power} = I^2R_L = \left(\frac{5}{2029}\right)^2 1166 = 0.00708\ \text{W}.$$

SELF-ASSESSMENT EXAMPLE 2.25

A source with e.m.f. 10 V and internal impedance (3 + j4) Ω is connected to a load through a section of line with inductive reactance +j5 Ω and negligible resistance. (a) Calculate the value of load impedance to give maximum power transfer (i) if the load is purely resistive, (ii) if the load has a power factor of 0.9 lagging. (b) for (a) (i) and (ii), calculate the value of power developed in the load.

FURTHER PROBLEMS

2.26 With respect to the terminals AB, determine the equivalent (a) Thevenin voltage, (b) Norton current. With what internal resistance is each of these generators associated. (Fig. 2.62.)

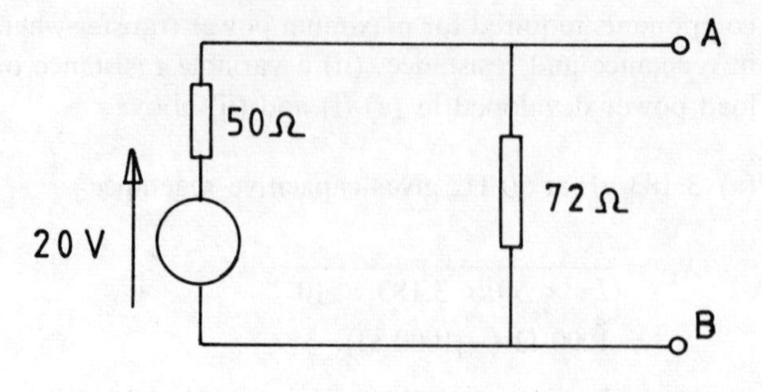

Fig. 2.62

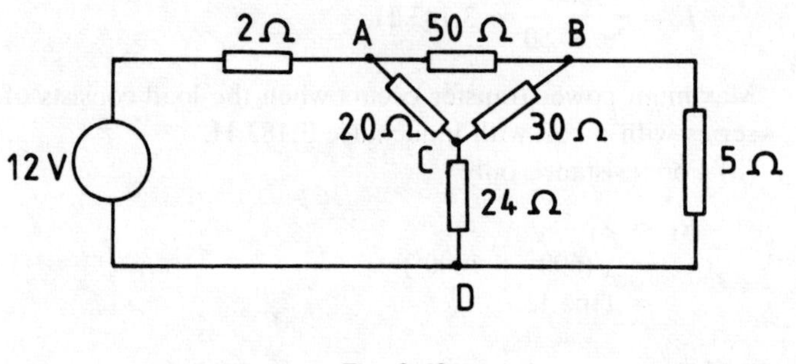

Fig. 2.63

2.27 Replace the delta ABC in Fig. 2.63 with an equivalent star. Hence determine the power dissipated in the 5 Ω resistor.

2.28 Using the delta/star transformation in delta XYZ, followed by an application of Thevenin's theorem, determine the value of a component to be connected to terminals AB which will cause maximum power to be transferred into it. (Fig. 2.64).

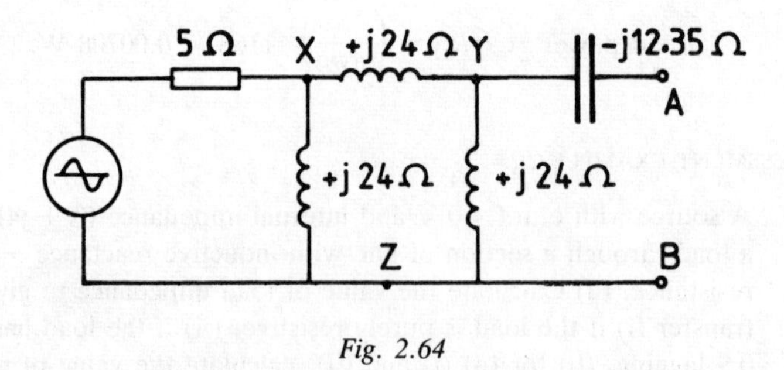

Fig. 2.64

2.29 Using the delta/star transformation and Thevenin's theorem determine the value of current in the 5 Ω resistor in Fig. 2.65.

2.30 Use Thevenin's theorem to determine the current in the 4.58 Ω resistor in Fig. 2.66.

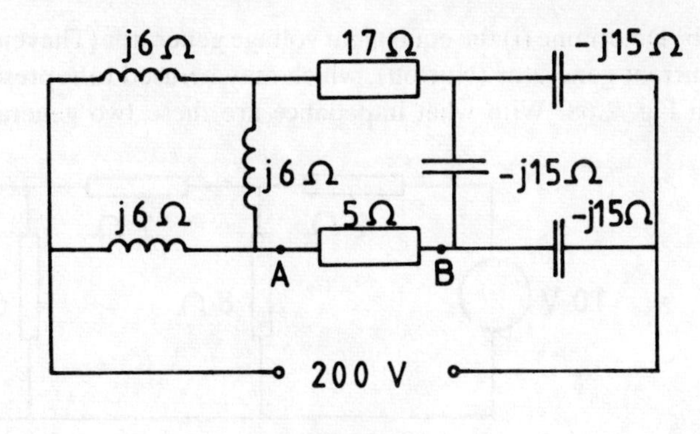

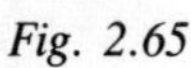

Fig. 2.65

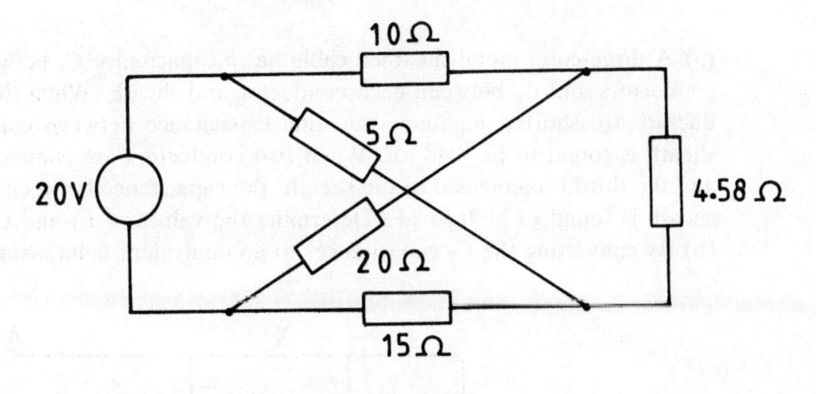

Fig. 2.66

2.31 Determine the value of current in the 3 Ω resistor when terminals X and Y in Fig. 2.67 are connected directly together.

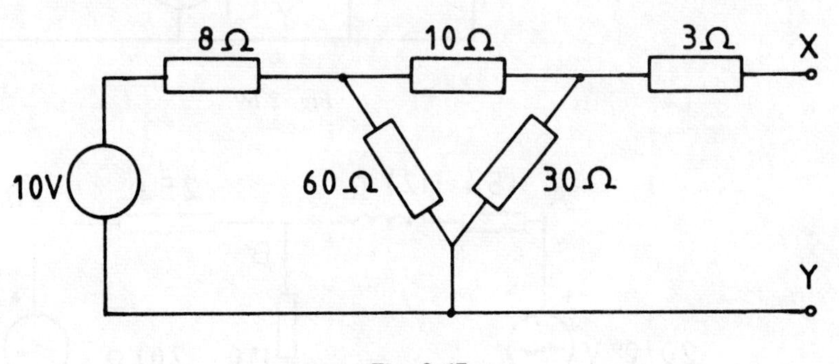

Fig. 2.67

2.32 (a) A load of (R + j20) Ω is fed from a generator with an e.m.f. of 100 V and internal impedance (25 + j50) Ω. (i) What value of R in the load would cause maximum power transfer to be accomplished? (ii) What is the value of the power in (i)?

(b) Determine (i) the equivalent voltage generator (Thevenin), (ii) the equivalent current generator (Norton), which may be used to represent the network given in Fig. 2.68. With what impedance are these two generators associated?

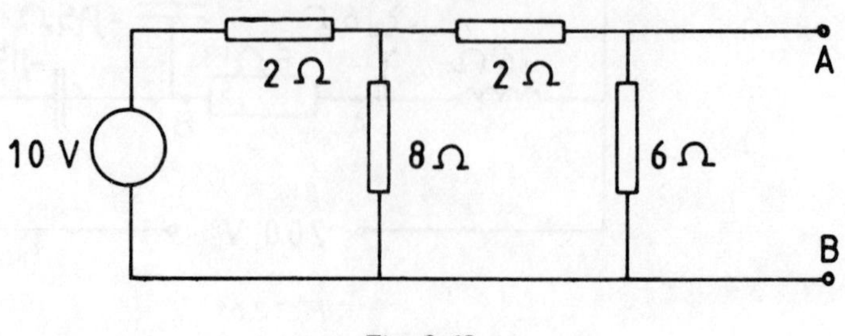

Fig. 2.68

2.33 (a) A three-core, metal-sheathed cable has a capacitance C_1 between any two conductors and C_2 between each conductor and sheath. When the three conductors are shorted together, the total capacitance between conductors and sheath is found to be 0.48 μF. When two conductors are connected together and the third is connected to the sheath, the capacitance between the pair and sheath is found to be 0.54 μF. Determine the values of C_1 and C_2.
(b) By converting the C_2 capacitances to an equivalent delta arrangement find

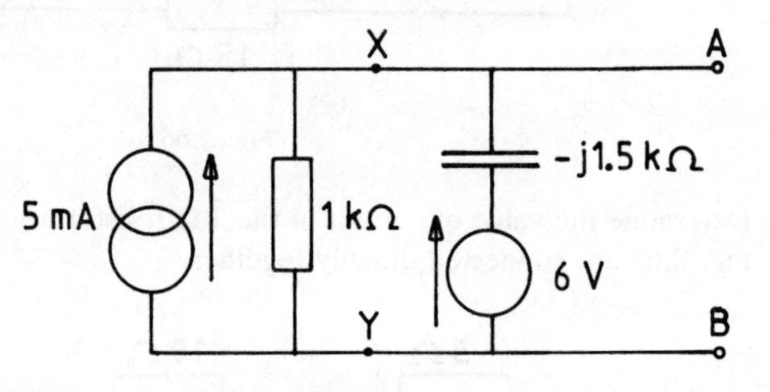

Fig. 2.69

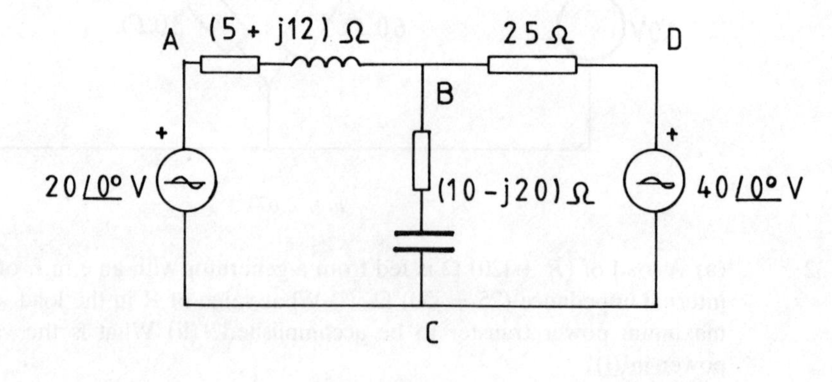

Fig. 2.70

the total capacitance between any two cores, and hence or otherwise determine the total effective capacitance between each core and sheath.

2.34 Convert the current source and its impedance to the left of XY in Fig. 2.69 into an equivalent voltage circuit. Hence determine the value of an inductive reactance with power factor 0.8 lagging which should be connected to terminals AB to get maximum power transfer into that impedance. What is the value of this power?

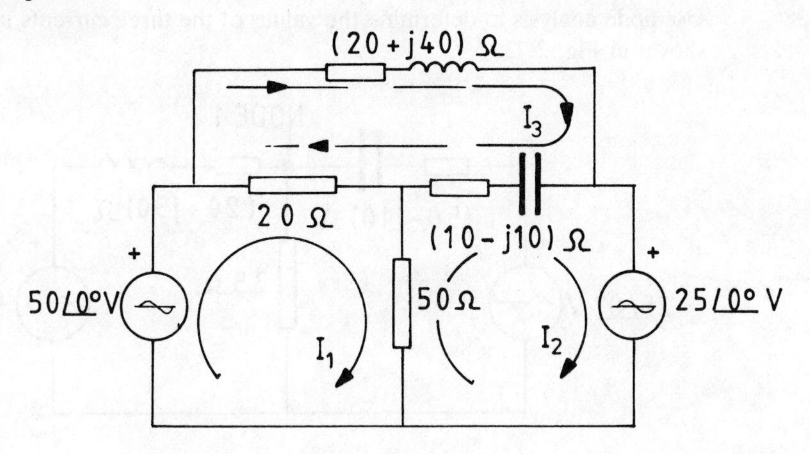

Fig. 2.71

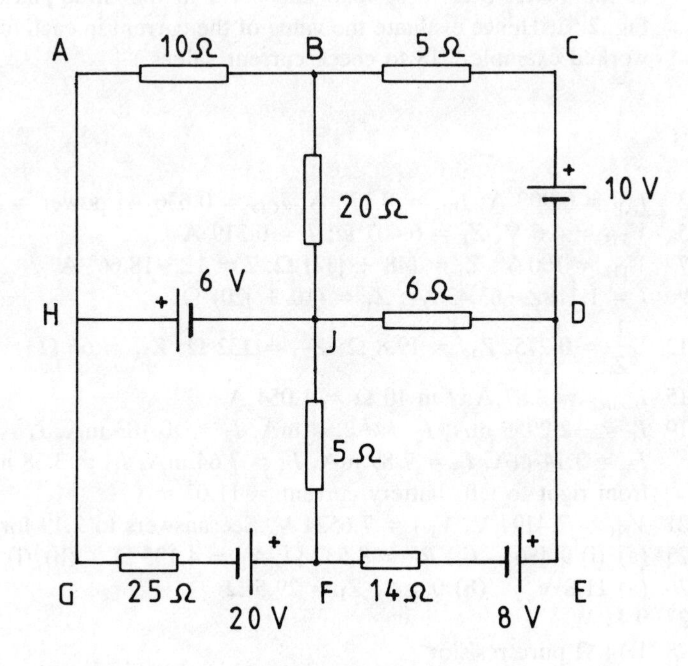

Fig. 2.72

2.35 For the network shown in Fig. 2.70, using loop analysis, determine each of the branch currents.

2.36 For the network shown in Fig. 2.71, using loop analysis, determine each of the branch currents.

2.37 For the network shown in Fig. 2.72, using loop analysis, determine each of the branch currents.

2.38 Use node analysis to determine the values of the three currents in the network shown in Fig. 2.73.

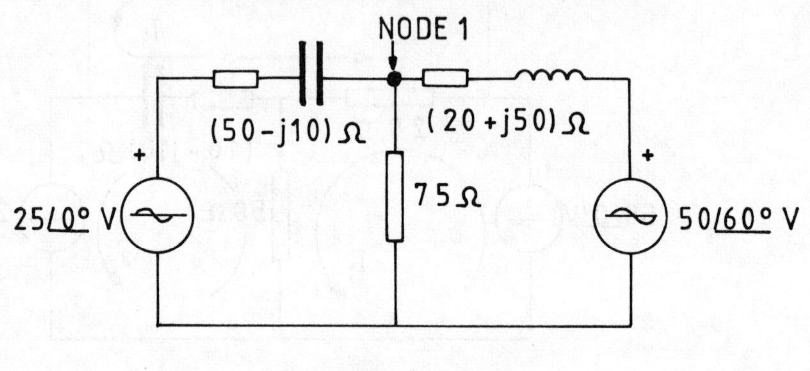

Fig. 2.73

2.39 Use node analysis to determine the potential difference between the star point of the 10 Ω, 8 Ω, 6 Ω load and earth in the three-phase system shown in Fig. 2.52. Hence evaluate the value of the current in each of the resistors. (See worked example 2.18 to check current values.)

Answers

2.2 I_{AF} = 0.908 A; I_{BE} = 0.272 A; I_{CD} = 0.636 A; power = 3.24 W
2.5 V_{TH} = 4.6 V; Z_I = 6.407 Ω; I = 0.319 A
2.7 V_{TH} = 100 V; Z_I = (48 + j32) Ω; $I = 1\angle -18.66°$ A
2.9 $I = 1.118\angle -63.43°$ A; Z_I = (10 + j20) Ω
2.12 $\Sigma\frac{1}{Z}$ = 0.275; Z_{12} = 19.8 Ω; Z_{23} = 132 Ω; Z_{31} = 66 Ω
2.15 I_{source} = 2.87 A; I in 10 Ω = 1.054 A
2.19 I_1 = −2.2959 mA, I_2 = −2.54 mA, I_3 = 10.183 mA, I_4 = 840.4 μA.
I_A = 3.14 mA, I_B = 7.89 mA, I_C = 7.64 mA, I_D = 3.38 mA, I_G = 0.24 mA from right to left; battery current = 11.02 mA
2.21 V_{N1} = 7.4194 V, V_{N2} = 7.6524 V. See answers to 2.19 for branch currents.
2.25 (a) (i) 9.486 (ii) R_L = 8.537 Ω; X_L = 4.135 Ω (b) (i) 4 W (ii) 2.79 W
2.26 (a) 11.8 V (b) 0.4 A; Z_I = 29.5 Ω
2.27 0.45 W
2.28 1.14 Ω pure resistor
2.29 V_{TH} (V_{AB}) = $191.5\angle 9.44°$ V; Z_I = (0.47 − j5.834) Ω; $I = 23.94\angle 56.3°$ A

2.30 0.111 A
2.31 0.405 A
2.32 (a) (i) 74.3 Ω (ii) 50.4 W (b) V_{TH} = 5 V; I_{Norton} = 2.22 A; Z_I = 2.25 Ω
2.33 (a) Joining three together results in $3C_2$. A common pair to sheath = $2C_2 + 2C_1$. C_2 = 0.16 μF; C_1 = 0.11 μF
(b) Between cores: 0.1633 μF. Core/sheath: 0.49 μF
2.34 R_L = 665.6 Ω. X_L = j499.2 Ω. Power = 0.01025 W
2.35 $I_{AB} = 0.71\angle 66.7°$ A, $I_{BD} = 0.6\angle 26°$ A, $I_{BC} = 1.22\angle 49.32°$ A
2.36 $I_1 = 1.29\angle -17.2°$ A, $I_2 = 0.71\angle -28.2°$ A, $I_3 = 0.559\angle -63.44°$ A
$I_{20\Omega} = 0.99\angle 6.65°$ A, $I_{(10 - j10)\Omega} = 0.41\angle 23.8°$ A; $I_{50\Omega} = 0.61\angle -4.1°$ A, $I_{(20 + j40)\Omega} = I_3$
2.37 I_{BA} = 0.174 A, I_{CB} = 0.56 A, I_{DJ} = 0.089 A, I_{FE} = 0.649 A, I_{FJ} = 0.324 A, I_{JH} = 0.8 A, I_{GF} = 0.97 A
2.38 Node 1 voltage = $33.6\angle +17.26°$. $I_{75\Omega} = 0.448\angle +17.26°$ A.
$I_{(50 - j10)\Omega} = 0.24\angle -114.1°$ A, $I_{(20 + j50)\Omega} = 0.633\angle 33.8°$ A both leaving their respective voltages sources
2.39 Star point voltage V_{N0} (say) = $35.74\angle -141.79°$ V

Chapter 3
Complex Waves

So far it has been assumed that alternating voltages and currents have been sinusoidal in form and that each could therefore be represented by a single phasor and the required calculations carried out using complex numbers.

The waveforms met in practice, however, often differ from the sinusoidal form, and even when this difference is relatively small ignoring it could result in quite large errors when calculating, for example, values of circuit current and power.

All such non-sinusoidal waveforms met in electrical engineering under steady-state conditions repeat regularly, in other words each positive half cycle is identical and each negative half cycle is identical. Positive and negative half cycles may however be different in shape. Such waves are called 'complex' and cannot be represented by a single phasor.

Two typical complex waves are shown in Fig. 3.1.

3.1 Synthesis of complex waves

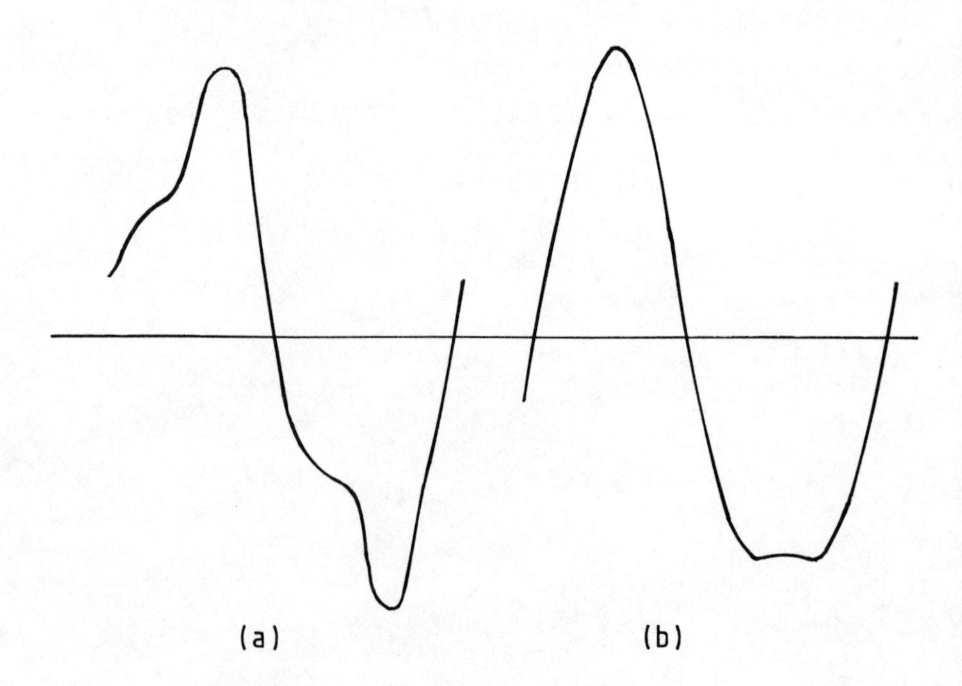

Fig. 3.1

All complex waves are made up of a number of pure sine waves of different frequencies and amplitudes. The first sine wave has a frequency equal to that of the complex wave and is known as the fundamental. The other sine waves have frequencies which are integral multiples of the fundamental, e.g. twice, four times, six times, etc., which are known as *even* harmonics or three times, five times, seven times, etc., which are known as *odd* harmonics.

A complex voltage waveform may be described by the equation

$$v = V_{m(1)} \sin(\omega t \pm \phi_1) + V_{m(2)} \sin(2\omega t \pm \phi_2) + V_{m(3)} \sin(3\omega t \pm \phi_3) + \ldots + V_{m(n)} \sin(n\omega t \pm \phi_n)$$

where $V_{m(1)}$, $V_{m(2)}$, ..., $V_{m(n)}$ are the maximum values of the various sine waves and ϕ_1, ϕ_2, ..., ϕ_n are their phase angles. The fundamental has angular velocity ω rad/s, the second harmonic, 2ω rad/s, and the nth harmonic, $n\omega$ rad/s.

Generally as the frequency of the harmonic increases, its amplitude decreases. Not all harmonics are present in a single complex wave, most being made up of a fundamental plus odd harmonics only or a fundamental plus even harmonics only. The shape of the complex waveform is affected by the amplitude of the harmonics present and their respective phase angles.

It is possible to resolve any complex wave into its fundamental and harmonics by the use of the Fourier analysis. This analysis yields both the magnitude and the phase of all components of the wave. However the number of different wave shapes met in electrical engineering is limited and here synthesis of waves (making complex waves by adding individual sine waves) will be considered rather than analysis. Practically, waveform analysis is often carried out using a waveform analyser which accepts the complex wave and gives a direct readout of all the waves present.

Two cases of a fundamental plus a third harmonic are now to be considered.

(a) $v = 20 \sin \omega t + 10 \sin 3\omega t$ V
(b) $v = 20 \sin \omega t + 10 \sin (3\omega t + 180°)$ V.

In both cases the fundamentals are the same size and the third harmonics are the same size. However in (b) the third harmonic has a phase change of 180°.

(a) $v = 20 \sin \omega t + 10 \sin 3\omega t$

In Fig. 3.2 the phasors for the fundamental and third harmonic are shown for $t = 0$.

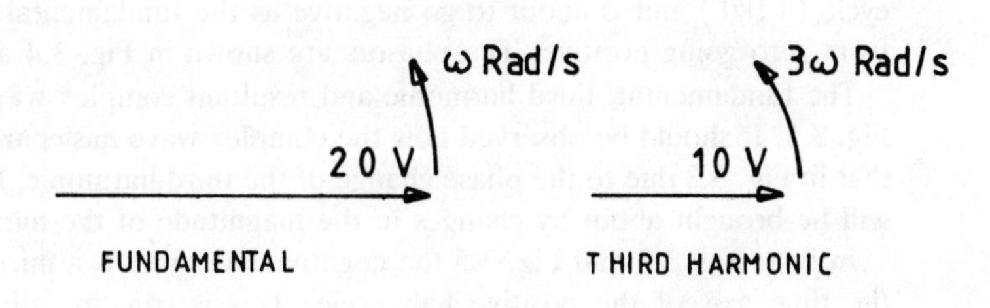

Fig. 3.2

At time $t = 0$, both waves start at zero and are travelling in a positive sense. For each complete cycle of the fundamental the third harmonic will complete three cycles. The waves are plotted to scale in Fig. 3.3. The fundamental and third harmonic are added together at regular intervals. For example, when the third harmonic reaches its first maximum of 10 V, the fundamental has reached 10 V also. Adding the two instantaneous values gives 20 V. Repeating the process at each zero and each maximum of the third harmonic, the complex wave can be plotted as shown in Fig. 3.3.

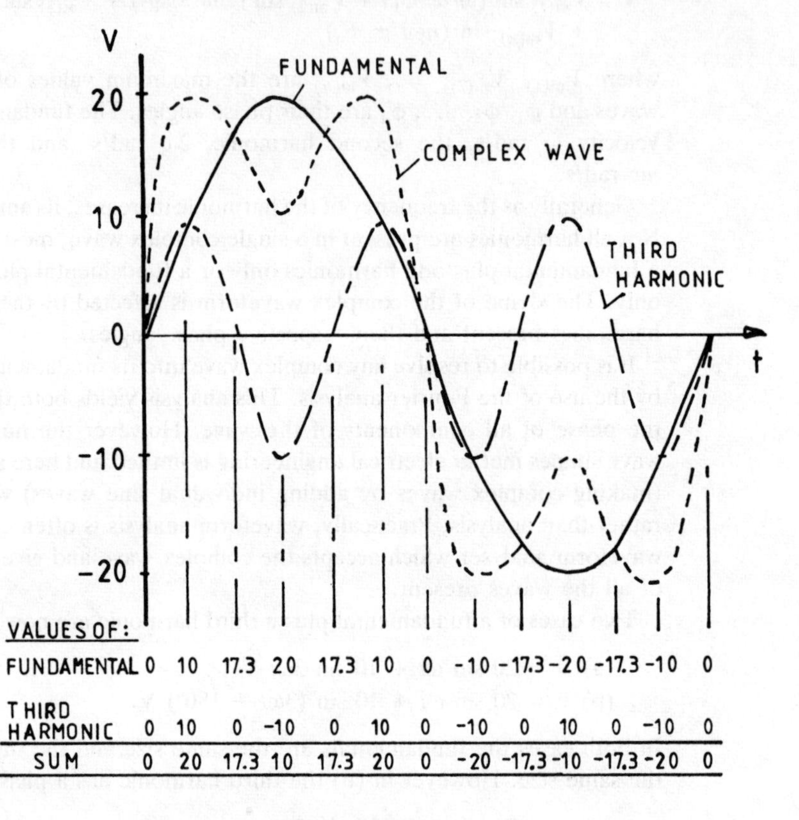

VALUES OF:													
FUNDAMENTAL	0	10	17.3	20	17.3	10	0	-10	-17.3	-20	-17.3	-10	0
THIRD HARMONIC	0	10	0	-10	0	10	0	-10	0	10	0	-10	0
SUM	0	20	17.3	10	17.3	20	0	-20	-17.3	-10	-17.3	-20	0

Fig. 3.3

(b) $v = 20 \sin \omega t + 10 \sin (3\omega t + 180°)$

In this case the third harmonic voltage has already completed one half of a cycle (+180°) and is about to go negative as the fundamental is about to rise from zero going positive. The phasors are shown in Fig. 3.4 at $t = 0$.

The fundamental, third harmonic and resultant complex wave are shown in Fig. 3.5. It should be observed how the complex wave has changed shape from that in Fig. 3.3 due to the phase change of the third harmonic. Further changes will be brought about by changes in the magnitude of the harmonic.

In both Fig. 3.3 and Fig. 3.5 the negative half cycle is a mirror image about the time axis of the positive half cycle. This is true for all complex waves consisting of a fundamental plus odd harmonics only. Looking back to Fig.

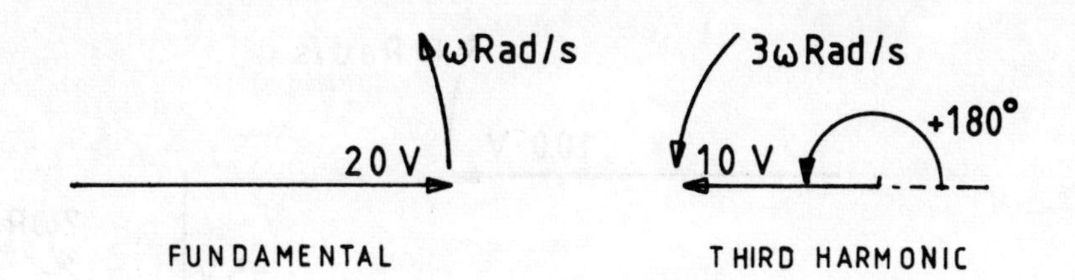

Fig. 3.4

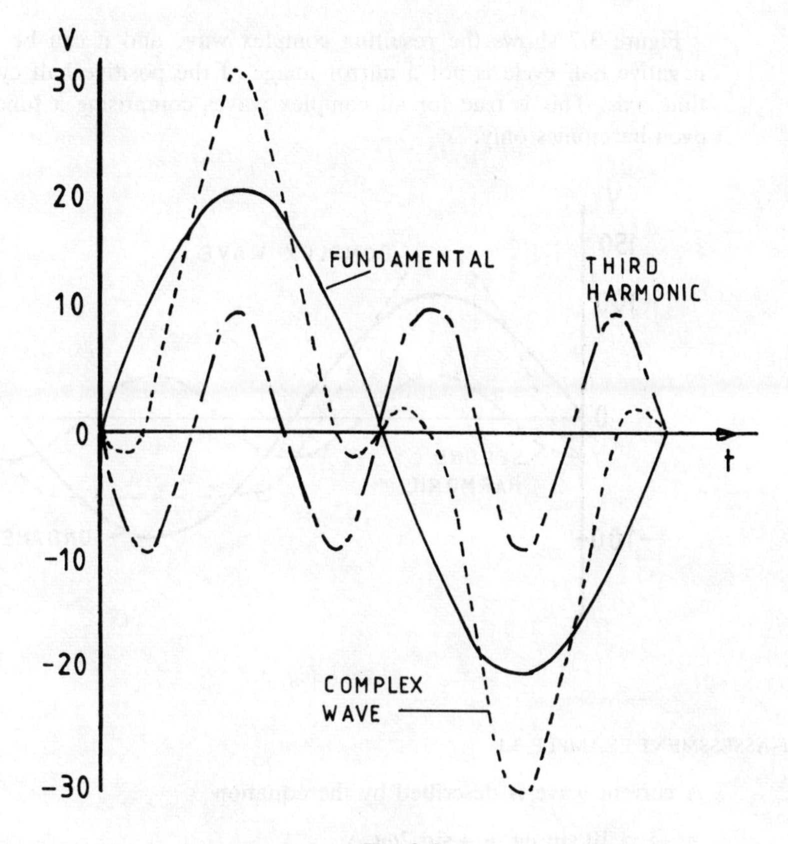

Fig. 3.5

3.1(a), it will be observed that this condition is fulfilled so that the wave contains only odd harmonics.

Now consider the complex wave described by

$$v = 100 \sin \omega t + 35 \sin (2\omega t - 90°) \text{ V}.$$

The phasors at $t = 0$ are shown in Fig. 3.6. It should be observed that at this instant the fundamental is at zero about to go positive while the second harmonic is at a maximum negative.

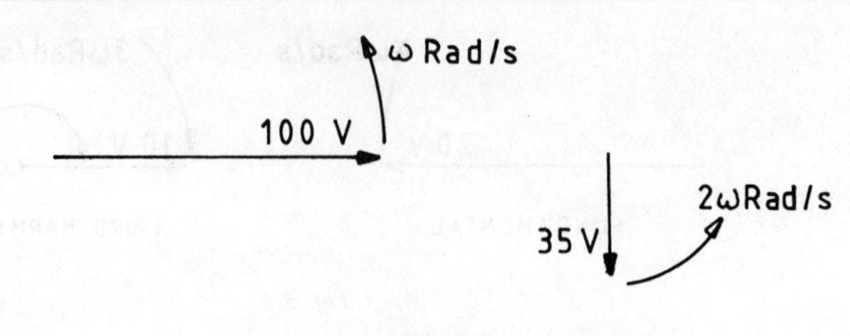

Fig. 3.6

Figure 3.7 shows the resulting complex wave and it can be seen that the negative half cycle is not a mirror image of the positive half cycle about the time axis. This is true for all complex waves comprising a fundamental plus even harmonics only.

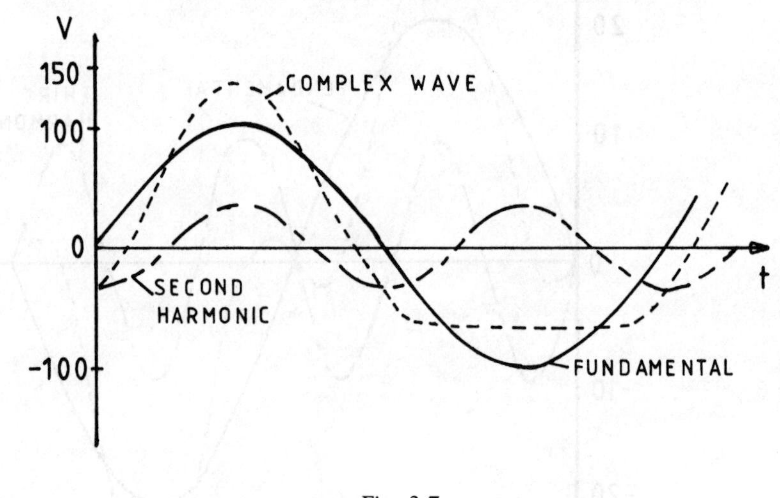

Fig. 3.7

SELF-ASSESSMENT EXAMPLE 3.1

A current wave is described by the equation

$$i = 10 \sin \omega t + 4 \sin 2\omega t \text{ A.}$$

Draw the waveform of (a) the fundamental for $1\frac{1}{2}$ complete cycles, (b) the second harmonic for the same time interval, (c) the resultant complex wave. Comment on the comparative shapes of the positive and negative half cycles.

In some circuits a fundamental, harmonics and a direct quantity or bias are present. Consider the effect of adding a direct voltage of +65 V to the complex wave in Fig. 3.7. The whole wave will be moved up by 65 V with respect to the time axis, resulting in the waveform shown in Fig. 3.8. This is the voltage waveform one would expect from a half-wave rectifier. The equation of this wave is $v = 65 + 100 \sin \omega t + 35 \sin (2\omega t - 90°)$ V.

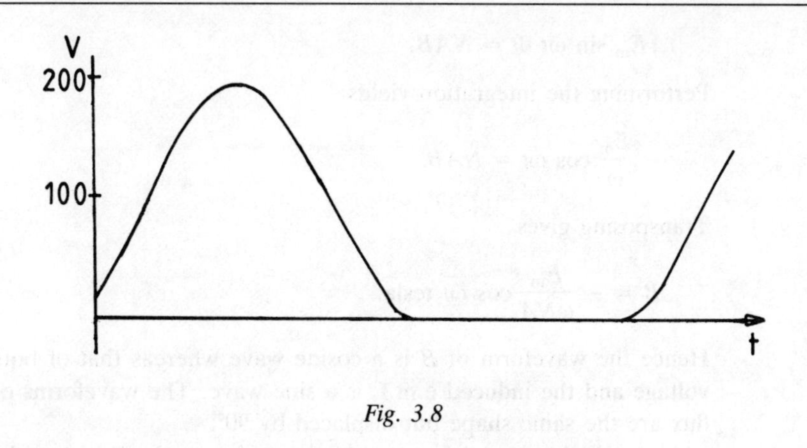

Fig. 3.8

Waves may be given a positive or a negative bias which shifts the waves up or down with respect to the time axis.

3.2 Production of harmonics

Harmonics are generated in devices which have a non-linear response to their inputs. Principally these are

(a) circuits containing iron,
(b) circuits containing transistors, rectifiers or valves.

Alternators generate voltages which contain odd harmonics due to inequalities in the flux distribution within the machine.

3.2.1 *Free magnetisation of ferrous material*

When a sinusoidal alternating voltage is applied to a coil wound on a ferrous core, a magnetising current flows which creates a flux in the core. Suppose the coil has N turns and a cross-sectional area of A m^2. From Faraday

$$e \propto \frac{N\mathrm{d}\Phi}{\mathrm{d}t} \quad \text{and in SI units} \quad e = \frac{N\mathrm{d}\Phi}{\mathrm{d}t}\ \mathrm{V}.$$

Now $\Phi = BA$ webers, where B = flux density in tesla (Wb/m^2), hence

$$e = \frac{NA\mathrm{d}B}{\mathrm{d}t}.$$

Integrating both sides of the equation with respect to t gives

$$\int e\mathrm{d}t = NAB.$$

By Lenz's law, the induced e.m.f. (e volts) opposes the applied voltage (v volts) and neglecting the resistance of the coil, $e = v$.

If the applied voltage is of sinusoidal form then e will be of the same form when $e = E_m \sin \omega t$ volts. Hence

$$\int E_m \sin \omega t \, dt = NAB.$$

Performing the integration yields

$$-\frac{E_m}{\omega} \cos \omega t = NAB.$$

Transposing gives

$$B = -\frac{E_m}{\omega NA} \cos \omega t \text{ tesla.}$$

Hence the waveform of B is a cosine wave whereas that of both the applied voltage and the induced e.m.f. is a sine wave. The waveforms of voltage and flux are the same shape but displaced by 90°.

An applied voltage of sinusoidal form also results in sinusoidal variation in the flux in the core.

Now consider applying a sinusoidal voltage to a coil wound on a core of magnetic material with a hysteresis loop, as shown in Fig. 3.9.

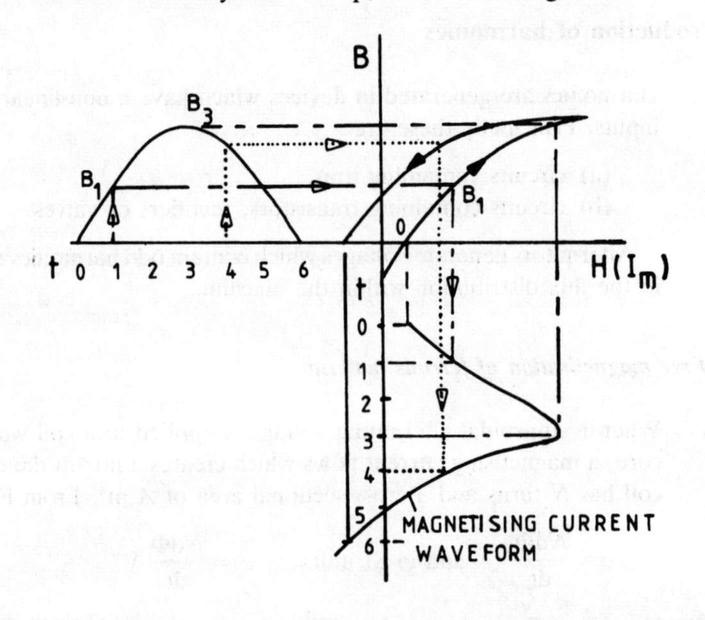

Fig. 3.9

In Fig. 3.9 there are four quadrants. In the top left-hand quadrant a sinusoidal wave representing flux density is drawn. In the top right-hand quadrant is the top part of a hysteresis loop for the core material. The horizontal axis (H) has dimensions amperes/metre so that it may be directly scaled as magnetising current (I_m) knowing the effective length of the core.

In the bottom right-hand quadrant is developed the magnetising current waveform in the following manner.

At $t(0)$, $B = 0$. In order for the flux density to be zero, H must have some positive value (the coercive force) to overcome the effect from the negative half cycle (point 0 on the hysteresis loop). This is projected down into the fourth quadrant of the diagram to meet horizontal $t(0)$.

At $t(1)$, the flux density has increased to value (B_1). This is projected on to the rising side of the hysteresis loop and then down to meet a horizontal drawn through $t(1)$ in the fourth quadrant. The process is repeated at regular intervals until we have seven values of I_m corresponding to the seven values of flux density, starting at zero and returning to zero on the top side of the hysteresis loop. At $t(6)$, negative coercive force is required so that I_m has become negative for the last point plotted.

The resulting magnetising current waveform is distinctly non-sinusoidal. Repeating the process for the negative half cycle of flux density results in an identical wave shape so that it may be deduced that the magnetising current contains a fundamental plus odd harmonics only.

3.2.2 *Forced magnetisation of ferrous material*

If the resistance of a circuit is high compared with its inductance and the applied voltage is of sinusoidal form, the effect of the resistance will be to cause the current to be of sinusoidal form. In three-phase, star-connected circuits where the supply has only three wires (no neutral) as, for example, with some transformers and induction motors, the line currents will be of sinusoidal form. Since it is necessary to have a current containing harmonics to produce a sinusoidal flux, it follows that if the current is of sinusoidal form the flux wave must be distorted. Further, if the flux wave is distorted then any voltages induced by this flux must also be distorted. In Fig. 3.10 it is assumed

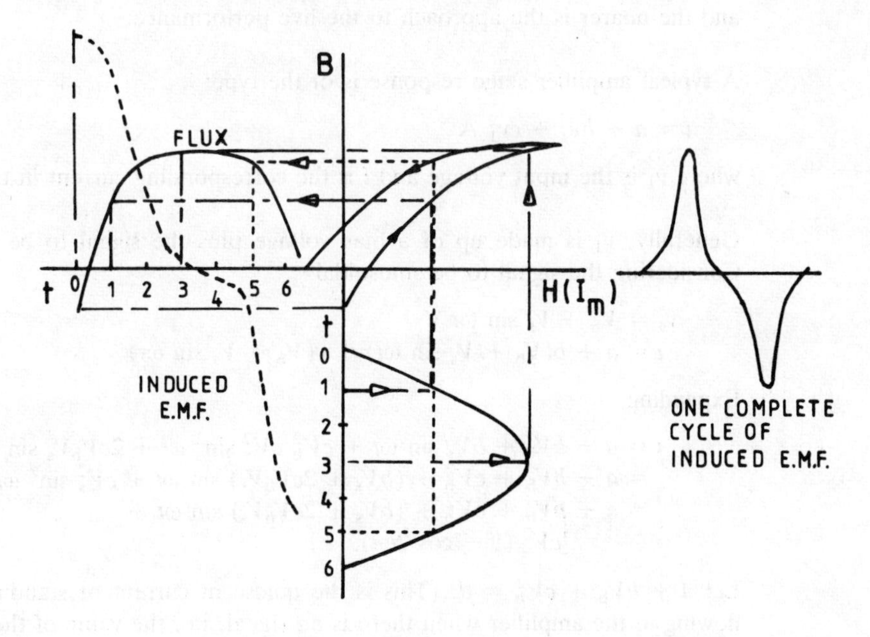

Fig. 3.10

that the current in the fourth quadrant is sinusoidal. Projecting values from this curve on to the hysteresis loop results in a flux wave which is distinctly flat topped.

Now since

$$e = NA\frac{dB}{dt}\text{ V}$$

the rate of change of flux with respect to time must be considered.

A tangent drawn to the flux wave at $t(0)$ has a large positive slope so that dB/dt is large and hence the induced e.m.f. will be large. At $t(3)$, there is virtually no change in flux with respect to time so the induced e.m.f. is near zero. At $t(5)$, dB/dt is again large and with negative slope which implies that the induced e.m.f. is large and negative. A complete cycle of induced e.m.f. is also shown in Fig. 3.10; it is distinctly non-sinusoidal and contains a substantial amount of third harmonic.

3.2.3 *Transistor and valve amplifiers*

All amplifying devices are non-linear to some extent in their response to an input. This means that with a sine wave input the positive half cycle is amplified by a different amount from the negative half cycle so that the two half cycles have different amplitudes. Being different in shape, even harmonic distortion is indicated. It is, in part, the presence of second harmonic distortion that informs the listener that it is an amplifier output that is being listened to and not a live performance. The better the equipment the less is this distortion and the nearer is the approach to the live performance.

A typical amplifier stage response is of the type:

$$i = a + bv_1 + cv_1^2 \text{ A}$$

where v_1 is the input voltage and i is the corresponding current in the output.

Generally, v_1 is made up of a bias voltage plus the signal to be amplified. Considering the signal to be sinusoidal

$$v_1 = V_b + V_s \sin \omega t \text{ V}$$

$$\therefore i = a + b(V_b + V_s \sin \omega t) + c(V_b + V_s \sin \omega t)^2$$

Expanding

$$\begin{aligned} i &= a + bV_b + bV_s \sin \omega t + cV_b^2\, cV_s^2 \sin^2 \omega t + 2cV_bV_s \sin \omega t \\ &= a + bV_b + cV_b^2 + (bV_s + 2cV_bV_s) \sin \omega t + cV_s^2 \sin^2 \omega t \\ &= a + bV_b + cV_b^2 + (bV_s + 2cV_bV_s) \sin \omega t + \\ &\quad \tfrac{1}{2}cV_s^2 (1 - \cos 2\omega t). \end{aligned} \qquad \text{(i)}$$

Let $a + bV_b + cV_b^2 = d$. (This is the quiescent current or standing current flowing in the amplifier when there is no signal, i.e. the value of the equation when $V_s = 0$.)

$$\tfrac{1}{2}cV_s^2 = f$$
$$(bV_s + 2cV_bV_s)\sin\omega t = e.$$

Substitute these values in equation (i) above

$$i = d + e\sin\omega t + f - f\cos 2\omega t.$$

Now $d + f$ = steady direct current output
e = the peak magnitude of the sine wave output which is what is required from the amplifier
f = the peak magnitude of a wave with twice the frequency of the input, the peak magnitude of the second harmonic distortion.

The ratio of second harmonic distortion to the fundamental $= \dfrac{f}{e}$ (ii)

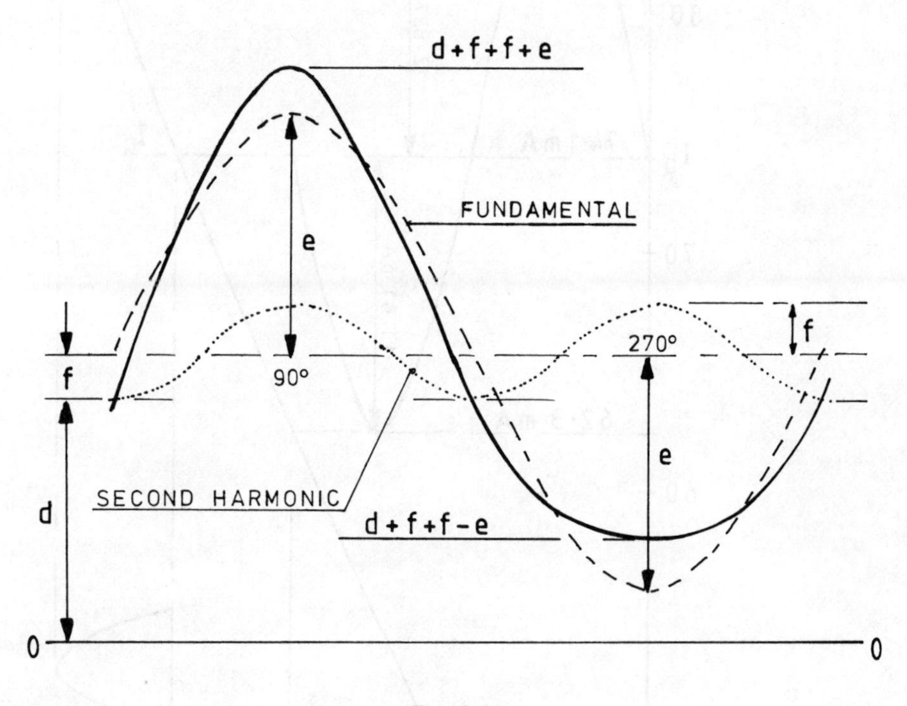

Fig. 3.11

The fundamental, second harmonic and complex waves are shown in Fig. 3.11.
Now at $\omega t = 90°$ the complex wave has its maximum value $= d + e + f + f$.
At $\omega t = 270°$ the wave has its minimum value $= d - e + f + f$.
Adding maximum and minimum values we get $d + d + e - e + f + f = 2d + 4f$.
Subtract $2d$, i.e. 2 times the quiescent current in the amplifier to give $4f$.
In words, we have: maximum value + minimum value − 2 times the quiescent current $= 4f$.
Also maximum value $(d + e + f + f)$ minus minimum value $(d - e + f + f) = 2e)$.
Multiplying by 2: 2(maximum value − minimum value) $= 4e$

$$\therefore \frac{\text{(maximum value + minimum value} - 2 \times \text{quiescent current}}{2\text{(maximum value} - \text{minimum value)}} = \frac{4f}{4e} = \frac{f}{e}$$

Which, from equation (ii) above, is the ratio of second harmonic distortion to the fundamental.

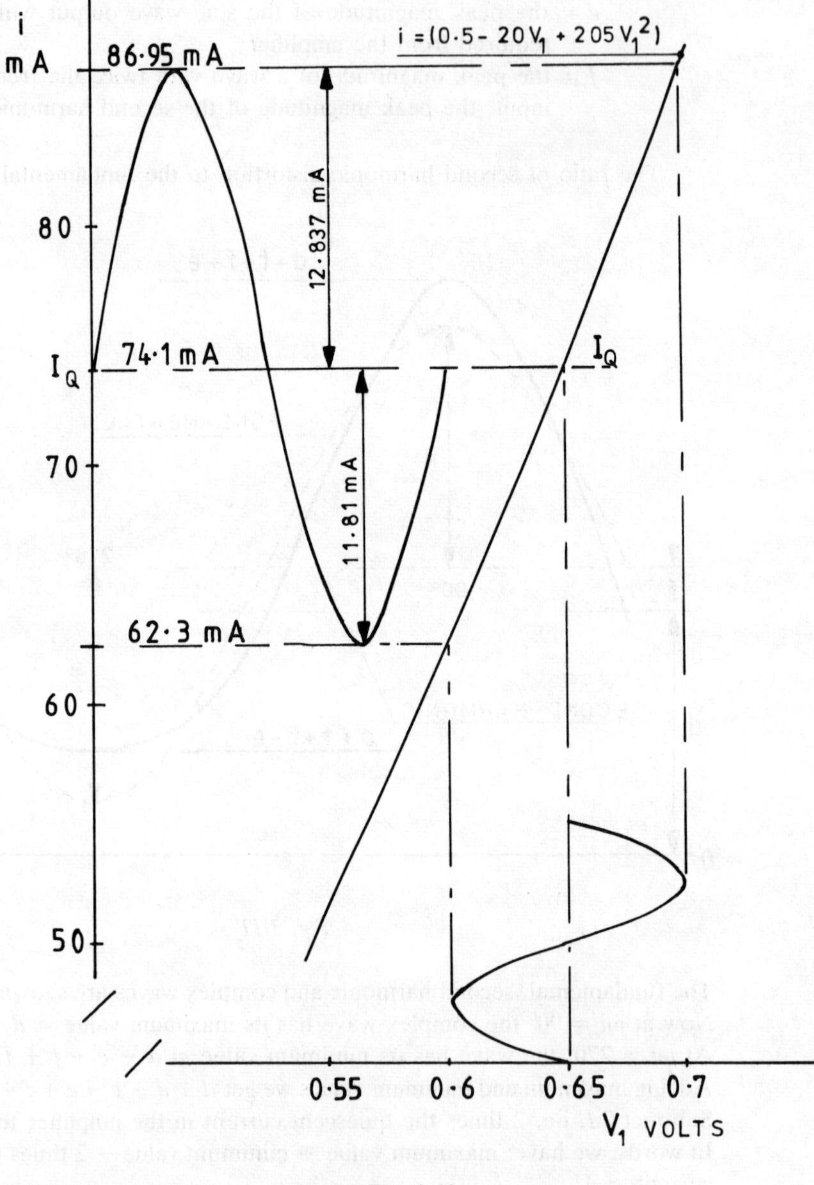

Fig. 3.12

Figure 3.12 shows part of the response curve of a transistor amplifier which follows the law

$$i = (0.5 - 20v_1 + 205v_1^2)\text{ mA} \qquad (b = -20, \quad c = 205).$$

The bias voltage is 0.65 V and the quiescent current = [0.5 −(20 × 0.65) + (205 × 0.65²)] mA = 74.1125 mA.

It is envisaged that the alternating input signal has a peak value of 50 mV so that at the peak of the signal, v_1 = 0.65 + 50 mV = 0.7 V. The current has risen to (0.5 − 20 × 0.7 + 205 × 0.7²) = 86.95 mA. At the instant of minimum signal v_1 = 0.6 V the corresponding current is 62.3 mA.

$$\frac{\text{maximum} + \text{minimum} - 2 \times \text{quiescent current}}{2(\text{maximum} - \text{minimum currents})} = \frac{86.95 + 62.3 - 2 \times 74.1125}{2(86.95 - 62.3)}$$
$$= 0.02079 \text{ or } 2.079\%.$$

The second harmonic distortion in the wave is 2.079%.

The same result may be achieved using the constants from the equation for the wave.

$$\frac{f}{e} = \frac{\frac{1}{2}cv_s^2}{bV_s + 2cV_bV_s} = \frac{\frac{1}{2} \times 205 \times (50 \times 10^{-3})^2}{-20 \times 50 \times 10^{-3} + 2 \times 205 \times 0.65 \times 50 \times 10^{-3}}$$
$$= 0.02079$$

SELF-ASSESSMENT EXAMPLE 3.2

Using the characteristic shown in Fig. 3.12 and the associated constants, b = −20 and c = 205; investigate the changes in output second harmonic content for the following conditions: (i) the bias is left at 0.65 V but the amplitude of the signal is reduced to 25 mV peak, (ii) the bias is changed to 0.68 V whilst the signal is maintained at 50 mV peak.

3.3 The rms value of a complex wave

Consider a voltage wave described by the equation

$$v = V_0 + V_{m(1)} \sin(\omega t + \phi_1) + V_{m(2)} \sin(2\omega t + \phi_2) + V_{m(3)} \sin(3\omega t + \phi_3) + \ldots + V_{m(n)} \sin(n\omega t + \phi_n)\text{ V}.$$

The rms value of a wave is the square root of the mean value of v^2 over a complete cycle.

The instantaneous value of the above wave is equal to the sum of the instantaneous values of the individual waves

$$v = V_0 + v_1 + v_2 + v_3 + \ldots + v_n.$$

Hence

rms value of wave = $\sqrt{\text{mean value of}}\ (V_0 + v_1 + v_2 + v_3 + \ldots + v_n)^2$ over a complete cycle.

Multiplying out the squared term gives

$$(V_0 + v_1 + v_2 + v_3 + \ldots + v_n)^2 = V_0^2 + V_0v_1 + V_0v_2 + V_0v_3 + \ldots + V_0v_n$$

$$+ v_1^2 + v_1v_2 + v_1v_3 + \ldots + v_1v_n$$
$$+ v_2^2 + v_2v_3 + \ldots + v_2v_n$$
$$\ldots + v_n^2$$

There are four distinct types of term present:

(a) V_0^2, (b) $V_0v_1, V_0v_2, \ldots, V_0v_n$, (c) $v_1^2, v_2^2, \ldots, v_n^2$, (d) v_1v_2, v_2v_3, v_3v_n.

By the superposition principle each such term may be considered separately and the total effect summed at the end.

Examining each type of term in turn:

(a) V_0^2. V_0 is a direct bias and so is constant in magnitude. The average value of V_0^2 over any period is V_0^2.

(b) V_0v_1, V_0v_n. These terms are all a constant V_0 multiplied by a voltage which is sinusoidally varying. For example, $v_n = 50 \sin (n\omega t + \phi_n)$ and $V_0 = 100$ V. Then $V_0v_n = 100 \times 50 \sin (n\omega t + \phi_n)$ V. The average value of a sine wave over any number of complete cycles is zero since there are equal areas above and below the time axis. The effect of V_0 is merely to increase the amplitude of the wave in both positive and negative half cycles so that the average value is still zero.

(c) v_1^2, v_n^2. Each of these terms is the square of a sinusoidally varying voltage. Considering the general case for v_n^2, the average value of v_n^2 over a complete cycle is the average value of $V_{m(n)}^2 \sin^2 (n\omega t + \phi_n)$ over that period. Now $\cos (A + B) = \cos A \cos B - \sin A \sin B$, so that if $A = B = (n\omega t + \phi_n)$

$$\cos 2(n\omega t + \phi_n) = \cos^2 (n\omega t + \phi_n) - \sin^2 (n\omega t + \phi_n). \quad (3.1)$$

Also $\cos (A - B) = \cos A \cos B + \sin A \sin B$, so for $A = B = (n\omega t + \phi_n)$

$$\cos 0 = \cos^2 (n\omega t + \phi_n) + \sin^2 (n\omega t + \phi_n) \qquad (\cos 0 = 1)$$

Transposing gives

$$\cos^2 (n\omega t + \phi_n) = 1 - \sin^2 (n\omega t + \phi_n) \quad (3.2)$$

Substituting equation (3.2) into equation (3.1) gives

$$\cos 2(n\omega t + \phi_n) = 1 - 2 \sin^2 (n\omega t + \phi_n)$$
$$\sin^2 (n\omega t + \phi_n) = \tfrac{1}{2}(1 - \cos 2(n\omega t + \phi_n))$$

So that

$$V_{m(n)}^2 \sin^2 (n\omega t + \phi_n) = \tfrac{1}{2}V_{m(n)}^2 (1 - \cos 2(n\omega t + \phi n))$$
$$= \tfrac{1}{2}V_{m(n)}^2 - \tfrac{1}{2}V_{m(n)} \cos 2(n\omega t + \phi_n).$$

This expression consists of a constant term and a cosine term. It is of the same form as that shown in Fig. 1.33(a). The mean value of the cosine term over a complete cycle is zero. The mean value of the expression is therefore $\tfrac{1}{2}V_{m(n)}^2$. Now

$$\frac{V_{m(n)}^2}{2} = \frac{V_{m(n)}}{\sqrt{2}} \times \frac{V_{m(n)}}{\sqrt{2}} = V_{(n)}^2$$

where V_n = rms value of wave.

For particular waves

The average value of $v_{(1)}^2$ over a complete cycle $= \dfrac{V_{m(1)}^2}{2} = V_{(1)}^2$

and of $v_{(2)}^2$ over a complete cycle $= \dfrac{V_{m(2)}^2}{2} = V_{(2)}^2$, etc.

(d) v_1v_2, v_2v_n. Each of these terms is the product of two sinusoidally varying voltages with different frequencies. Generally

$$V_{m(p)} \sin(p\omega t + \phi_p) \times V_{m(q)} \sin(q\omega t + \phi_q) \qquad \text{(where } p \neq q\text{)}$$

$$\cos(A + B) = \cos A \cos B - \sin A \sin B \tag{3.3}$$

$$\cos(A - B) = \cos A \cos B + \sin A \sin B \tag{3.4}$$

Subtracting equation (3.3) from equation (3.4) gives

$$\cos(A - B) - \cos(A + B) = 2 \sin A \sin B$$

and transposing gives

$$\sin A \sin B = \tfrac{1}{2}(\cos(A - B) - \cos(A + B)). \tag{3.5}$$

Letting $A = (p\omega t + \phi_p)$ and $B = (q\omega t + \phi_q)$ in equation (3.5)

$$\begin{aligned} V_{m(p)} &\sin(p\omega t + \phi_p) \times V_{m(q)} \sin(q\omega t + \phi_q) \\ &= \tfrac{1}{2}(\cos((p\omega t + \phi_p) - (q\omega t + \phi_q)) - \cos((p\omega t + \phi_p) + (q\omega t + \phi_q))) \\ &= \tfrac{1}{2}(\cos((p - q)\omega t + (\phi_p - \phi_q)) - \cos((p + q)\omega t + (\phi_p + \phi_q))). \end{aligned}$$

Both terms represent sinusoidally varying waves, each having a different frequency. Once again, the average value of a sine wave over any number of complete cycles is zero so that the average value of the complete expression is zero.

The mean value of the product of two sine waves of different frequencies is zero.

To sum up the preceding results:
The average value of v^2 is the sum of all the non-zero terms: V_0^2, $V_{m(1)}^2/2$, $V_{m(2)}^2/2, \ldots, V_{m(n)}^2/2$.
The rms value is the square root of this sum. Hence

$$V = \sqrt{\left(V_0^2 + \frac{V_{m(1)}^2}{2} + \frac{V_{m(2)}^2}{2} + \ldots + \frac{V_{m(n)}^2}{2}\right)} \tag{3.6}$$

Similarly for current

$$I = \sqrt{\left(I_0^2 + \frac{I_{m(1)}^2}{2} + \frac{I_{m(2)}^2}{2} + \ldots + \frac{I_{m(n)}^2}{2}\right)} \tag{3.7}$$

Now, since, $V_m/\sqrt{2} = V$, where V is the rms value of voltage, $V_m^2/2 = V_2$.
Equation (3.6) may therefore be expressed as

$$V = \sqrt{(V_0^2 + V_1^2 + V_2^2 + \ldots + V_n^2)}$$

using rms values throughout.

Similarly equation (3.7) may be rewritten

$$I = \sqrt{(I_0^2 + I_1^2 + I_2^2 + \ldots + I_n^2)}.$$

WORKED EXAMPLE 3.3

Determine the rms value of a complex voltage wave described by the equation

$$v = 10 + 50 \sin(\omega t + 30°) + 10 \sin(3\omega t + 15°) + 5 \sin(5\omega t - 90°) \text{ V}.$$

Using equation (3.6)

$$V = \sqrt{\left(10^2 + \frac{50^2}{2} + \frac{10^2}{2} + \frac{5^2}{2}\right)} = \sqrt{(100 + 1250 + 50 + 12.5)}$$

$$= 37.58 \text{ V}.$$

SELF-ASSESSMENT EXAMPLE 3.4

Determine the rms value of a complex current wave described by the equation

$$i = -5 + 20 \sin(\omega t - 10°) + 4 \sin(2\omega t + 30°) + 0.6 \sin(4\omega t + 25°) \text{ A}.$$

SELF-ASSESSMENT EXAMPLE 3.5

A complex voltage wave consisting of a fundamental and third harmonic only has a rms value of 39.53 V. The rms value of the fundamental is $50/\sqrt{2}$ V. Determine the rms value of the third harmonic.

3.4 Power conveyed by complex waves

Consider a wave described by the equation

$$v = V_0 + V_{m(1)} \sin(\omega t + \phi_a) + V_{m(2)} \sin(2\omega t + \phi_b) + \ldots + V_{m(n)} \sin(n\omega t + \phi_x) \text{ V}.$$

It is applied to a circuit and a current flows given by the equation

$$i = I_0 + I_{m(1)} \sin(\omega t + \phi_a - \phi_1) + I_{m(2)} \sin(2\omega t + \phi_b - \phi_2) + \ldots + I_{m(n)} \sin(n\omega t + \phi_x - \phi_n) \text{ A}$$

where $\phi_1, \phi_2, \ldots, \phi_n$ are phase angles determined by the circuit components.
The instantaneous power in the circuit is vi W.
The average power over a complete cycle is the average value of vi over this period.
As in section 3.3, the instantaneous value of the voltage wave may be written as the sum of the instantaneous values of the constituent voltages

$$v = V_0 + v_1 + v_2 + \ldots + v_n \text{ V}.$$

Similarly the instantaneous value of the current is

$$i = I_0 + i_1 + i_2 + \ldots + i_n \text{ A}.$$

Hence

$$v \times i = V_0 I_0 \\ + V_0 i_1 + V_0 i_2 + \ldots + V_0 i_n \\ + v_1 i_1 + v_1 i_2 + \ldots + v_1 i_n \\ + v_2 i_2 + \ldots + v_2 i_n, \text{ etc.}$$

There are four different types of terms here, as in section 3.3, from which the rms value has to be determined:

(a) $V_0 I_0$, (b) $V_0 i_1$, $V_0 i_2$, ..., $V_0 i_n$, (c) $v_1 i_1$, $V_2 i_2$, $v_n i_n$, (d) $v_1 i_2$, $v_2 i_3$, $v_n i_m$ where $n \neq m$.

Dealing with each of these types in turn.

(a) $V_0 I_0$. Since these are direct components, the power contributed to the circuit is $V_0 I_0$ W.

(b) $V_0 i_1$, $V_0 i_n$. These terms are all the product of a constant term multiplied by a sinusoidally varying quantity. The average value over a complete cycle is zero as already reasoned in section 3.3 (part **b**).

(c) $v_1 i_1$, $v_n i_n$. The average value of power in the circuit is the average value of $V_{m(n)} \sin(n\omega t + \phi_x) \times I_{m(n)} \sin(n\omega t + \phi_x - \phi_n)$ over a complete cycle.

Using equation (3.5) $\sin A \sin B = \frac{1}{2}(\cos(A - B) - \cos(A + B))$.
Let $A = (n\omega t + \phi_x)$ and $B = (n\omega t + \phi_x - \phi_n)$, which gives

$$V_{m(n)} \sin(n\omega t + \phi_x) \times I_{m(n)} \sin(n\omega t + \phi_x - \phi_n)$$

$$= \frac{V_{m(n)} I_{m(n)}}{2} (\cos((n\omega t + \phi_x) - (n\omega t + \phi_x - \phi_n)) - \cos((n\omega t + \phi_x) + (n\omega t + \phi_x - \phi_n)))$$

$$= \frac{V_{m(n)} I_{m(n)}}{2} (\cos \phi_n - \cos(2(n\omega t + \phi_x) - \phi_n)) \text{ W}.$$

ϕ_n is determined at the particular frequency by the circuit components and is therefore a constant at that frequency.

Looking back to Fig. 1.33 it will be seen that this expression for power reduces to

$$P = \frac{V_{m(n)} I_{m(n)}}{2} \cos \phi_n \text{ W}$$

(again the average value of the time-varying cosine term is zero).
Using rms values of voltage and current, $P = V_n I_n \cos \phi_n$ W.
(d) $v_1 i_2$, all terms like $v_p i_q$, where $p \neq q$.
The average value of $v_p i_q$ is the average value over a complete cycle of

$$V_{m(p)} \sin(p\omega t + \phi_p) \times I_{m(q)} \sin(q\omega t + \phi_p - \phi_q).$$

This is the product of two sine terms with different frequencies which has already been examined in section 3.3 (part **d**), the average value of which was found to be zero.

Summing the non-zero terms, the power supplied by complex voltages and currents is therefore given by

$$P = V_0 I_0 + V_{(1)} I_{(1)} \cos \phi_1 + V_{(2)} I_{(2)} \cos \phi_2 + \ldots + V_{(n)} I_{(n)} \cos \phi_n \text{ W.} \quad (3.8)$$

Thus, the total circuit power = sum of the powers due to the individual harmonics.

3.5 Power factor

The power factor of an a.c. circuit $= \dfrac{\text{circuit power in watts}}{\text{circuit volt-amperes}}$

$$= \frac{V_0 I_0 + \Sigma V_n I_n \cos \phi_n}{VI}. \quad \text{(equation 1.7)}$$

Notice that whereas power factor is equal to cosine ϕ when a sinusoidal voltage and current are involved, this may not be used where complex waveforms are present because each harmonic may involve a different phase angle and, in addition, there may be direct power present.

Where the circuit series resistance is known the power may be found by using the formula

$$P = I^2 R \text{ W.}$$

3.6 Percentage harmonic content

The size of the nth harmonic is sometimes expressed by comparing it with the size of the fundamental.

$\dfrac{V_{m(n)}}{V_{m(1)}} \times 100\%$ is defined as the percentage nth harmonic content of the wave.

For example, for $V_{m(2)} = 20$ V, and $V_{m(1)} = 80$ V

$$\frac{20}{80} \times 100 = 25\%.$$

The complex wave contains 25% second harmonic.

Since power is only contributed to a circuit by a current and voltage of the same frequency, a dynamometer wattmeter may be employed as a simple complex wave analyser at power frequencies giving information as to the number of harmonics present and their magnitudes but not their phase with respect to the fundamental.

The complex wave to be analysed is fed to the current coil of the wattmeter while the voltage coil is fed at constant voltage from an oscillator which delivers a sine-wave output at varying frequency. By altering the oscillator frequency over a range from the fundamental value upward, the particular frequencies at which the wattmeter is deflected are noted. Whenever there is a deflection, the current in the coil must have a component which is of the same frequency as that presently applied to the voltage coil. All frequencies present in the current wave are therefore detected as the oscillator frequency is varied. The magnitude of the wattmeter deflection at a particular harmonic frequency is a measure of the relative size of that harmonic.

$$\frac{\text{wattmeter deflection for the } n\text{th harmonic}}{\text{wattmeter deflection at fundamental frequency}} \times 100\% = \begin{array}{l}\text{percentage } n\text{th}\\ \text{harmonic content.}\end{array}$$

WORKED EXAMPLE 3.6

A voltage given by

$$v = 100 \sin \omega t + 50 \sin (3\omega t - 30°) + 10 \sin (5\omega t + 120°) \text{ V}$$

is applied to a circuit and the resultant current is found to be

$$i = 7.07 \sin (\omega t - 45°) + 1.58 \sin (3\omega t - 101.5°) + 0.196 \sin (5\omega t + 41.31°) \text{ A.}$$

Determine (a) total power supplied, (b) the overall power factor.

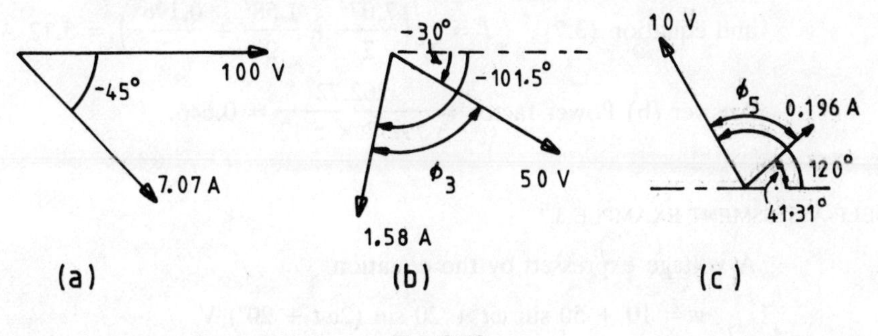

Fig. 3.13

From equation (3.8) it should be observed that only voltages and currents of the same frequency contribute power to a circuit. In this case there is no direct component.

Consider the fundamental voltage and current (Fig. 3.13(a))

$V_{m(1)} = 100$ V $\quad I_{m(1)} = 7.07$ A. $\quad$ The angle between them is 45°.

$$\text{Power} = \frac{V_m I_m}{2} \cos \phi \text{ W}$$

$$= \frac{100 \times 7.07}{2} \cos 45° = 250 \text{ W}$$

(or convert both quantities to rms values and use $VI \cos \phi$).

Consider the third harmonic voltage and current (Fig. 3.13(b))

$V_{m(3)} = 50$ V $\quad I_{m(3)} = 1.58$ A

The angle between them is $\phi_3 = (101.5 - 30)°$.

$$\text{Power} = \frac{50 \times 1.58}{2} \cos 71.5° = 12.53 \text{ W.}$$

Finally, consider the fifth harmonic voltage and current (Fig. 3.13(c))

$V_{m(5)} = 10$ V $\quad I_{m(5)} = 0.196$ A. $\quad$ The angle between them is $\phi_5 = (120 - 41.31)°$.

$$\text{Power} = \frac{10 \times 0.196}{2} \cos 78.69° = 0.192 \text{ W.}$$

Answer (a) Total power supplied $= \Sigma VI \cos \phi = 250 + 12.53 + 0.192 = 262.72$ W.

From equation (1.7)

$$\text{power factor} = \frac{\text{power}}{\text{volt-amperes}}.$$

Both the rms values of voltage and current are needed.

Using equation (3.6), $V = \sqrt{\left(\frac{100^2}{2} + \frac{50^2}{2} + \frac{10^2}{2}\right)} = 79.37$ V

and equation (3.7) $\quad I = \sqrt{\left(\frac{7.07^2}{2} + \frac{1.58^2}{2} + \frac{0.196^2}{2}\right)} = 5.12$ A.

Answer (b) Power factor $= \frac{262.72}{79.37 \times 5.12} = 0.646.$

SELF-ASSESSMENT EXAMPLE 3.7

A voltage expressed by the equation

$$v = 10 + 50 \sin \omega t + 20 \sin (2\omega t + 20°) \text{ V}$$

is applied to an inductive circuit when the resulting current was found to be of the form

$$i = 10 + 4.47 \sin (\omega t - 26.6°) + 1.414 \sin (2\omega t - 25°) \text{ A.}$$

Determine (a) the rms value of the voltage, (b) the rms value of the current, (c) the circuit power, (d) the overall power factor, (e) the resistance of the circuit.

3.7 Harmonics in single-phase circuits

Consider a voltage given by the equation

$$v = V_{m(1)} \sin \omega t + V_{m(2)} \sin 2\omega t + \ldots + V_{m(n)} \sin n\omega t \text{ V.}$$

The effect will be considered of applying such a wave to linear circuit elements which have (a) resistance, (b) inductance, and (c) capacitance.

3.7.1 Pure resistance

The value of current in a resistor is expressed by Ohm's law ($I = V/R$ A).

The current is in phase with the voltage, and the resistance value is independent of frequency. Hence the current will be expressed by the equation

$$i = \frac{V_{m(1)}}{R} \sin \omega t + \frac{V_{m(2)}}{R} \sin 2\omega t + \ldots + \frac{V_{m(n)}}{R} \sin n\omega t \text{ A.}$$

The current and voltage waveforms are therefore identical in shape and the percentage nth harmonic current is the same as the percentage nth harmonic voltage (section 3.6).

3.7.2 *Pure inductance*

The reactance of an inductor with inductance L henry is given by $X_L = \omega L\ \Omega$, where ω rad/s is the angular velocity of the fundamental. For the nth harmonic the reactance will be $n\omega L\ \Omega$.

The inductive reactance is therefore proportional to the angular velocity and to frequency. The current in a pure inductor lags the driving voltage by 90° and this will be true for each constituent voltage of the complex wave. The general expression for the current wave in an inductor will therefore be

$$i = \frac{V_{m(1)}}{\omega L} \sin (\omega t - 90°) + \frac{V_{m(2)}}{2\omega L} \sin (2\omega t - 90°) + \ldots$$

$$+ \frac{V_{m(n)}}{n\omega L} \sin (n\omega t - 90°) \text{ A.}$$

The nth harmonic current expressed as a percentage of the fundamental current is

$$\left(\frac{V_{m(n)}/n\omega L}{V_{m(1)}/\omega L}\right) \times 100 = \left(\frac{V_{m(n)}\ \omega L}{V_{m(1)}\ n\omega L}\right) \times 100 = \frac{1}{n} \times \begin{array}{l}\text{percentage } n\text{th} \\ \text{harmonic content} \\ \text{of voltage wave.}\end{array}$$

This reduction in magnitude of successive harmonics in addition to the 90° phase shift causes the current wave shape to differ considerably from that of the voltage wave.

3.7.3 *Pure capacitance*

The capacitive reactance of a capacitor with capacitance C farads is given by $X_C = (1/\omega C)\ \Omega$, where ω is the angular velocity of the fundamental. For the nth harmonic, $X_C = (1/n\omega C)\ \Omega$.

The current in a pure capacitor leads the voltage by 90°.

The general expression for the current wave will therefore be

$$i = \frac{V_{m(1)}}{1/\omega C} \sin (\omega t + 90°) + \frac{V_{m(2)}}{1/2\omega C} \sin (2\omega t + 90°) + \ldots$$

$$+ \frac{V_{m(n)}}{1/n\omega C} \sin (n\omega t + 90°) \text{ A}$$

$$= V_{m(1)}\ \omega C \sin (\omega t + 90°) + V_{m(2)}\ 2\omega C \sin (2\omega t + 90°) + \ldots$$
$$+ V_{m(n)}\ n\omega C \sin (n\omega t + 90°) \text{ A.}$$

The nth harmonic current expressed as a percentage of the fundamental current is

$$\left(\frac{V_{m(n)}\ n\omega C}{V_{m(1)}\ \omega C}\right) \times 100 = n \times n\text{th percentage harmonic content of voltage wave.}$$

Once again the current wave shape will be different from that of the voltage wave due this time to a magnification by a factor n of the voltage content and to the phase advance of 90°.

Having seen how individual linear circuit elements affect the harmonic content of the current wave, series combinations of resistance and inductance and resistance and capacitance will now be examined.

3.7.4 *Resistance and inductance in series*

WORKED EXAMPLE 3.8

A coil with resistance 10 Ω and inductance 10 mH is connected to a power source that provides a voltage given by the equation

$$v = 200 \sin \omega t + 50 \sin (3\omega t + 30°) + 10 \sin (5\omega t - 90°) \text{ V}$$
$$(\omega = 400 \text{ rad/s}).$$

Derive an expression for the instantaneous value of the circuit current.

Each of the voltages has to be considered separately since the impedance of the coil will be different at each of the frequencies present. Consider the fundamental

$$Z_1 = 10 + j400 \times 10 \times 10^{-3} = (10 + j4) = 10.77\angle 21.8° \ \Omega.$$

$$I_{m(1)} = \frac{V_{m(1)}}{Z_1} = \frac{200}{10.77\angle 21.8°} = 18.57\angle -21.8° \text{ A}.$$

The fundamental current has a peak value of 18.57 A and it lags the fundamental voltage by 21.8°. Hence

$$i_1 = 18.57 \sin (\omega t - 21.8°) \text{ A}.$$

Consider the third harmonic

$$Z_3 = 10 + j(3 \times 400) \times 10 \times 10^{-3} = 10 + j12 = 15.62\angle 50.19° \ \Omega.$$

$$I_{m(3)} = \frac{V_{m(3)}}{Z_3} = \frac{50}{15.62\angle 50.19°} = 3.2\angle -50.19° \text{ A}.$$

The third harmonic current has a peak value of 3.2 A and it lags the third harmonic voltage by 50.19°. Hence

$$i_3 = 3.2 \sin (3\omega t + 30° - 50.19°) = 3.2 \sin (3\omega t - 20.19°) \text{ A}.$$

An alternative method, if preferred, is to introduce the 30° lead of the voltage wave as follows

$$I_{m(3)} = \frac{V_{m(3)}}{Z_3} = \frac{50\angle+30°}{15.62\angle 50.19°} = 3.2\angle-20.19° \text{ A.}$$

Finally consider the fifth harmonic

$$Z_5 = 10 + j(5 \times 400) \times 10 \times 10^{-3} = 10 + j20 = 22.36\angle 63.43° \ \Omega.$$

$$I_{m(5)} = \frac{10}{22.36\angle 63.43°} = 0.447\angle-63.43° \text{ A.}$$

$I_{m(5)}$ lags the fifth harmonic voltage by 63.43°. Hence

$$i_5 = 0.447 \sin(5\omega t - 90 - 63.43) = 0.447 \sin(5\omega t - 153.43°) \text{ A.}$$

The required expression for the current is

$$i = 18.57 \sin(\omega t - 21.8°) + 3.2 \sin(3\omega t - 20.19°) + 0.447 \sin(5\omega t - 153.43°) \text{ A.}$$

The phasor diagrams are shown in Fig. 3.14.

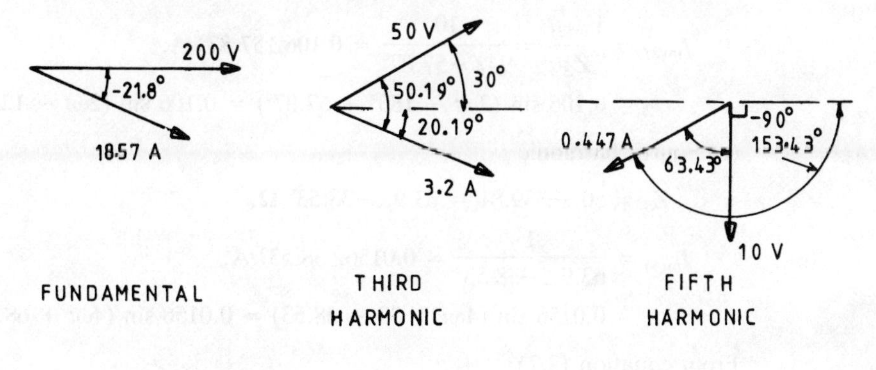

Fig. 3.14

SELF-ASSESSMENT EXAMPLE 3.9

A voltage described by the equation

$$v = 250 \sin \omega t + 50 \sin(3\omega t + 60°) + 20 \sin(5\omega t + 150°) \text{ V}$$

is applied to a series circuit consisting of a 15 Ω pure resistor and an inductor which has a resistance of 5 Ω and an inductance of 0.05 H. (ω = 314 rad/s.)

Derive an expression for the instantaneous current in the inductor and hence determine (a) the rms value of the current, (b) the power lost in the inductor.

3.7.5 Resistance and capacitance in series

WORKED EXAMPLE 3.10

A voltage described by the equation

$$v = 100 \sin \omega t + 10 \sin(2\omega t - 180°) + 1 \sin(4\omega t + 30°) \text{ V}$$

is applied to a circuit consisting of a 50 Ω resistor in series with a 10 μF capacitor. Determine the rms value of the circuit current for $\omega = 628$ rad/s.

As with the inductive circuit in worked example 3.8, it is necessary to consider each constituent voltage separately.

(a) Fundamental

$$Z_1 = 50 + \frac{1}{\text{j}628 \times 10 \times 10^{-6}} = 50 - \text{j}159.2 = 166.9\angle{-72.57^\circ}\ \Omega.$$

$$I_{\text{m}(1)} = \frac{V_{\text{m}(1)}}{Z_1} = \frac{100}{166.9\angle{-72.57^\circ}} = 0.6\angle 72.57^\circ\ \text{A}.$$

$$i_1 = 0.6 \sin(\omega t + 72.57^\circ)\ \text{A}.$$

(b) Second harmonic

$$Z_2 = 50 + \frac{1}{\text{j}(2 \times 628) \times 10 \times 10^{-6}} = 50 - \text{j}79.62 = 94\angle{-57.87^\circ}\ \Omega.$$

$$I_{\text{m}(2)} = \frac{V_{\text{m}(2)}}{Z_2} = \frac{10}{94\angle{-57.87^\circ}} = 0.106\angle 57.87^\circ\ \text{A}.$$

$$i_2 = 0.106 \sin(2\omega t - 180^\circ + 57.87^\circ) = 0.106 \sin(2\omega t - 122.13^\circ)\ \text{A}.$$

(c) Fourth harmonic

$$Z_3 = 50 - \text{j}39.81 = 63.9\angle{-38.53^\circ}\ \Omega.$$

$$I_{\text{m}(4)} = \frac{1}{63.9\angle{-38.53^\circ}} = 0.0156\angle 38.53^\circ\ \text{A}.$$

$$i_4 = 0.0156 \sin(4\omega t + 30 + 38.53) = 0.0156 \sin(4\omega t + 68.53^\circ)\ \text{A}.$$

From equation (3.7)

$$I = \sqrt{\left(\frac{0.6^2}{2} + \frac{0.106^2}{2} + \frac{0.0156^2}{2}\right)}$$

$$= 0.431\ \text{A}.$$

SELF-ASSESSMENT EXAMPLE 3.11

A circuit consisting of an 18 Ω resistor in series with a capacitor of value 0.884 μF is connected to a source of e.m.f. when the current in the circuit is described by the equation

$$i = 0.982 \sin(\omega t + 45^\circ) + 0.527 \sin(3\omega t + 108.43^\circ) + 0.1089 \sin(5\omega t - 38.69^\circ)\ \text{A}$$

$(\omega = 2\pi \times 10000$ rad/s).

Derive an expression for the instantaneous voltage applied to the circuit and hence determine (a) the rms value of the circuit voltage, (b) the power developed in the resistor, (c) the circuit power factor.

3.8 Selective resonance

The impedance of a circuit comprising resistance, inductance and capacitance in series is given by

$$Z = R + \mathrm{j}\left(\omega L - \frac{1}{\omega C}\right)\ \Omega.$$

At one particular frequency, $\omega L = \dfrac{1}{\omega C}$, when $Z = R\ \Omega$.

The circuit appears to be purely resistive, the impedance is a minimum and the circuit current will therefore be at a maximum value. The circuit is said to exhibit series resonance.

In the case of a parallel combination of a resistive inductor and a capacitor, at a particular frequency the circuit appears to be purely resistive. The value of this apparent resistance is known as the dynamic impedance. The dynamic impedance is the largest impedance that the particular circuit can possess. The circuit current is at a minimum and the circuit is said to exhibit parallel resonance.

If a supply voltage containing harmonics is applied to either circuit, resonance may occur, not at the fundamental frequency but at that of one of the harmonics. Such an occurrence is termed selective or harmonic resonance.

In the series circuit a large current at the particular harmonic frequency will flow giving rise to large voltages across the reactive components, i.e. voltage magnification. In the parallel circuit, current magnification at the harmonic frequency takes place, the currents in the reactive components at that frequency being much larger than that taken from the supply.

Selective resonance is one reason why it is undesirable to have non-sinusoidal supply voltages except for special purposes. Alternating current generators frequently generate waveforms with appreciable odd harmonic content. This is eliminated in the generator transformer, sometimes wound with a delta-connected tertiary, or by the use of tuned saturable reactors.

WORKED EXAMPLE 3.12

A 50 Hz, three-phase alternator has an internal impedance of $(1 + \mathrm{j}10)\ \Omega$ per phase at the fundamental frequency. It is connected to a cable, each core of which has an effective total capacitance to earth of 1.884 μF. Each phase of the alternator delivers a voltage described by the equation

$$v = 26\,944 \sin 314t + 50 \sin (13 \times 314)t\ \mathrm{V}.$$

Calculate the value of the maximum voltage impressed on each core of the cable to earth while no load is connected. (The resistance and inductance of the cable may be ignored.)

Consider one phase of the symmetrical system as shown in Fig. 3.15.

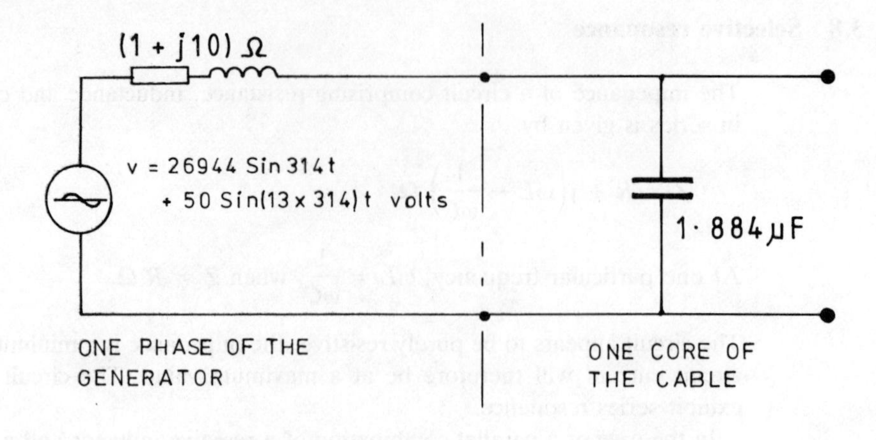

Fig. 3.15

At the fundamental frequency
Resistance = 1 Ω. Inductive reactance = +j10 Ω.

$$\text{Capacitive reactance of the cable, } X_C = \frac{1}{314 \times 1.884 \times 10^{-6}}$$
$$= 1690.4\ (-\text{j}1690.4\ \Omega).$$

With no load on the cable the effective impedance is that of a series circuit consisting of R, L and C.

$$Z = R + \text{j}(X_L - X_C) = 1 + \text{j}(10 - 1690.4) = -\text{j}1680.4\ \Omega \text{ (ignoring } R\text{)}.$$

$$\text{Peak value of fundamental current } I_{m(1)} = \frac{V_{m(1)}}{Z_1} = \frac{26\,944}{-\text{j}1680.4} = \text{j}16.03\text{ A}.$$

Peak voltage impressed on the cable = peak voltage across the capacitor

$$= I_{m(1)} \times -\text{j}X_C$$
$$= \text{j}16.03 \times (-\text{j}1690.4) = 27\,104\text{ V (in phase with the supply voltage).}$$

At the 13th harmonic frequency
Resistance = 1 Ω. Inductive reactance = 13 × 10 = 130 Ω.

$$\text{Capacitive reactance of the cable} = \frac{1}{13 \times 314 \times 1.884 \times 10^{-6}}$$
$$= 130\ (-\text{j}130\ \Omega).$$

Circuit impedance = 1 + j(130 − 130) = 1 Ω.

$$\text{Peak value of 13th harmonic current} = \frac{V_{m(13)}}{Z_{13}} = \frac{50}{1} = 50\text{ A}.$$

$$\text{Peak 13th harmonic voltage} = 50 \times -\text{j}130$$
$$= -\text{j}6500\text{ V } (6500\angle -90^\circ\text{ V}).$$

Total voltage on the cable = $27\,104 \sin 314t + 6500 \sin (13 \times 314t - 90^\circ)$ V.

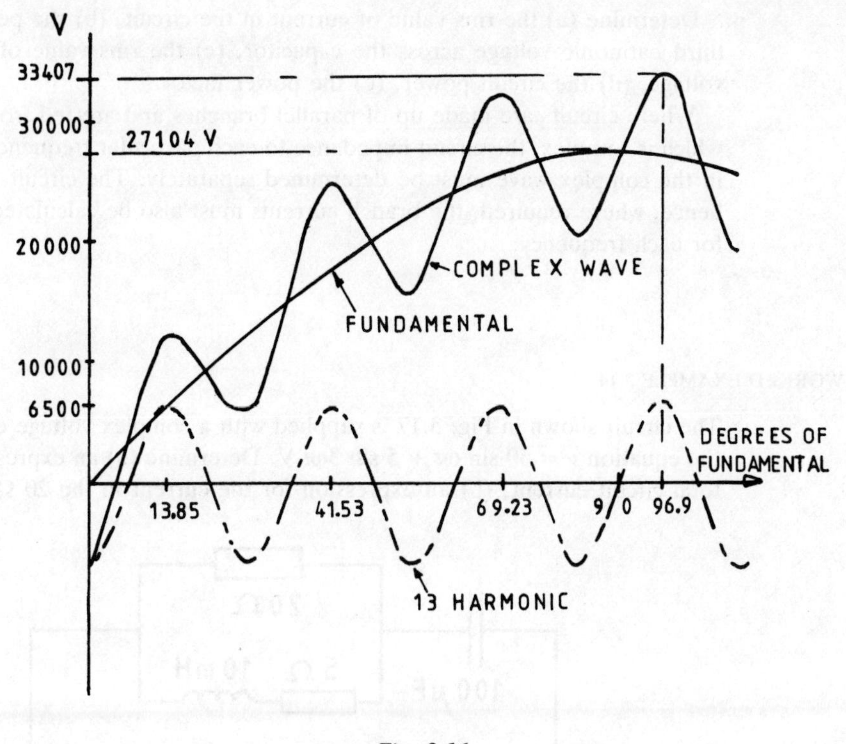

Fig. 3.16

Each complete cycle of the 13th harmonic takes the same time as 360/13 = 27.69° of the fundamental. In Fig. 3.16 the horizontal scale is marked in degrees of the fundamental. The 13th harmonic starts at a maximum negative (−90° at harmonic frequency or 90/13 = 6.92° on the fundamental scale). During the first 90° of the fundamental the 13th harmonic completes $3\frac{1}{4}$ cycles. Adding the fundamental and harmonic waves together gives the complex wave which is observed to have a maximum value at 96.9°. Here, the fundamental has a value 27 104 sin 96.9° = 26 907 V. Adding the peak of the harmonic, 6500 V, gives a value of 33 407 V.

This is the required answer. It is not necessary to plot the negative half cycle of the fundamental voltage since an odd harmonic, which is known to produce identical positive and negative half cycles, is being added.

SELF-ASSESSMENT EXAMPLE 3.13

A voltage described by the equation

$$v = 2000 \sin \omega t + 400 \sin 3\omega t + 100 \sin 5\omega t \text{ V}$$

is applied to a series circuit comprising a resistance of 10 Ω, a capacitance of 30 μF, and an inductance of 0.04114 H. (ω = 300 rad/s.)

Determine (a) the rms value of current in the circuit, (b) the peak value of third harmonic voltage across the capacitor, (c) the rms value of the supply voltage, (d) the circuit power, (e) the power factor.

Where circuits are made up of parallel branches and are fed from a voltage which is complex, the circuit impedance to each particular frequency contained in the complex wave must be determined separately. The circuit current and hence, where required, the branch currents must also be calculated separately for each frequency.

WORKED EXAMPLE 3.14

The circuit shown in Fig. 3.17 is supplied with a complex voltage described by the equation $v = 50 \sin \omega t + 5 \sin 3\omega t$ V. Determine (a) an expression for the total circuit current, (b) an expression for the current in the 20 Ω resistor.

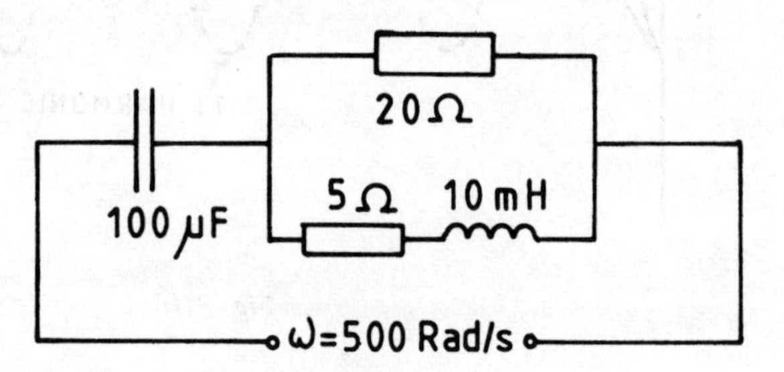

Fig. 3.17

At the fundamental frequency, $X_C = \dfrac{1}{500 \times 100 \times 10^{-6}} = 20\ \Omega$.

$$Z_{coil} = 5 + j500 \times 10 \times 10^{-3} = (5 + j5)\ \Omega.$$

$$\text{Equivalent impedance of the parallel branches} = \frac{20(5 + j5)}{20 + 5 + j5}$$

$$= (4.62 + j3.08)\ \Omega.$$

Adding the reactance of the capacitor

$$\text{Total impedance} = 4.62 + j3.08 - j20$$
$$= (4.62 - j16.92)\ \Omega.$$

Total circuit current at the fundamental frequency

$$= \frac{50}{(4.62 - j16.92)}\ \text{A}.$$

Rationalising and converting to polar form yields

$$I = 2.85\angle 74.72^\circ\ \text{A}.$$
$$i_1 = 2.85 \sin (\omega t + 74.72^\circ)\ \text{A}.$$

Potential difference across the parallel arms = $I_1 \times$ equivalent impedance $= 2.85\angle 74.72° \times (4.62 + j3.08) = 2.85\angle 74.72° \times 5.55\angle 33.69°$ $= 15.82\angle 108.4°$ V.

Therefore

$$\text{peak current in the } 20\ \Omega \text{ resistor} = \frac{15.82\angle 108.4°}{20} = 0.79\angle 108.4° \text{ A.}$$

$$i_{(20\ \Omega)} = 0.79 \sin(\omega t + 108.4°) \text{ A.}$$

At the third harmonic frequency

$$X_C = \frac{1}{3 \times 500 \times 100 \times 10^{-6}} = 6.667\ \Omega.$$

$$Z_{coil} = 5 + j3 \times 500 \times 10 \times 10^{-3} = (5 + j15)\ \Omega.$$

$$\text{Impedance of the parallel branches} = \frac{20(5 + j15)}{20 + 5 + j15}$$

$$= (8.25 + j7.05)\ \Omega.$$

$$\text{Total impedance} = 8.25 + j7.05 - j.6.667 = (8.25 + j0.386)\ \Omega.$$

$$\text{Total circuit current at third harmonic frequency} = \frac{5}{(8.25 + j0.386)}$$

$$= 0.605\angle -2.71° \text{ A.}$$

$$i_3 = 0.605 \sin(3\omega t - 2.71°) \text{ A.}$$

Potential difference across the parallel arms = $I_3 \times$ equivalent impedance $= 0.605\angle -2.71° \times (8.25 + j7.05) = 0.605\angle -2.71° \times 10.85\angle 40.51° =$ $6.57\angle 37.8°$ V.

Hence

$$\text{peak current in the } 20\ \Omega \text{ resistor} = \frac{6.57\angle 37.8°}{20} = 0.328\angle 37.8° \text{ A.}$$

$$i_{(20\ \Omega)} = 0.328 \sin(3\omega t + 37.8°) \text{ A.}$$

(a) The total current is expressed as

$$i_{(T)} = 2.85 \sin(\omega t + 74.2°) + 0.605 \sin(3\omega t - 2.71°) \text{ A.}$$

(b) The current in the 20 Ω resistor is expressed as

$$i_{(20\ \Omega)} = 0.79 \sin(\omega t + 108.4°) + 0.382 \sin(3\omega t + 37.8°) \text{ A.}$$

3.9 Errors in the measurement of impedance created by the presence of harmonics

Considerable error may occur when measuring the impedance of a circuit containing reactive components if any harmonic content of the supply voltage is not allowed for. For example, if the reactance of a coil is to be determined, using instruments which indicate rms values, the impedance is given by

$$Z = \frac{V}{I}\ \Omega.$$

Also

$$Z = \sqrt{(R^2 + X_L^2)}\ \Omega.$$

For small resistance values the impedance is very nearly the same as the reactance. Now if sinusoidal waveforms are assumed, only one frequency is present, and the apparent inductance of the coil is found from

$$X_L = 2\pi f L\ \Omega.$$

Hence

$$L = \frac{X_L}{2\pi f}\ \text{H}.$$

However, let us consider the true effect when a complex voltage is applied to a coil with inductance L henry and a resistance small enough to be ignored.

Let the voltage wave be expressed as

$$v = V_{m(1)} \sin \omega t + V_{m(2)} \sin 2\omega t + \ldots + V_{m(n)} \sin n\omega t\ \text{V}.$$

At the fundamental frequency $i_1 = \dfrac{V_{m(1)}}{\omega L} \sin(\omega t - 90°)$ A.

At the second harmonic frequency $i_2 = \dfrac{V_{m(2)}}{2\omega L} \sin(2\omega t - 90°)$ A

and so on, up to the nth harmonic $i_n = \dfrac{V_{m(n)}}{n\omega L} \sin(n\omega t - 90°)$ A.

The rms value of the current (equation 3.7) gives

$$I = \sqrt{\left[\frac{1}{2}\left(\frac{V_{m(1)}}{\omega L}\right)^2 + \frac{1}{2}\left(\frac{V_{m(2)}}{2\omega L}\right)^2 + \ldots + \frac{1}{2}\left(\frac{V_{m(n)}}{n\omega L}\right)^2\right]}$$

$$= \frac{1}{\sqrt{2}\omega L}\sqrt{\left[V_{m(1)}^2 + \left(\frac{V_{m(2)}}{2}\right)^2 + \ldots + \left(\frac{V_{m(n)}}{n}\right)^2\right]}\ \text{A}.$$

The rms value of the voltage (equation 3.6) gives

$$V = \sqrt{\left[\frac{(V_{m(1)})^2}{2} + \frac{(V_{m(2)})^2}{2} + \ldots + \frac{(V_{m(n)})^2}{2}\right]}$$

$$= \frac{1}{\sqrt{2}}\sqrt{(V_{m(1)}^2 + V_{m(2)}^2 + \ldots + V_{m(n)}^2)}.$$

The impedance of the circuit as $\dfrac{V}{I}$

$$= \frac{\dfrac{1}{\sqrt{2}}\sqrt{(V_{m(1)}^2 + V_{m(2)}^2 + \ldots + V_{m(n)}^2)}}{\dfrac{1}{\sqrt{2}\omega L}\sqrt{\left[V_{m(1)}^2 + \left(\dfrac{V_{m(2)}}{2}\right)^2 + \ldots + \left(\dfrac{V_{m(n)}}{n}\right)^2\right]}}$$

$$= \omega L \sqrt{\left[\frac{V_{m(1)}^2 + V_{m(2)}^2 + \ldots + V_{m(n)}^2}{V_{m(1)}^2 + \frac{(V_{m(2)})^2}{4} + \ldots + \frac{(V_{m(n)})^2}{n^2}}\right]}. \quad (3.9)$$

Now ωL is the reactance of the coil at the fundamental frequency. When harmonics are present the value of the reactance calculated as V/I is too large by the factor

$$\sqrt{\left[\frac{V_{m(1)}^2 + V_{m(2)}^2 + \ldots + V_{m(n)}^2}{V_{m(1)}^2 + \frac{(V_{m(2)})^2}{4} + \ldots + \frac{(V_{m(n)})^2}{n^2}}\right]}.$$

Where a pure capacitor is concerned a similar proof yields the result

$$\frac{V}{I} = \frac{1}{\omega C} \sqrt{\left[\frac{V_{m(1)}^2 + V_{m(2)}^2 + \ldots + V_{m(n)}^2}{V_{m(1)}^2 + 4(V_{m(2)})^2 + \ldots + n^2(V_{m(n)})^2}\right]}. \quad (3.10)$$

The calculated value of capacitive reactance (V/I) is less than the true value since in this case the value of the square-root term is always less than unity.

For circuits containing resistance and a reactive component an error will still be present but it will be reduced by the presence of the resistance.

WORKED EXAMPLE 3.15

A coil with negligible resistance is supplied with an e.m.f. of the form

$v = 28.24 \sin \omega t + 33.88 \sin 3\omega t + 35.3 \sin 5\omega t$ V
($\omega = 2\pi \times 50$ rad/s).

The rms value of the current in the coil is measured using a moving-iron instrument and is found to be 15.64 A.

Determine the value of the coil inductance and the percentage error when calculating it as V/I.

Equation (3.6)

$$V = \sqrt{\left[\frac{28.24^2}{2} + \frac{33.88^2}{2} + \frac{35.3^2}{2}\right)} = 39.95 \text{ V}.$$

$$\frac{V}{I} = \frac{39.95}{15.64} = 2.55\ \Omega.$$

$$\text{Apparent inductance} = \frac{2.55}{2\pi \times 50} = 8.12 \text{ mH}.$$

Using equation (3.9)

$$\frac{V}{I} = \omega L \sqrt{\left[\frac{28.24^2 + 33.88^2 + 35.3^2}{28.24^2 + \frac{33.88^2}{9} + \frac{35.3^2}{25}}\right]} = \sqrt{\left(\frac{797.5 + 1147.8 + 1246.1}{797.5 + 127.54 + 49.48}\right)}$$

$$2.55 = \omega L \sqrt{(3.27)} = 1.809\ \omega L. \qquad (V/I = 2.55\ \Omega, \text{ from above})$$

$$\text{Therefore } L = \frac{2.55}{2\pi \times 50 \times 1.809} = 4.49 \text{ mH.}$$

$$\text{Error} = 8.12 - 4.49 = 3.63 \text{ mH.}$$

As a percentage of the true value this is

$$\frac{3.63}{4.49} \times 100\% = 80.85\% \text{ error.}$$

SELF-ASSESSMENT EXAMPLE 3.16

A pure capacitor with capacitance 10 μF is fed with a complex e.m.f. The supply current is given by the equation

$$i = 3 \sin(\omega t + 90°) + 4.3 \sin(3\omega t + 150°) \text{ A}$$
$$(\omega = 2\pi \times 5000 \text{ rad/s}).$$

Determine (a) the equation for the voltage wave, (b) the apparent value of the capacitance calculated using the rms values of voltage and current ($V/I = X_C = 1/\omega C\ \Omega$).

FURTHER PROBLEMS

3.17 What are the fundamental and harmonic frequencies of a voltage wave described by the equation

$$v = 100 \sin 314.2t + 50 \sin(942.5t + 60°) + 10 \sin(1571t - 45°) \text{ V?}$$

3.18 Determine the rms value of a complex voltage whose instantaneous value is given by the equation

$$v = 100 \sin(\omega t + 30°) + 20 \sin(3\omega t - 90°) + 10 \sin(5\omega t + 15°) \text{ V.}$$

3.19 (a) Explain what 'harmonics' of a non-sinusoidal, periodic alternating wave are. (b) A series circuit, consisting of a resistor with a resistance of 10 Ω, an inductor with value 0.0375 H and negligible resistance and a capacitor, is supplied from a power source with an output voltage described by the equation

$$v = 2000 \sin \omega t + 600 \sin 3\omega t + 400 \sin 5\omega t \text{ V}$$
$$(\omega = 2\pi \times 50 \text{ rad/s}).$$

Determine (i) the required value of capacitance to give resonance at the third harmonic frequency, then with this value of capacitance in circuit (ii) the rms value of current, (iii) the circuit power, (iv) the peak voltage at 3rd and 5th harmonic frequencies developed across the capacitor.

3.20 A complex voltage waveform is described by the equation

$$v = 200 \sin \omega t + 40 \sin 3\omega t \text{ V}$$

($\omega = 100\pi$ rad/s). It is applied to a coil of inductance 200 mH and resistance 50 Ω.

Calculate (a) the rms value of the voltage, (b) the rms value of the current, (c) the power, (d) the power factor.

3.21 (a) Explain what is meant by the term 'selective resonance'. (b) A coil with resistance 10 Ω and inductance 50 mH is connected in parallel with a capacitor across a supply voltage given by the equation

$$v = 0.5 \sin \omega t + 0.1 \sin (3\omega t + 30°) + 0.05 \sin (5\omega t + 90°) \text{ V}$$

($\omega = 2\pi \times 7500$ rad/s).

Calculate the value of capacitance necessary to give resonance at the 5th harmonic frequency.

With this value of capacitance in circuit, deduce an expression for the instantaneous current in the capacitor.

3.22 Sketch the shapes of the following complex waves over a complete cycle of the fundamental (a) $10 + 30 \sin \omega t + 10 \sin (2\omega t - 90°)$, (b) $5 \sin \omega t + 2 \sin (3\omega t + 180°)$, (c) $2 + 4 \sin \omega t + 1.5 \sin (3\omega t + 90°)$.

3.23 The generator in Fig. 3.18 produces a complex e.m.f. with a fundamental having an rms value of 20 V and a 3rd harmonic of 10 V rms. The capacitor has a reactance of 6 Ω at the fundamental frequency. Use Thevenin's theorem to find the current in the 3 Ω resistor due to the fundamental and third harmonic voltages separately and hence the rms value of the current due to these currents combined.

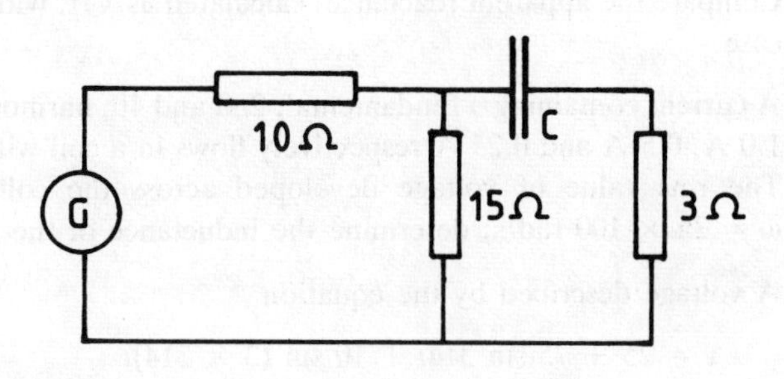

Fig. 3.18

3.24 (a) Explain briefly with the aid of a diagram why the magnetising current of a single-phase transformer contains harmonics. (b) A current containing a fundamental, 3rd and 5th harmonics of peak amplitudes 5 A, 2.45 A and 1 A respectively flows in an inductive circuit. The power input to the circuit is 48 W. The inductance of the coil is 0.01 H.

Determine (i) the rms value of potential difference developed across the coil, (ii) the circuit power factor. (The fundamental frequency is 100 Hz.)

3.25 A coil with inductance L henry and resistance R Ω is supplied with a complex e.m.f. given by the equation

$$v = 100 \sin \omega t + V_{m(2)} \sin (2\omega t + 60°) + V_{m(4)} \sin (4\omega t - 10°) \text{ V}.$$

The resulting current in the coil is described

$$i = 2.685 \sin(\omega t - 57.51°) + 0.758 \sin(2\omega t - 12.34°) + 0.039 \sin(4\omega t - 90.95°) \text{ A}$$

($\omega = 200\pi$ rad/s).

Calculate (a) the impedance of the circuit at the fundamental frequency and hence the values of R and L, (b) the value of $V_{m(2)}$, (c) the value of $V_{m(4)}$, (d) the circuit power, (e) the power factor.

3.26 A voltage given by the equation

$$v = 50 + 120 \sin(\omega t + 30°) + 25 \sin(3\omega t - 45°) \text{ V}$$

is applied to the circuit in Fig. 3.19 ($\omega = 2\pi \times 100$ rad/s). Derive an expression for the current in the branch AB.

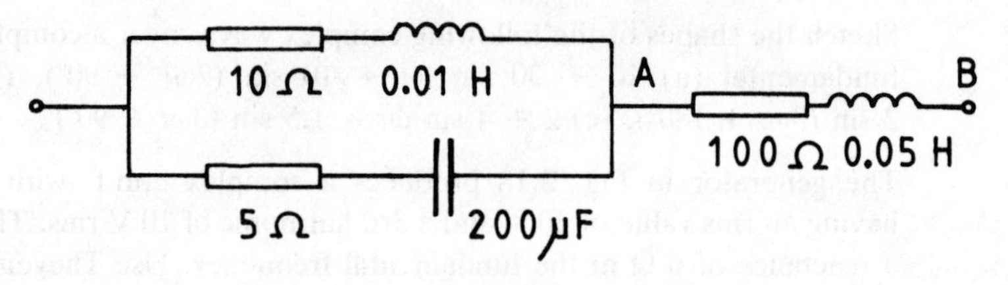

Fig. 3.19

3.27 A voltage wave containing a fundamental and 20% third harmonic is applied to (a) an inductor with negligible resistance, (b) a pure capacitor. Compare the apparent reactance, calculated as V/I, with the true value in each case.

3.28 A current containing a fundamental, 2nd and 4th harmonics of peak amplitudes 1.0 A, 0.5 A and 0.25 A respectively flows in a coil with negligible resistance. The rms value of voltage developed across the coil is 15 V. Given that $\omega = 2\pi \times 100$ rad/s, determine the inductance of the coil.

3.29 A voltage described by the equation

$$v = 25 + 75 \sin 314t + 10 \sin(3 \times 314)t$$

is applied to the circuit shown in Fig. 3.20. Determine (a) the rms value of current in each branch, (b) the circuit power.

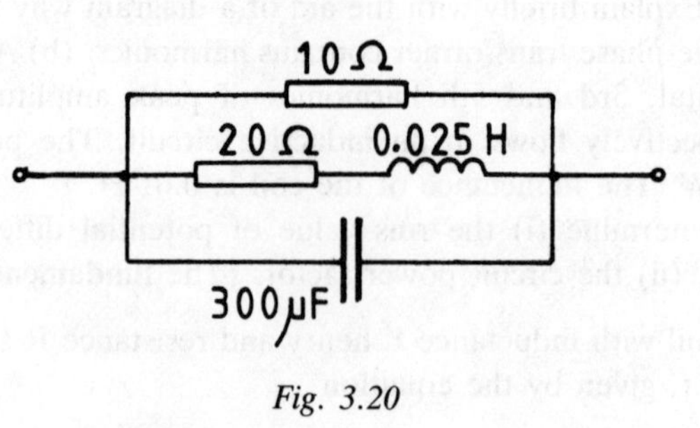

Fig. 3.20

3.30 The dynamic current/voltage characteristic of an amplifier with a resistive load may be represented by the equation

$i = a + bv_1 + cv_1$ A.

(a) Deduce the ratio of second harmonic current to the fundamental in the output. (b) If the amplifier has a steady, no-signal current of 60 mA and the application of a sinusoidal voltage causes the output to vary between 105 mA and 25 mA, calculate the percentage second harmonic content in the output current. (c) What would be the effect on the harmonic content of the output of reducing the magnitude of the input signal?

Answers

3.1 See Fig. 3.21.

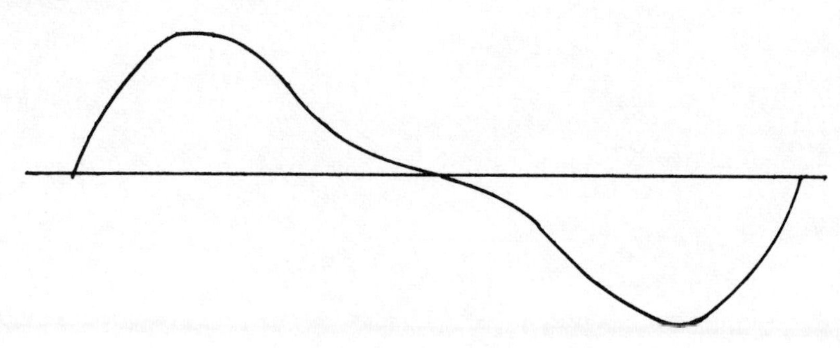

Fig. 3.21

3.2 Second harmonic content reduced to (i) 1.04%, (ii) 1.98%.
3.4 15.27 A
3.5 $25/\sqrt{2}$ V
3.7 (a) 39.37 V (b) 10.54 A (c) 209.92 W (d) 0.506 (e) 10Ω
3.9 $i = 9.84 \sin(\omega t - 38.1°) + 0.98 \sin(3\omega t - 7°) + 0.247 \sin(5\omega t + 74.3°)$; $I = 6.99$ A; power = 244.6 W
3.11 $v = 25 \sin \omega t + 10 \sin(3\omega t - 90°) + 2 \sin(5\omega t - 50°)$ V (a) $V = 19.1$ V (b) 11.28 W (c) 0.746
3.13 (a) 31.72 A (b) 1481 V (c) 1445 V (d) 10069 W (e) 0.218
3.16 $v = 9.55 \sin \omega t + 4.56 \sin(3\omega t + 60°)$ V; $V = 7.48$ V; $I = 3.707$ A; X_C 2.018 Ω; apparent $C = 15.77$ μF
3.17 50 Hz; 150 Hz; 250 Hz
3.18 72.46 V
3.19 (b) (i) 30 μF (ii) 45.53 A (iii) 20.73 kW (iv) 10.26 A: $X_C = 21.22$ Ω, 217.7 V; 60 A: $X_C = 35.4$ Ω, 2122 V
3.20 (a) 144.2 V (b) 1.77 A (c) 156 W (d) 0.613
3.21 (b) 360.25 pF; $i = 8.488 \sin(\omega t + 90°) + 5.093 \sin(3\omega t + 120°) + 4.244 \sin(5\omega t + 180°)$ μA
3.23 Fundamental: (0.922 + j0.616) A; third: (0.635 + j0.141) A; rms = 1.29 A
3.24 (b) $R = 3$ Ω (i) 46.83 V rms (ii) 0.256

3.25 (a) $R = 20\ \Omega$; $X_L = 31.41\ \Omega$; $L = 50$ mH (b) 50 V (c) 5 V
(d) 77.85 W (e) 0.499
3.26 $i = 4.55 + 1.083 \sin(\omega t + 14.84°) + 0.178 \sin(3\omega t - 86.44°)$ A
3.27 (a) 1.0175 × true value (b) 0.874 × true value
3.28 19.5 mH
3.29 (a) Top: 5.905 A; middle: 2.776 A; bottom: 5.38 A (b) 503 W
3.30 (b) 6.25% (c) a reduction

Chapter 4
Electrostatics

4.1 Coulomb's law

The basic model of the atom consists of a positively charged nucleus and orbiting electrons which in total carry the same charge as the nucleus but of opposite polarity. Benjamin Franklin discovered that when certain insulating substances are rubbed vigorously with fabric they become electrically charged. Glass rubbed with silk becomes positively charged since the silk removes electrons from the glass leaving some atoms with insufficient electrons to neutralise the positive charge on the nucleus. Ebonite rubbed with fur becomes negatively charged since the fur donates electrons to the ebonite.

It must be said that the terms 'positive' and 'negative' were chosen quite arbitrarily since the reasons for the phenomena were not understood at the time. All our conventions with respect to currents, forces, charges and the like have been established on the basis of that original choice.

It was found that two charged glass rods repelled one another and two charged ebonite rods repelled one another. However a charged glass rod attracted a charged ebonite rod. This implies that like charges of either polarity repel one another while unlike charges attract one another.

Coulomb experimented with such charges and concluded that the force between two charges q_1 and q_2 could be expressed in the form:

$$F \propto \frac{q_1 q_2}{d^2} \qquad \text{or introducing a constant k, } F = \frac{\mathrm{k} q_1 q_2}{d^2} \qquad (4.1)$$

where F = force and d = distance between the charges, both in a particular system of units.

4.2 Lines of electric force

A charge of q_1 coulombs at a point will have an effect all round it. Another charge q_2 of like polarity brought near to q_1 from any direction will be repelled along a line drawn through the two charges. From equation (4.1) it will be deduced that the effect will be identical for any particular distance d between the charges whatever the direction of approach of q_2.

Suppose charge q_1 is positive and is situated at the centre of an imaginary sphere with radius r metres. For charge $+q_2$ at any point on the surface of that sphere the values of the forces will all be the same. The directions of the forces are all normal (at right angles) to the surface of the sphere. For a straight conductor carrying charge $+q_1$ coulombs per metre length, a charge $+q_2$ brought near to it from any direction will be repelled along a radius drawn normal to the surface of the conductor. Again, for any particular distance r from the conductor, all the forces will be equal in magnitude. The force directions will be reversed for q_1 negative. A plan showing the directions of forces acting on a positive charge $(+q_2)$ is known as an electrostatic field plot and the line along which the charge moves is a line of electric force. Figures 4.1 and 4.2 show the two cases discussed.

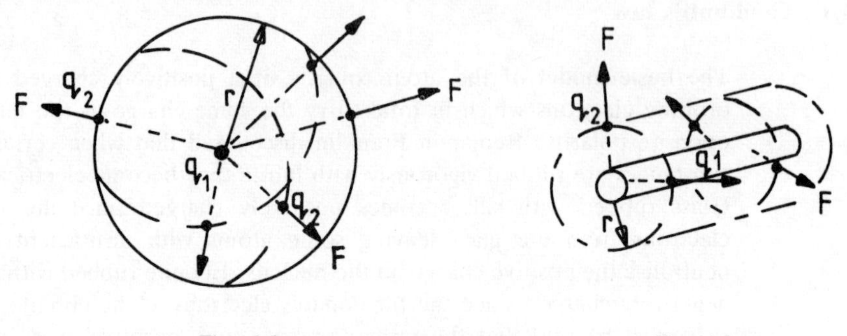

Fig. 4.1 *Fig. 4.2*

For a parallel-plate capacitor the field plot will be as shown in Fig. 4.3.

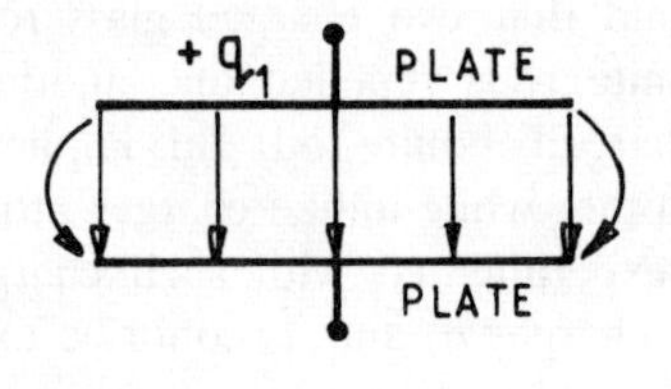

Fig. 4.3

4.3 Charge or flux density

The convention used comes from an electrostatic theorem due to Gauss. This states that the total outward flux normal to the enclosing surface is numerically equal to the charge enclosed.

$$D = \frac{q}{A} \text{ C/m}^2 \tag{4.2}$$

where D = charge or flux density, q = charge (in coulombs) enclosed, A = area (in m^2) across which the flux crosses normally. As an example, consider Fig. 4.1. The charge enclosed is q_1 and the area across which the lines of force and hence electrostatic flux pass is the surface area of the sphere, i.e. $4\pi r^2$ m^2. Hence at any radius r m

$$D = \frac{q_1}{4\pi r^2} \text{ C/m}^2 \tag{4.3}$$

Similarly, for the straight conductor in Fig. 4.2, if a 1 m length of conductor is considered upon which the charge is q_1 C, at any radius r m from the conductor, the surface area of the enclosing cylinder is $2\pi r$ m^2. Hence

$$D = \frac{q_1}{2\pi r} \text{ C/m}^2. \tag{4.4}$$

Finally, for the parallel-plate capacitor in Fig. 4.3, neglecting the curved lines or fringing at the edges of the plates, all lines cross the space between the plates, i.e. the dielectric, normally so that for the dielectric area A m^2

$$D = \frac{q_1}{A} \text{ C/m}^2. \tag{4.5}$$

4.4 Electric force or potential gradient *E*

The electric force or potential gradient in a material is expressed in volts per metre.

$$E = \frac{\text{potential difference in volts}}{\text{distance in metres}}. \tag{4.6}$$

This is analogous to magnetising force (A/m) in electromagnetism.

For the conductor shown in Fig. 4.4, the potential difference between the ends of the conductor is IR volts. This is on one metre length so that $E = IR$ V/m. In a dielectric material the resistance will be much higher and values of potential gradient will be extremely high in consequence. As an example, for the parallel-plate capacitor in Fig. 4.4 (b), d is 2 mm and the applied voltage is 4000 V, so that

$$E = \frac{4000}{2} = 2000 \text{ V/mm } (= 2 \times 10^6 \text{ V/m}).$$

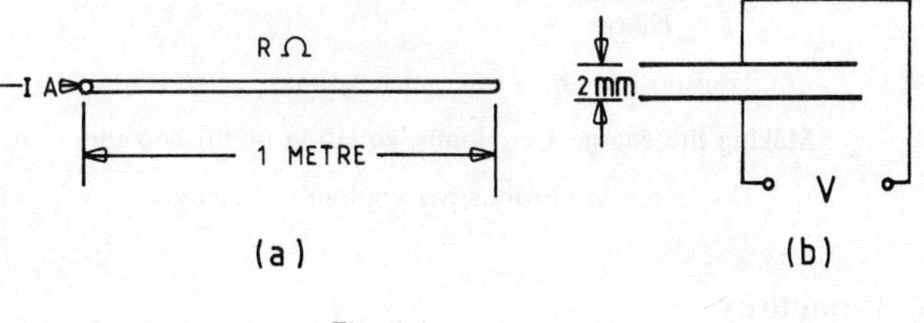

Fig. 4.4

4.5 Force on a charge in an electrostatic field

The energy (work done) in an electric circuit is given by

$$W = VIt = Vq \text{ J}. \tag{4.7}$$

where W = work done, I = current in amperes, t = time in seconds and V = potential difference.

Charge in coulombs $q = It$.

Transposing equation (4.7) gives

$$\frac{W}{q} = V. \tag{4.8}$$

In words, equation (4.8) states 'potential difference is equal to work done per coulomb of charge moved in a circuit'.

This is a useful concept when it comes to the absolute potential of a point in space. The absolute potential may be expressed in terms of work done in bringing unit charge (1 coulomb) from an infinite distance at which there is no force felt, up to the point at which the potential is to be determined.

Absolute potential at a point = energy required to bring the unit positive charge to this point from one at which no force is felt.

Transposing equation (4.6) gives

$$\text{p.d.} = E \times \text{distance}.$$

From equation (4.8)

$$\text{p.d.} = \frac{\text{work done}}{\text{charge}}.$$

Therefore

$$E \times \text{distance} = \frac{\text{work done}}{\text{charge}} = \frac{\text{force (N)} \times \text{distance}}{\text{charge}}.$$

Cancelling distance on both sides of the equation gives

$$E = \frac{\text{force (N)}}{\text{charge}}. \tag{4.9}$$

Transposing Eq = force in newtons. (4.10)

Making the charge 1 coulomb, equation (4.10) becomes

E = force in newtons per coulomb of charge. (4.11)

4.6 Permittivity

For any arrangement of conductors and dielectric the relationship between electric flux density D and potential gradient E is expressed as

$$\frac{D}{E} = \varepsilon$$

where ε is the permittivity of the dielectric.

For vacuum and air the permittivity of free space ε_0 (= 8.85×10^{-12} F/m) is used. For other materials a relative number must be included. This expresses how much better than vacuum the particular material is and is called the relative permittivity and is denoted by ε_r.

$$\frac{D}{E} = \varepsilon_o\varepsilon_r \quad \text{or} \quad E = \frac{D}{\varepsilon_o\varepsilon_r} \tag{4.12}$$

4.7 Electric potential in terms of *E*

In Fig. 4.5, a charge $+Q$ is located at P. A charge $+q$ is presently situated at A and may be moved along the x axis.

Between the two charges there will be a force of repulsion since they are of like polarity. At point A an electric field of strength E_x is present due to charge $+Q$ at P.

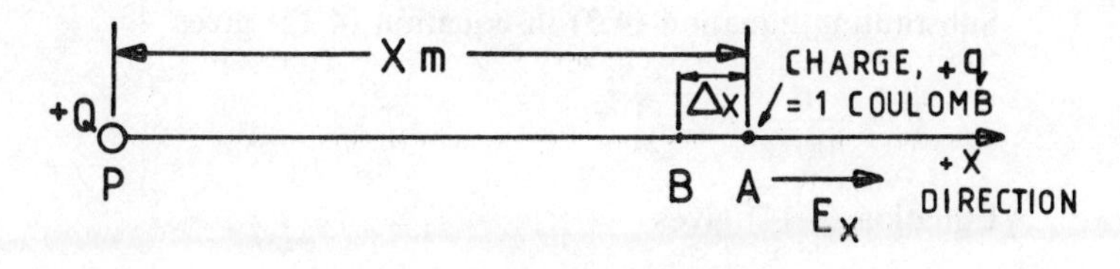

Fig. 4.5

To move the unit charge $+q$ a distance Δx to point B will require work to be done.

$$\text{Work done} = \text{force} \times \text{distance} = F\,\Delta x \text{ Nm.} \tag{4.13}$$

But from equation (4.11), E = force in newtons per coulomb of charge. Hence, $E = F$ and work done $= E\,\Delta x$ Nm (from equation 4.13).

From equation (4.8), the p.d. between two points is expressed in newton metres of work done per coulomb of charge displaced. Therefore

$$\text{p.d. between A and B} = \text{work done on 1 coulomb} = E\,\Delta x.$$

Coulomb's law states (equation 4.1) that force is inversely proportional to the square of the distance between the charges, so that as x decreases in Fig. 4.5, i.e. as the unit charge is moved in the $-x$ direction, more work has to be done due to the increasing force. This means that the potential is increasing as the charge is moved to the left, i.e. in the $-x$ direction.

To take this into account when required the following is usually written

$$\text{p.d.}_{(AB)} = -E\,\Delta x.$$

In the limit as $\Delta x \to 0$

$$V_{BA} = -\int_a^b E\,dx \text{ V.} \tag{4.14}$$

V_{BA} is the potential difference between two points A and B which are situated at distances a and b respectively from the charge $+Q$. The potential difference between the points is a scalar quantity which depends only on the distance of each point from the charge and not on the distance between the two points. Thus any route may be taken between A and B as it gives the same result. From this can be deduced that moving a charge round any closed loop in a static electric field from A and back to A requires no energy expenditure.

4.8 Relationship between electric lines of force and equi-potential surfaces

Consider a point charge of q coulombs. At any radius r m from the charge

$$\text{flux density } D = \frac{q}{4\pi r^2} \text{ C/m}^2. \qquad \text{(equation 4.3)}$$

$$\text{Also} \qquad E = \frac{D}{\varepsilon} \text{ V/m.} \qquad \text{(equation 4.12)}$$

Substituting equation (4.3) in equation (4.12) gives

$$E = \frac{q}{4\pi r^2 \varepsilon}. \tag{4.15}$$

Equation (4.14) gives

$$V_{BA} = -\int_a^b E \, dx.$$

Therefore, substituting equation (4.15) in equation (4.14) gives

$$V_{BA} = -\frac{q}{\varepsilon}\int_a^b \frac{dr}{4\pi r^2} = \frac{q}{4\pi\varepsilon}\left(\frac{1}{r}\right)_a^b$$

$$= \frac{q}{4\pi\varepsilon}\left(\frac{1}{b} - \frac{1}{a}\right) \text{V.} \tag{4.16}$$

If a is an extremely large radius at which no effect is felt, the absolute potential of point B at radius b becomes

$$V_B = \frac{q}{4\pi\varepsilon}\left(\frac{1}{b}\right) \text{V.}$$

Since q, 4π and ε are constants it can be seen that all points at radius b have the same potential so that a sphere with radius b metres is an equi-potential surface. The equi-potential surface is at right angles to all the lines of force emanating from the charge (see Fig. 4.1). Equi-potential points may be joined by conductors without any effect since if both ends of the conductor are at the same potential no current will flow.

Any sphere with a point charge at its centre is an equi-potential surface so it would be possible to fabricate thin metal spheres and dispose them around such a charge without affecting the electrostatic field.

For a straight conductor

$$D = \frac{q}{2\pi r}\ \text{C/m}^2. \qquad \text{(equation 4.4)}$$

Hence

$$E = \frac{q}{2\pi r\varepsilon}\ \text{(from equation 4.12)}$$

$$V_{BA} = -\frac{q}{2\pi\varepsilon}\int_a^b \frac{1}{r}\,dr = \frac{q}{2\pi\varepsilon}\log_e \frac{a}{b}\ \text{V}.$$

Fixing either a or b as a datum point allows the potential difference between this and any other point to be determined. For example, suppose b is the radius r of the conductor itself, then the potential difference between the surface of the conductor and any point outside the conductor at a distance a normal to its surface is given by

$$V = \frac{q}{2\pi\varepsilon}\log_e \frac{a}{r}.$$

All such points lie on the surface of a cylinder with radius a surrounding the conductor.

Any concentric cylinder surrounding a conductor forms an equi-potential surface and cylindrical metal sheaths can be disposed within a solid dielectric with no effect on the field. In fact some bus-bars for power application and transformer bushings contain aluminium foil cylinders to smooth out any irregularity in the field which might be present due to non-uniformity within the dielectric, especially in the surface layers where paper, air and porcelain may be present in varying thicknesses.

WORKED EXAMPLE 4.1

In Fig. 4.6, conductors A and B are situated in air and carry charges of 10×10^{-9} C/m length with opposite polarity. The conductors are 0.7 m apart.

Calculate (a) the field strength at conductor B due to the charge carried on conductor A, (b) the force developed between the conductors.

(a) Using equation (4.4), the flux density D due to charge on conductor A at any radius r m is given by

$$D = \frac{q}{2\pi r} = \frac{10 \times 10^{-9}}{2\pi \times 0.7}.$$

From equation (4.12)

$$E = \frac{D}{\varepsilon_0}$$

$$= \frac{10 \times 10^{-9}}{2\pi \times 0.7 \times 8.85 \times 10^{-12}} = 256.9\ \text{V/m}.$$

(b) Force = Eq (equation 4.10). Conductor B, which carries a charge of 10×10^{-9} C/metre length, is situated in an electrostatic field of strength 256.9 V/m set up by the charge on conductor A. Therefore

$$\text{force} = 256.9 \times 10 \times 10^{-9} = 2.569\ \mu\text{N/m length}$$

(this is a force of attraction, since unlike charges attract one another).

SELF-ASSESSMENT EXAMPLE 4.2

Figure 4.6 shows the electrostatic field plot between a pair of long parallel conductors carrying opposite charges. Using the fact that equi-potential lines are at right angles to force lines, deduce the shape of the equi-potential surfaces between the conductors.

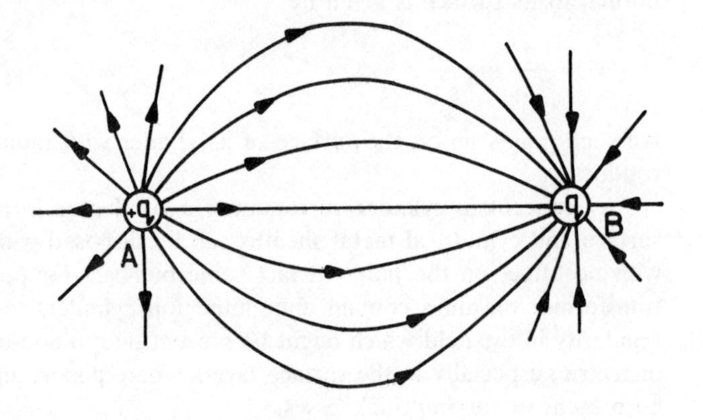

Fig. 4.6

4.9 Capacitance of a parallel-plate capacitor

The capacitance of a capacitor is measured in farads. By definition, a capacitor has a capacitance of 1 farad if, when a potential difference of 1 volt is applied between the plates, 1 coulomb of charge is stored. In symbols

$$q = CV.$$

where C = capacitance in farads.
For two parallel plates as shown in Fig. 4.3

$$E = \frac{D}{\varepsilon}. \qquad \text{(equation 4.12)}$$

When the voltage is applied to the capacitor a charge q coulombs is displaced, being taken from one plate and stored on the other.

$$\text{The charge density, } D = \frac{q}{A}\ \text{C/m}^2. \qquad \text{(equation 4.5)}$$

Substituting equation (4.5) into equation (4.12) gives

$$E = \frac{q}{A\varepsilon}$$

$$\text{but } V = -\int_a^b E \, dx \qquad \text{(equation 4.14)}$$

b and a are general limits giving the potential difference between two points A and B. In the parallel-plate capacitor the potential difference exists between two plates, d metres apart so that the limits become 0 and d.

$$V = -\int_0^d \left(\frac{q}{A\varepsilon}\right) dx = -\left(\frac{qx}{A\varepsilon}\right)_0^d = -\frac{qd}{A\varepsilon} \text{ V}. \qquad (4.17)$$

In this case the negative sign has no significance since either polarity can be obtained by reversing the connections to the plates.

Substituting $q = CV$ in equation (4.17) gives

$$V = \frac{CVd}{A\varepsilon}$$

$$\text{Hence } C = \frac{A}{d} \varepsilon \text{ F} \qquad (4.18)$$

4.10 Concentric cylinders (the co-axial cable)

4.10.1 Capacitance

Consider 1 metre length of co-axial cable as shown in Fig. 4.7. The centre conductor with radius r carries current to a load; the return being through the sheath which has inner radius R.

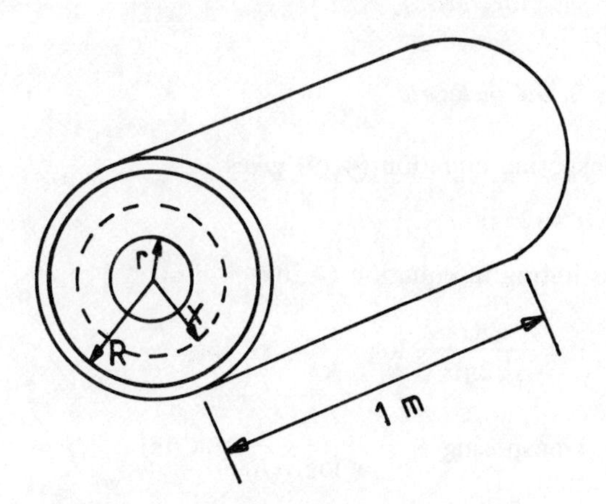

Fig. 4.7

The dielectric has absolute permittivity ε.
The surface area of a cylinder within the dielectric with radius x is $2\pi x$ m².

$$\text{The charge density at this radius} = \frac{q}{2\pi x}\ \text{C/m}^2. \qquad \text{(equation 4.4)}$$

$$\text{From equation (4.12), } E = \frac{D}{\varepsilon} = \frac{q}{2\pi x \varepsilon}\ \text{V/m}. \qquad (4.19)$$

$$\text{The potential difference between core and sheath, } V = -\int_a^b E\,\mathrm{d}x. \qquad \text{(equation 4.14)}$$

Now, if the potential of the core is higher than that of the sheath, the potential is highest at radius r and equation (4.14) becomes

$$V = -\int_R^r \left(\frac{q}{2\pi\varepsilon}\frac{1}{x}\,\mathrm{d}x\right).$$

Substituting equation (4.19) in equation (4.14)

$$V = \frac{q}{2\pi\varepsilon}(\log_e x)\frac{r}{R} = -\frac{q}{2\pi\varepsilon}\log_e\frac{r}{R}. \qquad (4.20)$$

Substituting $q = CV$ in equation (4.20) gives

$$V = -\frac{CV}{2\pi\varepsilon}\log_e\frac{r}{R}\ \text{V}.$$

Hence $C = \dfrac{-2\pi\varepsilon}{\log_e (r/R)}$ farads per metre length of the cable.

This may be rearranged to eliminate the negative sign

$$C = \frac{2\pi\varepsilon}{\log_e (R/r)}\ \text{F/m}. \qquad (4.21)$$

4.10.2 *Stresses in the dielectric*

Transposing equation (4.19) gives

$$q = 2\pi x\varepsilon\, E\ \text{C}.$$

Substituting in equation (4.20)

$$V = \frac{-2\pi x\varepsilon\, E}{2\pi\varepsilon}\log_e\frac{r}{R} = xE\log_e\frac{R}{r}.$$

$$\text{Transposing } E = \frac{V}{x\log_e (R/r)}\ \text{V/m}. \qquad (4.22)$$

Thus, the potential gradient in a co-axial cable has a maximum value when x has a minimum value. The smallest possible value for x within the dielectric is when $x = r$, the inner conductor radius. Thus

$$E_m = \frac{V}{r \log_e (R/r)} \text{ V/m.} \tag{4.23}$$

WORKED EXAMPLE 4.3

A concentric cable has inner sheath radius 5 cm and core radius 2 cm. The core potential is 53.7 kV rms with respect to the sheath. The relative permittivity of the dielectric is 2.5.

Calculate the capacitance of the cable per kilometre length.

Draw a graph showing the variation of potential gradient through the dielectric between core and sheath.

Substituting the values in equation (4.21) gives

$$C = \frac{2\pi \times 8.85 \times 10^{-12} \times 2.5}{\log_e (5/2)} \qquad (\varepsilon = \varepsilon_0\varepsilon_r)$$

$$= 151.7 \text{ pF/m} = 151.7 \text{ nF/km.} \quad (\times 1000)$$

The potential gradient throughout the dielectric is determined from equation (4.22)

$$53.7 \text{ kV rms} = 53\,700 \times \sqrt{2} = 75\,943.3 \text{ V peak.}$$

Substitute in equation (4.22)

$$E = \frac{75\,943.3}{x \log_e (5/2)} = \frac{75\,943.3}{x(0.916)} = \frac{82\,881.2}{x}.$$

For different values of x the corresponding values of E are calculated

x (cm)	2	3	4	5
E (V/cm)	41 440	27 627	20 720	16 576

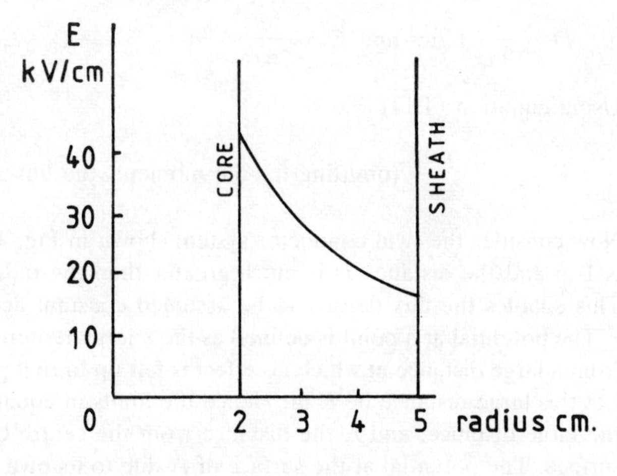

Fig. 4.8

These are the stresses at the various radii which are present at the peak of the voltage wave and are therefore the values which the dielectric must withstand if it is not to fail in service (Fig. 4.8).

SELF-ASSESSMENT EXAMPLE 4.4

Investigate the effect on the values of potential gradient throughout the dielectric in the concentric cable in worked example 4.3 if the core has a radius of 1.839 cm, all other parameters remaining unchanged.

It may be proved mathematically that the minimum value of stress at the core for a fixed sheath radius and working voltage occurs when R/r = natural e (2.718). In self-assessment example 4.4, $R/r = 5/1.839 = \text{e}$ and the stress has been reduced from that found in worked example 4.3.

4.11 Capacitance of a twin line not in close proximity to earth

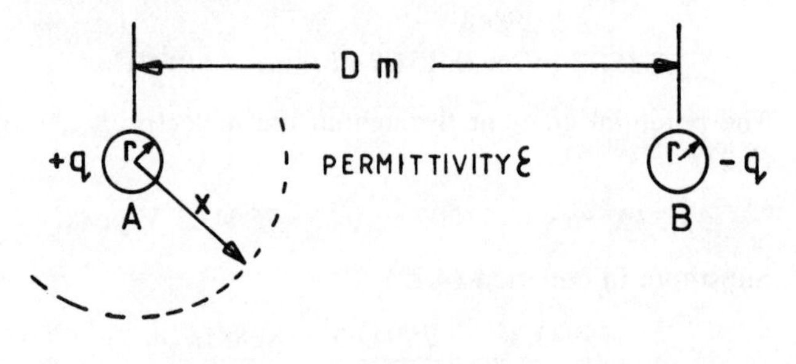

Fig. 4.9

Consider firstly the general case pertaining to a 1 metre length of a single conductor carrying a charge of q coulombs per metre length. At any radius x metres outside the conductor the surface area of the enclosing cylinder is $2\pi x$ m^2. Therefore

$$D = \frac{q}{2\pi x}\ \text{C/m}^2 \text{ and } E = \frac{q}{2\pi x \varepsilon}\ \text{V/m} \qquad \text{(equation 4.19)}$$

Using equation (4.14)

$$V = -\int \frac{q\,\text{d}x}{2\pi\varepsilon x} \text{ (omitting for the moment, the limits).} \qquad (4.24)$$

Now consider the twin conductor system shown in Fig. 4.9. The length of line is 1 m and the distance D is much greater than the radii r of the conductors. This enables the flux density to be assumed constant across a diameter.

The potential at a point is defined as the energy required to bring unit charge from a large distance at which no effect is felt up to that point (see section 4.5). Let this large distance be R m. Hence the limits in equation (4.24) become R, the large distance, and r, the distance from the centre of the conductor to its surface. The potential at the surface of A due to its own charge $+q$ is given by

$$V_A = -\int_R^r \frac{q\,dx}{2\pi\varepsilon x} = \frac{q}{2\pi\varepsilon}\log_e (R/r)\ \text{V (similar method to equation 4.20.)} \tag{4.25}$$

The potential at the surface of conductor B due to charge $+q$ on A is equal to the work done in bringing unit charge from R up to its surface, i.e. the distance D from conductor A. Hence

$$V_B = -\int_R^D \frac{q\,dx}{2\pi\varepsilon x} = \frac{q}{2\pi\varepsilon}\log_e (R/D)\ \text{V.} \tag{4.26}$$

Now the effect of charge $-q$ on B has to be considered. The method is exactly as above except that the sign of q has changed.

Potential at the surface of B due to its own charge $(-q)$ =

$$\frac{-q}{2\pi\varepsilon}\log_e \frac{R}{r}\ \text{V.} \tag{4.27}$$

Potential at the surface of A due to charge $(-q)$ on B =

$$\frac{-q}{2\pi\varepsilon}\log_e \frac{R}{D}\ \text{V.} \tag{4.28}$$

$$\text{Total potential at A} = \frac{q}{2\pi\varepsilon}\left(\log_e \frac{R}{r} - \log_e \frac{R}{D}\right)$$

(equation 4.25 + equation 4.28).

Subtracting logarithms is to perform the division function, hence

$$\text{Total potential at A} = \frac{q}{2\pi\varepsilon}\log_e\left(\frac{R/r}{R/D}\right) = \frac{q}{2\pi\varepsilon}\log_e \frac{D}{r}\ \text{V.}$$

$$\text{Total potential at B} = \frac{-q}{2\pi\varepsilon}\left(\log_e \frac{R}{r} - \log_e \frac{R}{D}\right)$$

(equation 4.27 + equation 4.26)

$$= \frac{-q}{2\pi\varepsilon}\log_e \frac{D}{r}\ \text{V.}$$

Total potential at A − total potential at B = p.d. V_{AB}

$$V_{AB} = \frac{q}{2\pi\varepsilon}\log_e \frac{D}{r} - \left(\frac{-q}{2\pi\varepsilon}\log_e \frac{D}{r}\right) = \frac{2q}{2\pi\varepsilon}\log_e \frac{D}{r} = \frac{q}{\pi\varepsilon}\log_e \frac{D}{r}\ \text{V.}$$

But $q = CV$, hence

$$V = \frac{CV}{\pi\varepsilon}\log_e \frac{D}{r}.$$

Cancelling V on both sides and transposing

$$C = \frac{\pi\varepsilon}{\log_e (D/r)}\ \text{farads per metre length.}$$

WORKED EXAMPLE 4.5

A long straight cylindrical wire of small diameter carrying a charge of q C/m length is suspended at a large distance from earth. Determine (a) the electric field strength at a distance x m from the conductor, (b) the charge per unit length if the potential difference between two points 1.5 cm and 8 cm from the wire is 125 V.

(a) $D = \dfrac{q}{2\pi x}$ C/m^2 considering 1 m length of the wire. (equation 4.5)

$$E = \frac{D}{\varepsilon_0} = \frac{q}{2\pi x \varepsilon_0} \text{ V/m.} \qquad \text{(equation 4.12)}$$

(b) Using equation (4.14)

$$V_{AB} = -\int_a^b E\,dx = -\int_8^{1.5} \frac{q}{2\pi\varepsilon_o}\frac{1}{x}\,dx = \frac{q}{2\pi\varepsilon_o}\log_e\frac{8}{1.5}.$$

But V_{AB} is given as 125 V, therefore

$$125 = \frac{q}{2\pi\varepsilon_o}\log_e\frac{8}{1.5}.$$

Transposing gives

$$q = \frac{2\pi\varepsilon_o \times 125}{\log_e(8/1.5)} = \frac{2\pi \times 8.85 \times 10^{-12} \times 125}{1.674}$$

$$= 4.15 \times 10^{-9} \text{ C/m.}$$

WORKED EXAMPLE 4.6

A parallel-plate capacitor has the following details:

area of plates 0.16 m^2
distance between the plates 0.8 mm
working voltage 3000 V (rms)
relative permittivity of the dielectric 3.3.

Determine (a) the peak electric stress in the dielectric assuming the applied voltage is of sine form, (b) the capacitance of the capacitor.

(a) 3000 V rms = $3000 \times \sqrt{2}$ = 4242.6 V peak.

$$\text{Peak stress} = \frac{\text{peak voltage}}{\text{distance between the plates}} = \frac{4242.6}{0.8} = 5303.3 \text{ V/mm.}$$

(b) $C = \dfrac{\varepsilon_o \varepsilon_r A}{d}$ F (equation 4.18)

$$= \frac{8.85 \times 10^{-12} \times 3.3 \times 0.16}{0.8 \times 10^{-3}} = 5840 \text{ pF.}$$

WORKED EXAMPLE 4.7

A concentric cable is required to have a capacitance of 160 pF/m length. The core diameter, which is determined by its current carrying capacity, is 1 cm. The relative permittivity of the dielectric is 4.

Determine the necessary inside diameter of the sheath.

$$\text{The capacitance of concentric cable per metre length} = \frac{2\pi\varepsilon_o\varepsilon_r}{\log_e (R/r)} \text{ F.}$$

(equation 4.21).

Hence

$$160 \times 10^{-12} = \frac{2\pi \times 8.85 \times 10^{-12} \times 4}{\log_e (R/r)}.$$

Transposing

$$\log_e (R/r) = \frac{2\pi \times 8.85 \times 10^{-12} \times 4}{160 \times 10^{-12}} = 1.39.$$

Therefore

$$e^{1.39} = R/r = 4.015$$

Now $D/d = R/r$, where D = inner diameter of sheath, d = core diameter. Hence

$$\frac{D}{d} = \frac{D}{1} = 4.015$$

Therefore $D = 4.015$ cm.

SELF-ASSESSMENT EXAMPLE 4.8

A single-phase circuit consists of two parallel conductors 0.6 cm in diameter spaced 1 m apart in air.

Determine for 1 km of this circuit (a) the capacitance, (b) the capacitive reactance, (c) the charging current if the potential difference between the conductors is 19052 V at 50 Hz. (Line charging current = normal capacitive current = V/X_C A.)

WORKED EXAMPLE 4.9

A concentric cable is to operated with an rms voltage to ground of 159 000 V. The maximum potential gradient within the cable is not to exceed 60 kV/cm. Determine (a) the radius of the core and the inner radius of the sheath for ideal operation, (b) the stress on the dielectric at the inner surface of the sheath.

(a) $V_m = \sqrt{2} \times 159\,000 = 224\,860$ V.

To give minimum stress at the core (ideal operation) $R/r = \mathrm{e}$.

Equation 4.23 gives

$$E_m = \frac{V}{r \log_e (R/r)}$$

and for ideal operation this becomes

$$E_m = \frac{V}{r \log_e \mathrm{e}} = \frac{V}{r}.$$

$$\text{Therefore } 60\,000 = \frac{224\,860}{r} \quad \text{or} \quad r = \frac{224\,860}{60\,000} = 3.75 \text{ cm.}$$

Outer sheath inner radius $= \mathrm{e} \times 3.75 = 10.2$ cm.

(b) At radius x, stress $= \dfrac{V}{x \log_e (R/r)}$ (equation 4.22)

Therefore, at 10.2 cm, $E = \dfrac{224\,860}{10.2} = 22\,045$ V/cm.

4.12 Concentric spheres

4.12.1 Capacitance

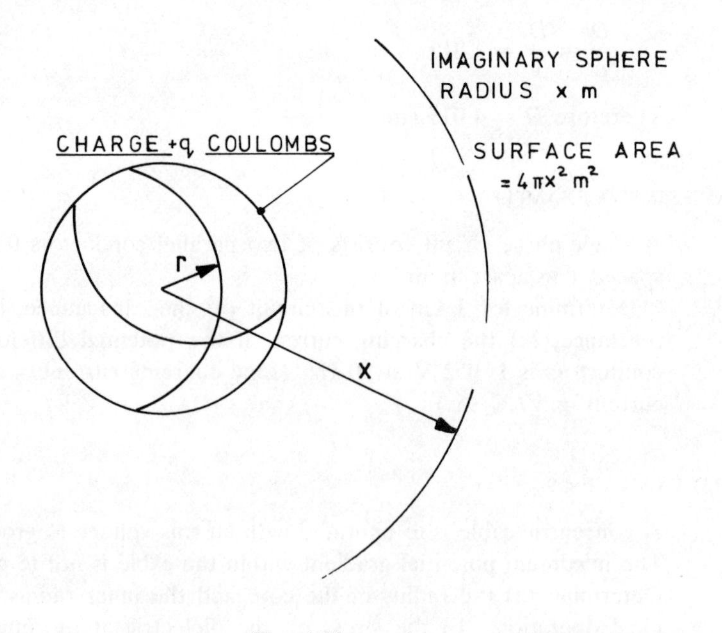

Fig. 4.10

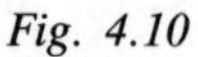

Figure 4.10 shows an isolated sphere carrying a charge of q coulombs.
Considering radii outside the sphere, we have from equation (4.2)

$$D \text{ at radius } x = \frac{q}{\text{Area}} = \frac{q}{4\pi x^2} \text{ C/m}^2.$$

Then, using equation (4.12)

$$E = \frac{q}{4\pi x^2 \varepsilon} \text{ V/m and from equation (4.14)}$$

$V = -\int E\mathrm{d}x$ between the limits r, the radius of the sphere, and a radius R large enough that the electrostatic field is zero. (See also section 4.11 which employs the same argument.)

$$V = -\frac{q}{4\pi\varepsilon}\int_R^r \frac{1}{x^2} = -\frac{q}{4\pi\varepsilon}\left(\frac{1}{x}\right)_R^r = \frac{q}{4\pi\varepsilon}\left(\frac{1}{r} - \frac{1}{R}\right) \text{V}.$$

Now since R is large, $1/R$ tends to zero and $V = \dfrac{q}{4\pi\varepsilon r}$ V.

But $q = CV$. Hence $V = CV/4\pi\varepsilon r$ V. Cancelling V and rearranging $C = 4\pi\varepsilon r$ F.

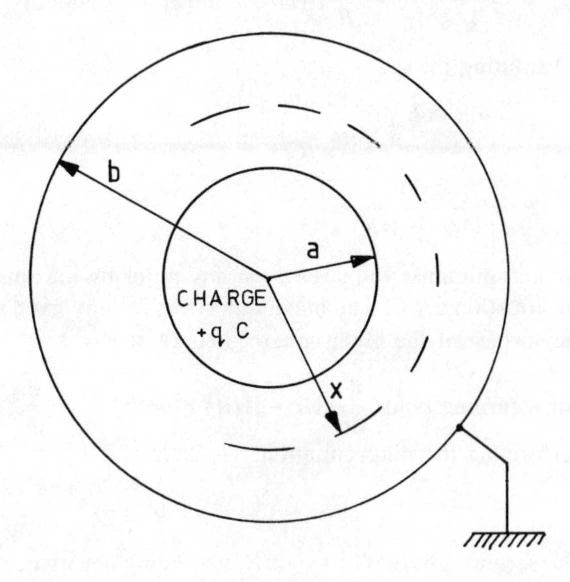

Fig. 4.11

Now extending the analysis to two concentric spheres as shown in Fig. 4.11. The outer sphere is earthed and the central sphere is at a potential of $+V$ volts with respect to earth. The limits of the electrostatic field are a and b.

$$\therefore \quad V = -\frac{q}{4\pi\varepsilon}\left(\frac{1}{x}\right)_b^a = \frac{q}{4\pi\varepsilon}\left(\frac{1}{a} - \frac{1}{b}\right) \text{V}.$$

$$C = \frac{4\pi\varepsilon}{\left(\dfrac{1}{a} - \dfrac{1}{b}\right)} \text{F}.$$

4.12.2 Stresses in the dielectric

WORKED EXAMPLE 4.10

A capacitor is formed from two concentric spheres with radii r (inner) and R (outer). A voltage V volts is applied to the centre sphere whilst the outer sphere is earthed. Demonstrate that if R is fixed and r may be varied that the maximum stress in the dielectric cannot be reduced below $4V/R$ V/m.
At any radius x between the two spheres

$$E = \frac{q}{4\pi x^2 \varepsilon} \text{ V/m.}$$

Transposing

$$q = 4\pi x^2 \varepsilon E \text{ coulombs.}$$

Now we have deduced that

$$V = \frac{q}{4\pi\varepsilon}\left(\frac{1}{r} - \frac{1}{R}\right) \quad (r = a \text{ and } R = b \text{ above})$$

Substituting for q:

$$V = \frac{4\pi x^2 \varepsilon E}{4\pi\varepsilon}\left(\frac{1}{r} - \frac{1}{R}\right)$$

$$E = \frac{V}{(1/r - 1/R)\, x^2}.$$

We can minimise the stresses at any point by maximising the denominator of the equation for E. The maximum stress for any fixed ratio of r and R occurs at the surface of the inner sphere, i.e. when $x = r$

For a turning point $\frac{\mathrm{d}}{\mathrm{d}r}(1/r - 1/R)\, r^2 = 0 \quad \therefore \frac{\mathrm{d}}{\mathrm{d}r}(r - r^2/R) = 0.$

Performing the differentiation: $1 - 2r/R = 0$

$$1 = 2r/R$$

$$R = 2r \qquad \text{or } r = \tfrac{1}{2}R$$

The second differential $= -2/R$, and being negative, a positive turning point is indicated as required.

Now since $E = \dfrac{V}{(1/r - 1/R)\, x^2}$ $\quad E_{max} = \dfrac{V}{(1/r - 1/R)\, r^2} = \dfrac{V}{(r - r^2/R)}$

and again, since $r = R/2$

$$E_{max} = \frac{V}{(R/2 - R^2/4R)} = \frac{V\,4R}{2R^2 - R^2} = \frac{4RV}{R^2}$$

$$= 4V/R \text{ V/m, which is the required result.}$$

Increasing r reduces the thickness of the dielectric between the spheres and increases the maximum dielectric stress. Reducing r increases the charge concentration on the surface of the inner sphere, again increasing the dielectric stress. The value $E = 4V/R$ is the minimum value achievable at the surface of the inner sphere.

SELF-ASSESSMENT EXAMPLE 4.11

Two concentric spheres with the outer one earthed have a potential difference between them of 100 kV. The outer sphere has inner radius 1 m. For three values of inner-sphere radius: (i) 0.2 m, (ii) 0.5 m, and (iii) 0.75 m, determine the maximum electric stress at the surface of the inner sphere and demonstrate that this is a minimum when $R = 2r$.

4.13 Energy stored in an electrostatic field

Consider a capacitive system initially charged such that the potential difference between the electrodes is v volts.

A small quantity of charge Δq coulombs is supplied to the system which causes the voltage to rise to $(v + \Delta v)$ V.

$$\text{The mean value of voltage during the change} = \left(v + \frac{\Delta v}{2}\right) \text{V}.$$

The energy supplied to an electric circuit is given as $W = Vq$ J (equation 4.7), hence the change in energy ΔW due to arrival of charge Δq will be

$$\Delta W = \left(v + \frac{\Delta v}{2}\right) \Delta q = v\,\Delta q + \frac{\Delta v\,\Delta q}{2} \text{ J}.$$

If the changes are made small, as Δq approaches zero, Δv approaches zero so that the product $\Delta v\,\Delta q$ is small enough to be ignored. In the limit, $\Delta v \to 0$, $\Delta q \to 0$ and $\mathrm{d}W = v\,\mathrm{d}q$.

The total energy supplied as the charge is increased from zero to a final value Q is obtained by integration

$$\int \mathrm{d}W = \int_0^Q v\,\mathrm{d}q.$$

But since $q = Cv$, $v = \dfrac{q}{C}$, which gives

$$\int \mathrm{d}W = \int_0^Q \frac{q\mathrm{d}q}{C}.$$

$$\text{Total energy } W = \left(\frac{q^2}{2C}\right)_0^Q = \frac{Q^2}{2C}.$$

Again, $Q = CV$, where V = final potential difference. Substituting gives

$$W = \frac{CV^2}{2} \text{ J}. \tag{4.29}$$

Now consider a very small element within the dielectric of the capacitive system. Whatever the arrangement of electrodes, i.e. concentric cable, parallel conductors, etc., with a small element the force lines may be considered parallel and the equi-potential lines may be considered parallel. In Fig. 4.12, AD is parallel to CB and DC is parallel to AB.

$DA = CB = \Delta x. \qquad DC = AB = \Delta y.$

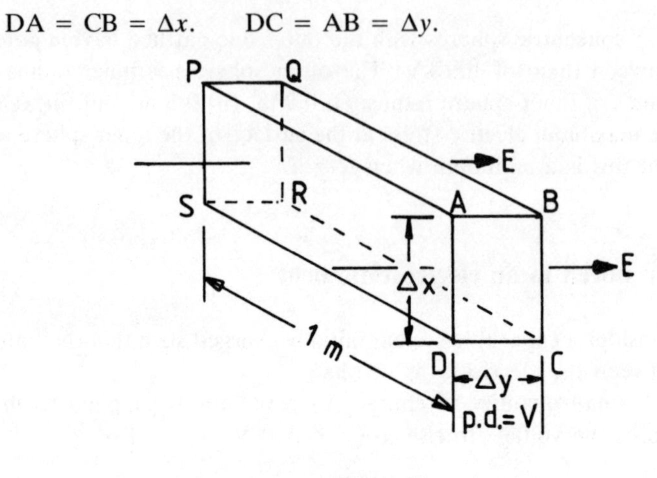

Fig. 4.12

A 1 m length of such an element is now considered. Surface DAPS is crossed by the electrostatic force lines and is an equi-potential surface. CBQR is also an equi-potential surface (see section 4.8). These surfaces can be made conducting with no change in the field and when in this form could form the plates of a parallel-plate capacitor.

$$\text{Capacitance } C = \varepsilon_o \varepsilon_r \frac{A}{d} \qquad \text{(equation 4.18)}$$

$$= \varepsilon_o \varepsilon_r \frac{\Delta x \times 1}{\Delta y} \text{ F.} \qquad (4.30)$$

The potential difference across Δy is V volts, therefore $E = V/\Delta y$ (equation 4.6).

Transposing gives

$$V = E\,\Delta y. \qquad (4.31)$$

$$\text{Energy stored} = \tfrac{1}{2}\,CV^2. \qquad \text{(equation 4.29)}$$

Substituting equations (4.30) and (4.31) into equation (4.29) gives

$$\text{energy stored} = \tfrac{1}{2}\,\varepsilon_o \varepsilon_r \frac{\Delta x \times 1}{\Delta y}\,(E\,\Delta y)^2$$

$$= \tfrac{1}{2}\,\varepsilon_o \varepsilon_r\,\Delta x\,E^2\,\Delta y \text{ J.}$$

Now, for 1 m length the volume of the dielectric $= 1 \times \Delta x \times \Delta y$ m^3.

Therefore, the energy stored per unit volume $= \frac{1}{2}\,\varepsilon_o\varepsilon_r\,E^2$ J. (4.32)

Substituting for E from equation (4.12) into equation (4.32) gives

$$\text{energy stored per unit volume} = \frac{1}{2}\,\varepsilon_o\varepsilon_r\left(\frac{D}{\varepsilon_o\varepsilon_r}\right)^2$$

$$= \frac{D^2}{2\varepsilon_o\varepsilon_r}\ \text{J/m}^3. \qquad (4.33)$$

4.14 Dielectric materials

A dielectric material is one which has a very high resistivity when compared with that of a conductor such as copper and aluminium, etc. Dielectrics are therefore used to maintain separation between conductors which are at different potentials, i.e. electric power lines from earth and from each other, capacitor plates, etc.

4.14.1 Electric strength of dielectrics

The breakdown voltage gradient E of an insulating material is dependent upon its thickness but not proportional to that thickness.

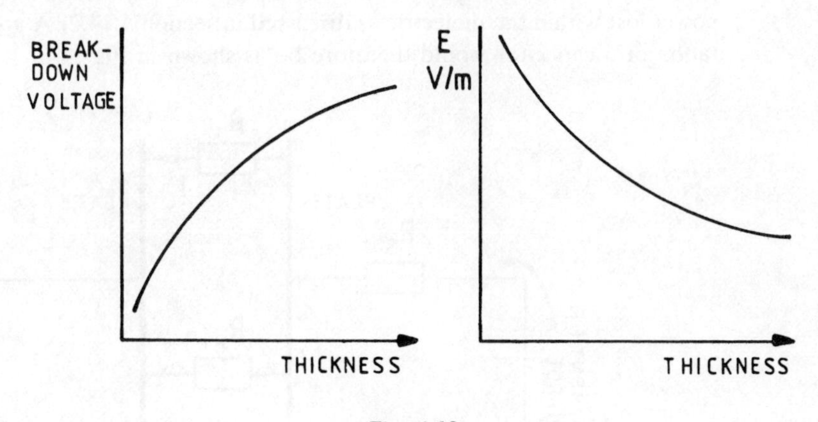

Fig. 4.13

Typical shapes for breakdown voltage versus thickness and potential gradient E versus thickness are shown in Fig. 4.13. The precise shapes of the curves depend on (i) the shape of the test electrodes (flat, cylindrical, pointed, etc.); (ii) the material being tested (air, glass, plastics); and (iii) the atmospheric pressure and humidity if the system is not sealed. Actual breakdown voltages quoted in reference works have to be treated with considerable care unless the precise conditions of the test are known.

4.14.2 Dielectric failure

Where a dielectric is subjected to alternating stresses, dielectric hysteresis has to be considered. This involves the outer electrons being attracted away from the nucleus in the direction of the electrostatic field. Provided that the force of attraction is not too great most electrons do not move very far through the material. However, work is done in this re-orienting of the electrons during each reversal of stress and heat is produced. Some electrons do finally end up on the electrode system and constitute a small charge movement through the dielectric. Such a movement of charge is a leakage current in phase with the voltage and again heat is produced.

Dielectrics become conducting if they are supplied with sufficient energy either in the form of applied potential difference or directly as heat. The insulation resistance of most dielectrics falls rapidly as the temperature is increased so that leakage current increases thus generating further heat and a condition known as thermal runaway or thermal avalanche may develop when the heat is generated faster than it can be dissipated to the surrounding environment. Under these conditions the dielectric fails due to burning.

4.14.3 Representation of an imperfect dielectric

As charge moves periodically from one electrode to the other in a capacitive system it passes through the resistance of the connecting wires, the capacitor plates, etc. There is a power loss due to this series resistance. There is also power lost within the dielectric as discussed in section 4.14.2. A good representation of a capacitor would therefore be as shown in Fig. 4.14.

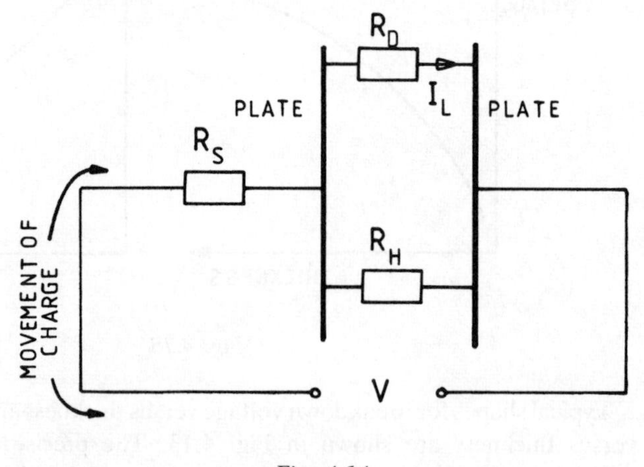

Fig. 4.14

R_s = series resistance, R_D = resistance of the dielectric through which the leakage current, I_L, flows. R_H is a resistance which represents the hysteresis loss.

However it is impossible to establish the individual value of such resistors: measurements using equipment such as the Schering bridge provide us with values of capacitance and effective resistance values for either series or parallel configurations.

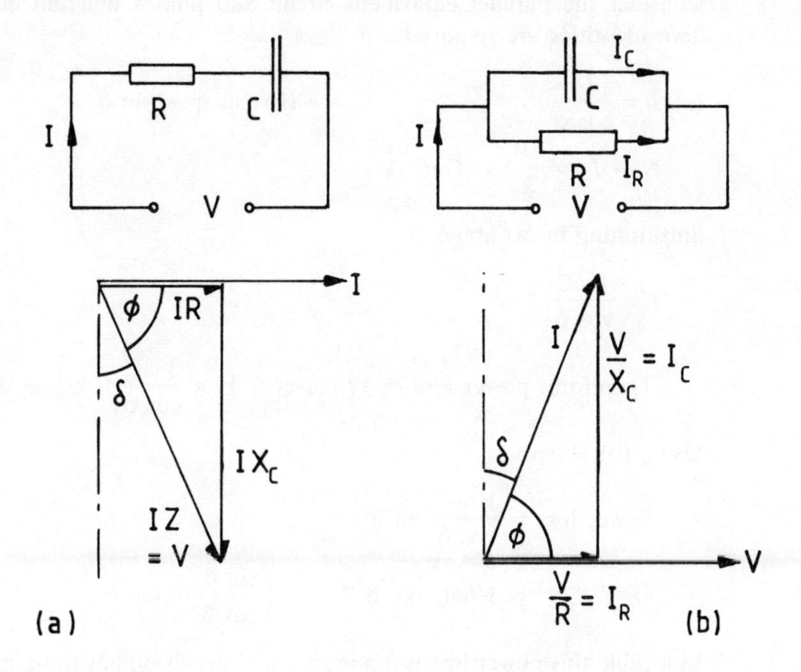

Fig. 4.15

The circuit and phasor diagrams for the series equivalent are shown in Fig. 4.15(a). The two voltage drops IR and IX_C sum phasorially to V, the supply voltage. The circuit phase angle is ϕ. Now if the resistance were zero, V would lag I by exactly 90°. The capacitor would then be perfect. The difference between the perfect and the practical case is the angle δ which is known as the loss angle. This will usually be a fraction of one degree and rarely more than two degrees.

Exactly the same result is achieved by considering the parallel equivalent shown in Fig. 4.15(b). The branch currents I_C and I_R sum to the total current which has the same value as in the series case. Notice that V/X_C is that value of current which would flow in a perfect capacitor. This is called the charging current in transmission line work.

Both values of capacitance C are in all practical cases identical, but the series and parallel resistance values will differ considerably. In the parallel case, since it is considered to be connected directly across the supply, it will be very high, being close to the straight insulation resistance of the system. In the series case, it will be considerably less, though in small capacitors it may be in the hundreds of kilohm range.

4.14.4 *Dielectric heating in terms of capacitance and loss angle*

Power loss in the dielectric = $VI \cos \phi$ W.

Consider the parallel equivalent circuit and phasor diagram in Fig. 4.15(b). Two identities are required

(a) $I = \dfrac{I_C}{\cos \delta}$ (b) $\cos \phi = \sin \delta$.

$$\text{Now } I_C = \frac{V}{X_C} = V\omega C \text{ A.}$$

Substituting in (a) above

$$I = \frac{V\omega C}{\cos \delta}.$$

$$\text{Therefore, power loss} = VI \cos \phi = V \times \frac{V\omega C}{\cos \delta} \times \cos \phi \text{ W.}$$

Using (b) above

$$\text{Power loss} = \frac{V^2\omega C}{\cos \delta} \sin \delta$$

$$= V^2\omega C \tan \delta \text{ W.} \qquad \left(\frac{\sin \delta}{\cos \delta} = \tan \delta\right). \qquad (4.34)$$

In a cable this power loss is a nuisance and provision has to be made to get rid of the heat produced. At extra-high voltages the V^2 term becomes very large and the dielectric loss can exceed the cable I^2R losses. In welding machines for joining plastics, the frequency is increased into the kilohertz or megahertz range and dielectric loss is used to melt the internal surfaces to be joined so that they fuse together under pressure.

SELF-ASSESSMENT EXAMPLE 4.12

A concentric cable has core radius 2 cm and inside sheath radius 5.5 cm. The relative permittivity of the dielectric is 2.3. The loss angle, $\delta = 0.16°$, and the working voltage $V = 194$ kV at 50 Hz.

Calculate, for a 1 km length of cable, (a) the capacitance, (b) the charging current, (c) the power loss.

4.14.5 *Properties of dielectrics*

Dielectrics used in capacitor manufacture need to have high permittivities in order to give the greatest possible capacitance within the smallest physical size. Paper/paraffin wax insulated capacitors are used at all voltages up to 150 kV in circuits where the loss angle is not too significant. They are not polarity sensitive and sizes from 250 pF to 10 μF are readily available. Electrolytic capacitors are used at voltages up to about 600 V where the applied potential is uni-directional. The dielectric is an extremely thin film of aluminium oxide which is chemically developed on aluminium foil. Electrical contact is made with the oxide through a conducting liquid, the electrolyte. The losses are greater than with the paper type but very large values of capacitance can be produced within a very small physical size.

Where the loss angle must be small, mica capacitors may be used. These comprise thin layers of mica and foil clamped together. These are not polarity-sensitive but are expensive. Sizes from 25 pF to 0.25 μF are available.

Capacitors using titanium oxide or polycarbonate have extremely high capacitance within a given physical size when used at low voltages, e.g. up to 10 V.

In power cable manufacture, high capacitance is a distinct disadvantage. This gives rise to system stability problems and a rising voltage on the system during periods of light loading. This is known as the Ferranti effect. In addition, as has already been seen in section 4.14.4 cable heating is directly proportional to capacitance (equation 4.34). The capacitance of oil-impregnated paper insulated cables is relatively large (equation 4.21). This capacitance could be reduced if the relative permittivity could be made small, ideally unity. Several cable models have been evaluated and one which uses polythene tapes pressurised with sulphur hexafluoride gas at a pressure of approximately 14 bar gives a reduction in capacitance of 40%, of thermal resistivity 30% and dielectric losses of over 95% compared with a corresponding paper/oil insulated cable. The costs of the two types of cable are comparable.

Table of properties of selected dielectrics

Material	ε_r	Approx. breakdown strength (kV/mm)	Resistivity (Ω cm)
Air	1.0006	3	
Transformer oil	2.3	13	7×10^{12}
Paraffin wax	2.2	7.5	0.9×10^{16}
Mica	7	180	2×10^{15}
Glass	6.5	5–7	2×10^{14}
Varnished paper	2	15	up to 10^{6}
Polythene	1.5	40	10^{17}
PTFE	2	15	10^{17}
Titanium oxide	70–80	Used in extremely thin layers in capacitors	–
Polycarbonate	2.8	– ditto –	$10^{17}+$

4.15 Field plotting

Where electrode systems are asymmetric, mathematical analysis is difficult and equi-potential lines may be plotted using models of the systems involved. Such plots may be used to determine capacitance. One method is to use demineralised water as the dielectric. A model of the system in section is made from copper or brass and is supported on insulators in a tank so that it is partly immersed in the demineralised water. In Fig. 4.16, a section of cable is shown in which the core is not concentric with the sheath. A Wheatstone bridge is used to determine points of equal potential in the water. For example, with $R_1 = R_2$, the probe is moved about in the water until the detector indicates a zero. At this point in the water the potential difference between the core and the probe is equal to that between the probe and the sheath. Many such points can be found and an equi-potential line drawn through them.

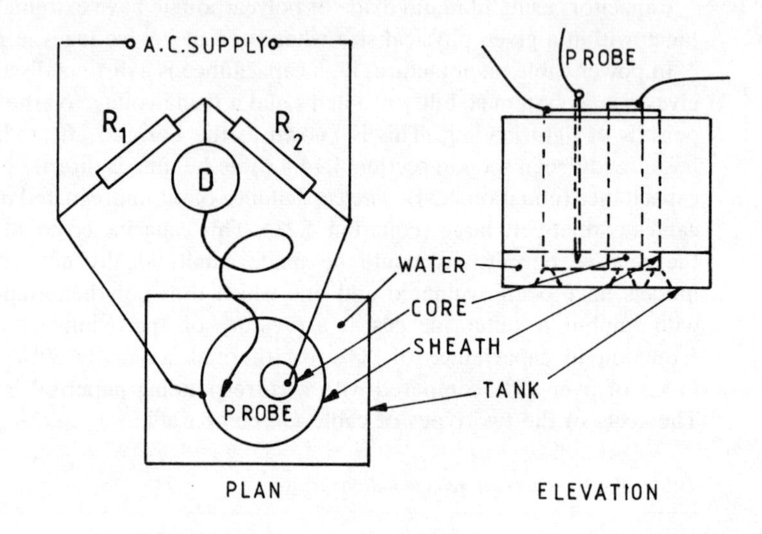

Fig. 4.16

Other equi-potential lines are drawn by the same method using different ratios of R_1 and R_2. A typical result is shown in Fig. 4.17.

With $R_2 = 5000\ \Omega$ and $R_1 = 15\,000\ \Omega$, the probe will detect the 25% equi-potential line, i.e. for 100 V on the core and the sheath earthed, the p.d. between the core and any point on this line is 25 V.

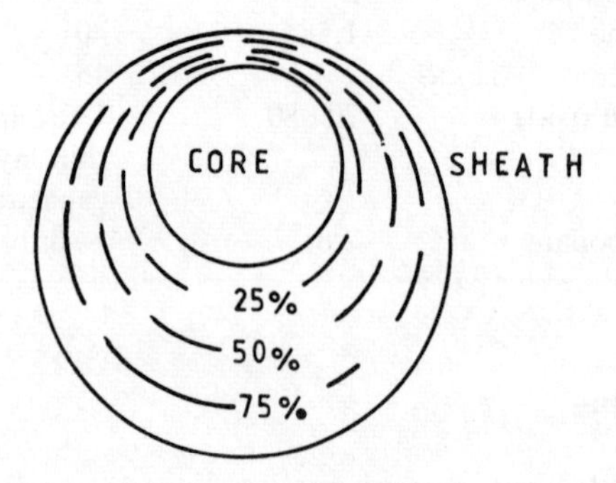

Fig. 4.17

For $R_2 = 15\,000\ \Omega$ and $R_1 = 5000\ \Omega$, the 75% equi-potential line may be drawn.

Since stress lines are at right angles to equi-potential lines, these may be drawn in by inspection. Similar results are achievable using Teledeltos paper. This has a surface which is lightly graphited making it slightly conducting. The electrode configuration is painted on the graphite in plan using highly conducting metal paint. The procedure is as with the demineralised water, using a probe to mark the paper at the required null points, between which the equi-potential lines are drawn.

4.16 Capacitance of systems deduced from a field plot

4.16.1 With stress and equi-potential lines making true squares

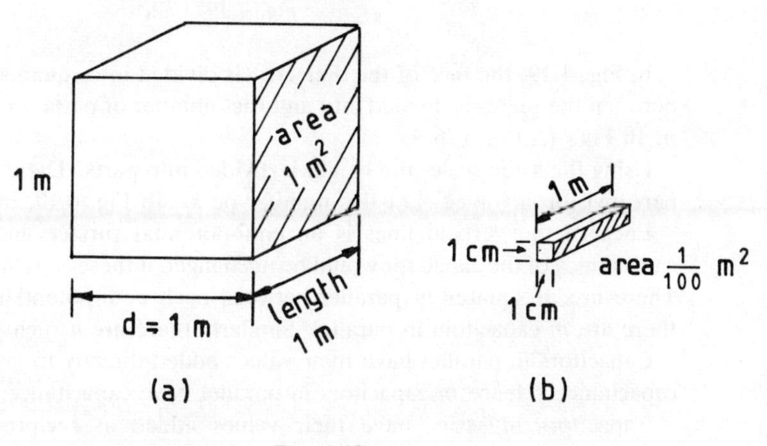

Fig. 4.18

Consider the capacitance of a parallel-plate capacitor formed using a 1 m cube of dielectric as shown in Fig. 4.18(a).

$$C = \frac{\varepsilon_o \varepsilon_r A}{d}\ \text{F}. \qquad \text{(equation 4.18)}$$

Now since $A = 1\ \text{m}^2$ and $d = 1$ m

$$C = \varepsilon_o \varepsilon_r\ \text{F}.$$

Reducing the area and distance between the plates in the same proportion, while leaving the length unchanged at 1 m, results in the same value of capacitance. The capacitor shown in Fig. 4.18(b) also has a capacitance of $\varepsilon_o \varepsilon_r$ F. A further reduction to 1 mm height and 1 mm between the plates, provided that the length is still 1 m, leaves the capacitance unchanged.

One method of determination of the capacitance of a parallel-plate capacitor would be to consider the dielectric to be split up into such squares.

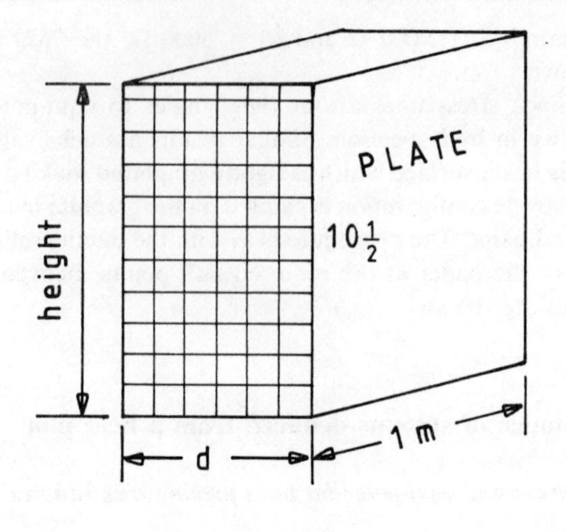

Fig. 4.19

In Fig. 4.19, the face of the dielectric is divided into squares. The distance d between the plates is divided into an exact number of parts. Let this number be n. In Fig. 4.19, $n = 6$.

Using the same scale, the height is divided into parts. There will usually be a part division involved. Let this number be m. In Fig. 4.19, m = 10.5.

Each of the vertical lines is an equi-potential surface and so the overall capacitance of the capacitor would be unchanged if these were made conducting. There are m squares in parallel between each equi-potential surface and so there are m capacitors in parallel. Similarly there are n such groups in series.

Capacitors in parallel have their values added directly to give an equivalent capacitance. Hence m capacitors in parallel have capacitance $m\varepsilon_o\varepsilon_r$ F.

Capacitors in series have their values added as reciprocals to give an equivalent capacitance. Therefore for n identical capacitors

$$\frac{1}{C_{eq}} = n\left(\frac{1}{C}\right).$$

Transposing

$$C_{eq} = \frac{C}{n}$$

where C is the value of each of the capacitors connected in series ($= m\varepsilon_o\varepsilon_r$ F).

Hence, the total capacitance of the arrangement in Fig. 4.19 is

$$\frac{m\varepsilon_o\varepsilon_r}{n}\text{ F.} \tag{4.35}$$

This will be the value for a capacitor 1 m in length. For other lengths the value above will be adjusted in proportion.

WORKED EXAMPLE 4.13

A parallel-plate capacitor has plates 10 cm × 15 cm. They are spaced 3 mm apart. The relative permittivity of the dielectric is 4.

Determine the capacitance of the capacitor (a) by the use of squares, (b) directly from equation (4.18).

It is of no significance whether 10 cm is considered as the height or the length of the plates. In Fig. 4.20, 10 cm is arbitrarily chosen as the height.

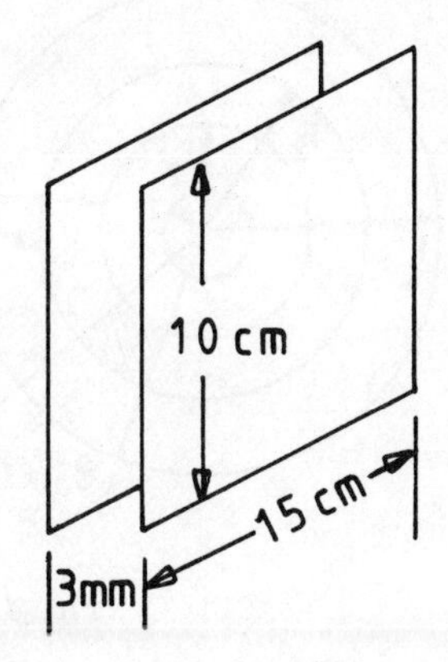

Fig. 4.20

(a) Any convenient size square may be used provided that the 3 mm distance between the plates is divided exactly. 3 mm squares, 1.5 mm squares or 1 mm squares, for example, could be used.
Using 1 mm squares, $n = 3$ and $m = 100$.

For a 1 m length, $C = \dfrac{m\varepsilon_o\varepsilon_r}{n}$ F (equation 4.35)

$$= \frac{100 \times 8.85 \times 10^{-12} \times 4}{3} = 1.18 \times 10^{-9} \text{ F.}$$

For a 15 cm length, $C = \dfrac{15}{100} \times 1.18 \times 10^{-9} = 177$ pF.

(The reader might care to verify that the same result is obtained however many squares are used.)

(b) Equation (4.18) gives

$$C = \frac{\varepsilon_o\varepsilon_r A}{d} = 8.85 \times 10^{-12} \times 4 \times \frac{10 \times 15 \times 10^{-4}}{3 \times 10^{-3}}$$

$$= 177 \text{ pF.}$$

4.16.2 *With stress and equi-potential lines forming curvilinear squares*

Consider a 1 m length of co-axial cable, a cross-section of which is shown in Fig. 4.21. The core radius is r_1 and the inner sheath radius is r_4. By the correct choice of spacing between stress lines and between equi-potential surfaces in a field plot, the 'near-squares' formed (one 'square' hatched in Fig. 4.21), or curvilinear squares as they are called, can be made to have the same capacitance as a true square in section 4.16.1.

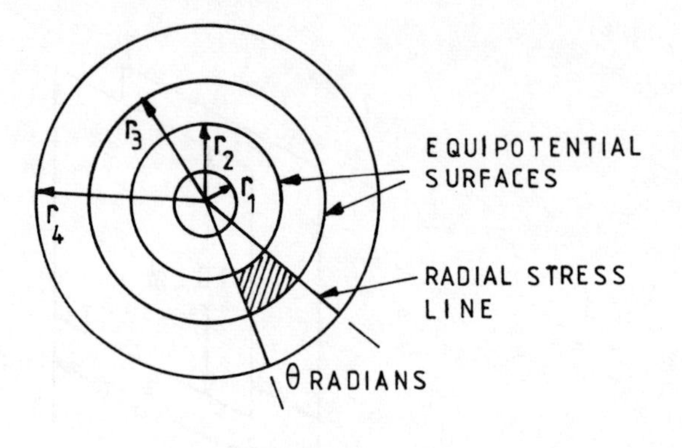

Fig. 4.21

From equation (4.21), the capacitance between cylindrical equi-potential lines at radii r_p and r_q respectively is given by

$$C = \frac{2\pi\varepsilon_o\varepsilon_r}{\log_e (r_q/r_p)} \text{ farads per metre length.}$$

Hence for a segment θ radians the capacitance will be

$$C_\theta = \frac{\theta}{2\pi} \times \frac{2\pi\varepsilon_o\varepsilon_r}{\log_e (r_q/r_p)} \text{ F/m}$$

$$= \frac{\theta\varepsilon_o\varepsilon_r}{\log_e (r_q/r_p)} \text{ F/m.} \quad (4.36)$$

Hence if $\theta = \log_e (r_q/r_p)$ (4.37), equation (4.36) becomes $C = \varepsilon_o\varepsilon_r$ F, which is the same as that for a perfect square.

The dielectric may be subdivided by as many equi-potential surfaces as desired, three or four usually being selected to give reasonably shaped curvilinear squares.

Now from equation (4.37)

$$\theta = \log_e (r_q/r_p) \text{ hence, } e^\theta = (r_q/r_p). \quad (4.38)$$

For core radius r_1 and equi-potential surfaces at r_2 and r_3 with sheath inner radius r_4, each ratio of successive radii must satisfy the condition in turn. Thus

$$e^\theta = \frac{r_2}{r_1}, \quad e^\theta = \frac{r_3}{r_2} \quad \text{and } e^\theta = \frac{r_4}{r_3}.$$

Multiplying together gives

$$e^{\theta} \times e^{\theta} \times e^{\theta} = \frac{r_2}{r_1} \times \frac{r_3}{r_2} \times \frac{r_4}{r_3}$$

$$e^{3\theta} = \frac{r_4}{r_1}. \tag{4.39}$$

This equation is used to determine the value of θ and hence the number of segments.

WORKED EXAMPLE 4.14

Draw a flux/equi-potential curvilinear square plot for a concentric cable with core radius 0.75 cm and inner sheath radius 4 cm. Determine from the plot, the capacitance of the cable per metre run if the relative permittivity of the dielectric is 3.5. Compare the result with that directly obtained from equation (4.21).

Using two equi-potential surfaces within the dielectric so forming three capacitors in series in each segment

$$e^{3\theta} = \frac{r_4}{r_1}. \qquad \text{(equation 4.39)}$$

Substituting in values gives

$$e^{3\theta} = \frac{4}{0.75} = 5.33.$$

Therefore

$$3\theta = \log_e 5.33 = 1.674$$

$$\theta = \frac{1.674}{3} = 0.558 \text{ radians } (31.9°).$$

Since 2π radians comprise a complete circle, with each segment having a value of 0.558 radians, there will be $2\pi/0.558 = 11.26$ segments in the plot.

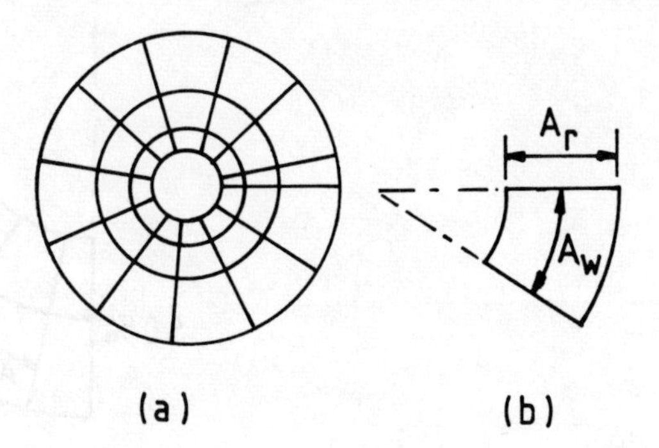

Fig. 4.22

The plot is shown in Fig. 4.22. If 1 metre length of the cable is considered, each segment has three capacitors in series between the core and sheath. There are 11.26 such groups of capacitors in parallel.
Each curvilinear square, 1 m in length, has a capacitance of $\varepsilon_o \varepsilon_r$ F.

$$\text{Therefore, total capacitance per metre length} = \varepsilon_o \varepsilon_r \times \frac{11.26}{3}$$

$$= 8.85 \times 10^{-12} \times 3.5 \times \frac{11.26}{3}$$

$$= 116.26 \text{ pF.}$$

Using equation (4.21)

$$C = \frac{2\pi\varepsilon_o\varepsilon_r}{\log_e (4/0.75)} = 116.26 \text{ pF.}$$

It should be observed from Fig. 4.22 that in any size of curvilinear square the mean width of the segment A_w is equal to its length measured along a radius A_r. This fact will be helpful with asymmetric electrode plots.

SELF-ASSESSMENT EXAMPLE 4.15

Using three equi-potential surfaces within the dielectric (four series capacitors), use the principle of curvilinear squares to determine the capacitance of a concentric cable with core radius 2 cm and inner sheath radius 5.5 cm. The relative permittivity of the dielectric is 2.3.

Considerable use is made of this method to determine the capacitance of asymmetric systems. Consider again Fig. 4.17. By selecting (say) the 50% equi-

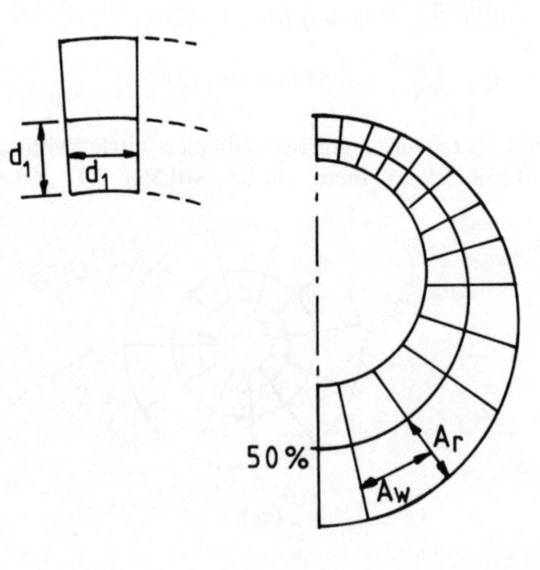

Fig. 4.23

potential line, the core and the sheath only, the dielectric space may be divided into curvilinear squares using dividers. The distance from the core to equi-potential line d_1 is stepped off midway between core and the equi-potential line circumferentially. Figure 4.23 should make this clear.

Resetting the dividers at each stage it is possible to create curvilinear squares progressing from the top to the bottom. In the half section shown in Fig. 4.23 there are 12.5 segments. The other half is symmetrical so there will be 25 sections in all. In each segment there are two capacitors in series between core and sheath.

$$\text{Capacitance per metre length} = \varepsilon_o \varepsilon_r \times \frac{25}{2}\ \text{F}.$$

SELF-ASSESSMENT EXAMPLE 4.16

What is the capacitance per metre length of the L-shaped air-insulated bus-bar system shown in Fig. 4.24?

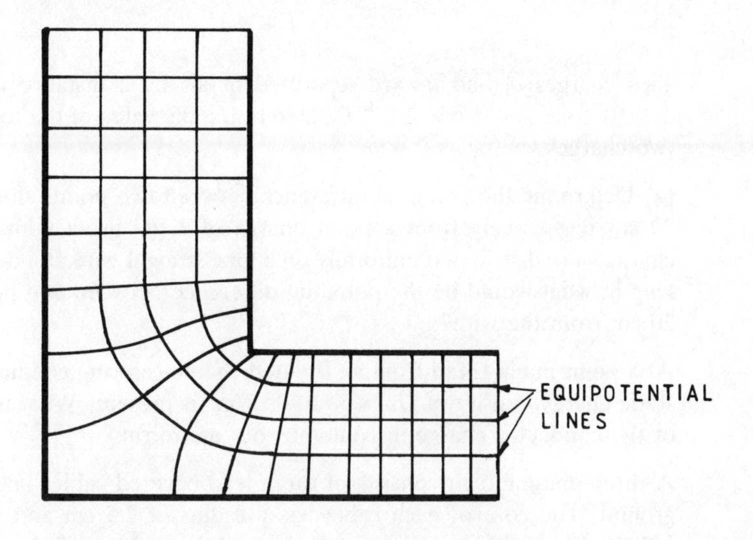

Fig. 4.24

(Notice that not all the curvilinear squares are perfectly regular especially those close to the corners. This makes the method a little imprecise but quite accurate enough for most applications. The alternative is a complicated calculation which itself is subject to error because of the simplifying assumptions made.)

FURTHER PROBLEMS

4.17 In Fig. 4.25, A and B are two long air-insulated conductors of small diameter in a plane normal to that of the paper. Each carries a charge of 5×10^{-8} C/m length.

Determine (a) the electric field strength at point P due to the charge on conductor A, (b) the electric field strength at point P due to the charge on conductor B, (c) the resultant of (a) and (b), taking into account their directions and adding as vectors, and (d) the resultant force on a positive charge of 1 mC placed at point P.

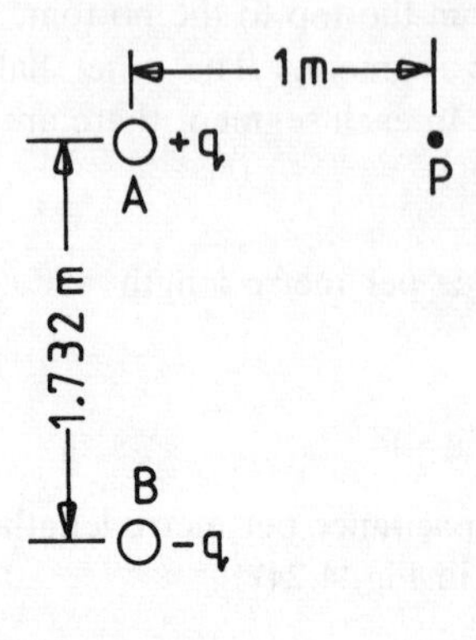

Fig. 4.25

4.18 Two charges q_1 and q_2 are separated in air by a distance of 1 mm. $q_1 = 2 \times 10^{-9}$ C, $q_2 = 1.6 \times 10^{-10}$ C. Determine the value of the force between the two charges.

4.19 (a) Determine the potential difference between two points situated 10 cm and 20 cm respectively from a point charge of 3 μC in vacuum. (b) Suppose a charge were distributed uniformly on a long straight wire at a density of 3 μC/m length, what would be the potential difference between two points 10 cm and 20 cm from the wire?

4.20 At a point in air 1.4 m from an isolated charge-carrying conductor, an electrostatic energy density of 26.28×10^{-15} J/m^3 is present. What is the magnitude of the conductor charge in coulombs per metre run?

4.21 A three-phase circuit consists of three lead covered cables laid parallel in the ground. The core of each cable has a radius of 2.5 cm and an inner sheath radius of 6 cm. The relative permittivity of the dielectric is 3 and the loss angle δ is 0.2°. The circuit operates at 132 kV (line) and 50 Hz.

Determine, for a 10 km length of circuit, (a) the capacitance of each cable to earth, (b) the charging current, (c) the total dielectric power loss of the circuit. (For the charging current you will need the voltage *to earth* of each conductor.)

4.22 Using the table of properties of dielectrics in section 4.14.5, determine the area and spacing between two plates which form a standard capacitor of value 10 pF when they are immersed in transformer oil. The working voltage is 15 kV rms (assume sinusoidal form). Allow a safety factor of 3 on the plate spacing.

4.23 Two metal plates separated by a sheet of glass form a parallel-plate capacitor. Each plate has an area of 0.36×10^{-3} m^2. The glass is 3 mm thick. The potential difference between the plates is 1.5 kV d.c. Using the information in the table of properties of dielectrics in section 4.14.5, calculate (a) the capacitance of the capacitor, (b) the charge displaced when the voltage is applied, starting from the uncharged state, (c) the minimum factor of safety at which the capacitor is operating (what voltage could the glass support before breakdown?), (d) the value of capacitance if the glass plate were removed.

4.24 An isolated twin line comprises two air-insulated conductors each with radius 0.5 cm, spaced 1.5 m apart. The potential difference between the lines is 6350 V at 50 Hz. Calculate, per kilometre length, (a) the line capacitance, (b) the rms value of charge carried by each wire, (c) the charging current.

4.25 A concentric cable has a capacitance of 2.5 μF and a dielectric power loss of 15.6 kW when operated at 66 kV and 50 Hz. Determine (a) the value of the loss angle, (b) the equivalent parallel resistance of the cable.

4.26 Determine the area of the plates of a parallel-plate capacitor and the spacing between them allowing a safety factor of 3, if the energy stored is to be 0.06 J at a steady voltage of 5000 V. The breakdown potential gradient of the dielectric is 35 kV/cm and it has a relative permittivity of 4.3.

4.27 A concentric cable has core radius 1.5 cm and sheath inner radius 3.5 cm. The permittivity of the dielectric is 2.7. Using the method of curvilinear squares, deduce the capacitance of the cable per metre length. (Use two equi-potential surfaces within the dielectric.)

4.28 A field plot between two metal plates is shown in Fig. 4.26. The relative permittivity of the dielectric is 2.5. What is the capacitance per metre length of the system?

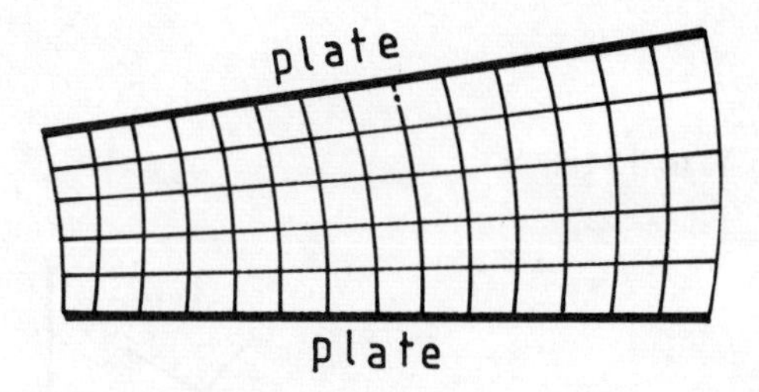

Fig. 4.26

4.29 A field plot between two copper bus-bars is shown in Fig. 4.27. They are air-insulated. Determine the capacitance of a 100 m length of the bus-bars using the method of curvilinear squares.

4.30 Determine the capacitance between two air-insulated concentric spheres, the outer one being earthed given that the inner sphere has radius 30 cm and the outer sphere has inner radius 50 cm. $\varepsilon_0 = 8.85 \times 10^{-12}$ F/m.

4.31 The capacitance between concentric, air-insulated spheres designed to give minimum electric stress in the dielectric is to be 102×10^{-12} F. (a) Determine the necessary radii r and R of the two spheres to accomplish this. (b) Given

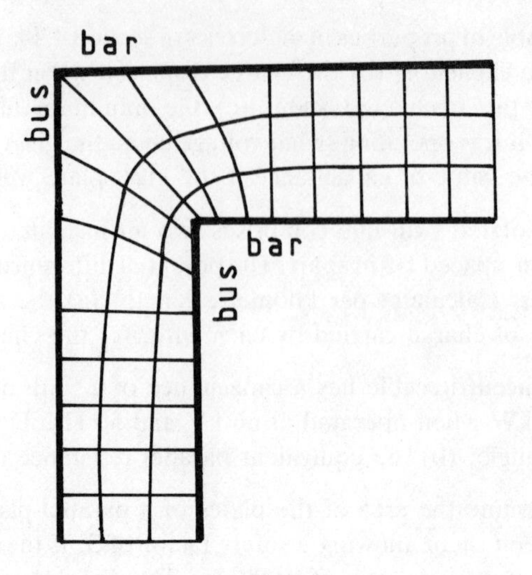

Fig. 4.27

that the maximum stress is not to exceed 3×10^6 V/m determine the maximum potential difference which can be applied between the two spheres.

4.32 Two air-insulated spheres, designed to give minimum electric stress have a breakdown voltage between them of 100 kV. The maximum stress is not to exceed 3×10^6 V/m. Determine the values of the radii of the inner and outer spheres respectively. What is the capacitance between the spheres?

Answers

4.2 *Fig. 4.28*
4.4 41 296 V to 15 188 V

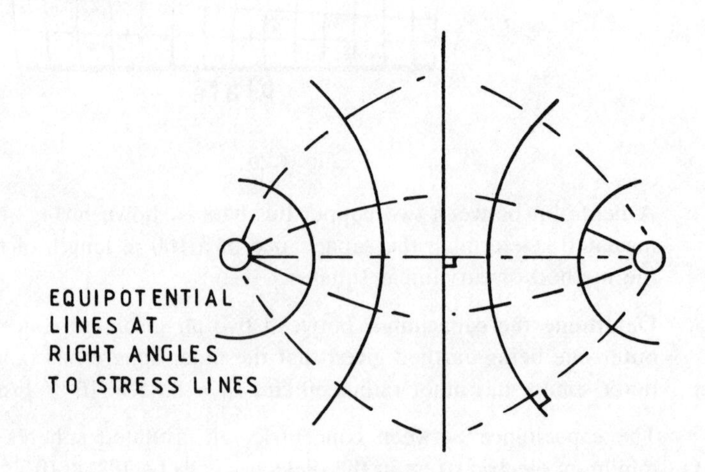

Fig. 4.28

4.8 (a) 4.786×10^{-9} F/km (b) $X_C = 665\,085$ (c) $I_c = 0.0286$ A
4.11 (i) 625 kV/m (ii) 400 kV/m (iii) 533.3 kV/m. Minimum value in (ii)
4.12 (a) 126.43×10^{-9} F (b) 7.705 A (c) 4174 W
4.15 $\theta = 14.49°$; 24.84 segments; $C = 126.4$ pF/m
4.16 37.6 pF/m
4.17 (a) 899.2 V/m from A to P (b) 449.6 V/m from P to B (c) 935 V/m (d) 0.935 N
4.18 2.88 mN
4.19 (a) 134.88 kV (b) 37.4 kV
4.20 6×10^{-12} C/m
4.21 (a) 1.905 μF (b) 45.6 A (c) 12.13 kW/core (= 36.4 kW total) (phase voltage (to earth) = $132/\sqrt{3}$ kV)
4.22 d = 4.9 mm; area = 24 cm^2
4.23 (a) 6.9 pF (b) 10.35×10^{-9} C (c) at 5 kV/mm, max. voltage would be 15 kV, actual 1.5 kV, safety factor 10 (d) 1.06 pF
4.24 (a) 4.87 nF (b) 30.9 μC (c) 9.72 mA
4.25 (a) 0.26° (b) R = 280 kΩ
4.26 $C = 4.8$ nF; $d = 0.428$ cm; area = 0.54 m^2
4.27 $\theta = 16.18°$; 22.25 segments; 177.2 pF/m
4.28 61.95 pF
4.29 4.72 nF
4.30 83.41×10^{-12} F
4.31 (a) $r = 0.459$ m, $R = 0.917$ m (b) 687 870 V
4.32 $R = 0.133$ m $r = 0.066$ m $C = 14.828$ pF

Chapter 5
Electromagnetism

5.1 Basic relationships between electric current, magnetic fields and forces

When a potential difference is created by an external e.m.f. between the ends of a conductor, an electric current is established. Such a current sets up a magnetic field within and surrounding the conductor.

When a magnetic field is established which links with a conductor, an e.m.f. is induced in that conductor during the period of time taken to establish the field. Lenz's law states that the induced e.m.f. is in such a direction as to oppose the establishment of the original current. During a field collapse, the induced e.m.f. will attempt to maintain the established current.

Associated with currents and magnetic fields, whether produced by those currents or provided externally, there are always mechanical forces.

In this section the principal relationships between the several quantities will be examined.

5.1.1 The electric and magnetic circuits. Relationships between (a) current density, electric force and conductivity, (b) flux density, magnetising force and permeability

For free space the permeability is μ_0 and this has a value of $4\pi \times 10^{-7}$ H/m. For other materials $\mu = \mu_0\mu_r$, where μ_r is the relative permeability.

Materials are divided into two classifications (1) paramagnetic, (2) diamagnetic. Paramagnetic materials have relative permeabilities greater than unity and when in the form of a needle and suspended in a magnetic field will align themselves with the magnetic field. Diamagnetic materials have relative permeabilities less than unity and will not orientate themselves with the field. Iron, nickel and cobalt are examples of ferro-magnetic materials which are very strongly paramagnetic. Air is considered to have a relative permeability of unity and therefore to perform as free space.

Now a single current-carrying conductor will be considered. The current I amperes sets up a magnetic field one line of which, at radius r metres, is $2\pi r$ metres long.

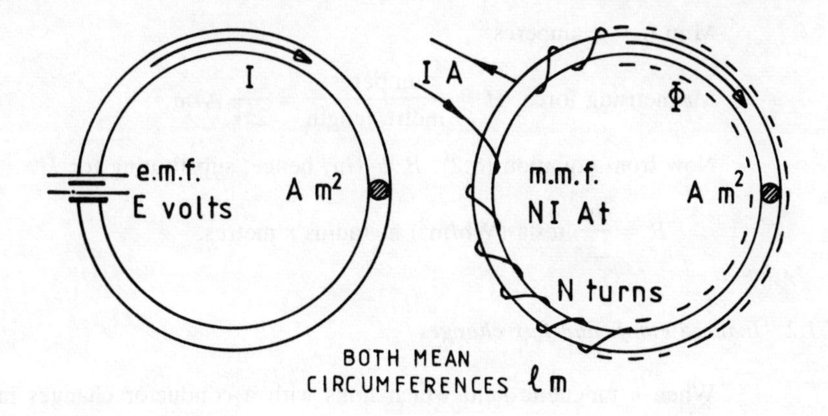

Fig. 5.1

Electric circuit	Magnetic circuit
Resistance, $R = \ell/\sigma A$	Reluctance $S = \ell/\mu A$
σ = conductivity of the material	μ = permeability of the material
e.m.f. = electromotive force (volts)	m.m.f. = magnetomotive force ($N \times I$ At)*
From Ohm's law: R = e.m.f./I Transposing: e.m.f. = IR or I = e.m.f./R	S = m.m.f./Φ m.m.f. = ΦS or Φ = m.m.f./S
Resistance has units volts per ampere $I = JA$ (J = current density, A/m^2) I = e.m.f./R can be rewritten as $JA = E/(\ell/\sigma A)$ (cancel A both sides) $J = (E/\ell)\ \sigma$	Reluctance has units magnetomotive force per weber $\Phi = BA$ (B = flux density, Wb/m^2) Φ = m.m.f./S can be rewritten as $BA = NI/(\ell/\mu A)$ (cancel A both sides) $B = (NI/\ell)\ \mu$
E/ℓ volts per metre length is called potential gradient or electric force, symbol E (see chapter 4, section 4.4) Therefore, $J = (E/\ell)\ \sigma$ may be written as $J = E\sigma$ (5.1)	NI/ℓ amperes per metre length is called the magnetising force, symbol H Therefore, $B = (NI/\ell)\ \mu$ may be written as $B = H\mu$ (5.2)
For resistors in series, the current is the same in each resistor. The total circuit e.m.f. = sum of the individual voltage drops $IR_1, IR_2, \ldots$ $E = IR_1 + IR_2 + IR_3 + \ldots$ $= I(R_1 + R_2 + R_3 + \ldots)$	For sections of different magnetic materials in series the flux is the same in each section. m.m.f.$_{(total)}$ = m.m.f.$_{(1)}$ + m.m.f.$_{(2)}$ + ... $= \Phi S_1 + \Phi S_2 + \Phi S_3 + \ldots$ $= \Phi(S_1 + S_2 + S_3 + \ldots)$

* A current I amperes flowing in N turns produces an m.m.f. of NI ampere-turns. Since turns have no dimensions the product $N \times I$ is expressed simply as amperes. Hence

$$H = \frac{NI}{\ell}\ \text{A/m}.$$

The same effect as I amperes in N turns could be achieved by NI amperes in one turn.

M.m.f. = I amperes.

$$\text{Magnetising force, } H = \frac{\text{amperes}}{\text{metre length}} = \frac{I}{2\pi r} \text{ A/m.}$$

Now from equation (5.2), $B = H\mu$, hence, substituting for H,

$$B = \frac{I\mu}{2\pi r} \text{ tesla (Wb/m}^2\text{) at radius } r \text{ metres.} \tag{5.3}$$

5.1.2 *Induced e.m.f. and flux changes*

When a magnetic field which links with a conductor changes in magnitude, during the period of the change an e.m.f. will be induced in the conductor.

From Faraday's original work the following relationships were obtained

induced e.m.f. $e \propto \Delta\Phi$

$e \propto N$ (N = number of turns linked by the flux)

$$e \propto \frac{1}{\Delta t}.$$

Combining these quantities gives

$$e \propto \frac{N\,\Delta\Phi}{\Delta t} \quad \text{or} \quad e = \frac{kN\,\Delta\Phi}{\Delta t}.$$

In S.I. units k = 1, and with Φ measured in webers and t measured in seconds

$$e = \frac{N\,\Delta\Phi}{\Delta t} \text{ V.}$$

In the limit as $\Delta\Phi \to 0$ and $\Delta t \to 0$

$$e = \frac{N\,\mathrm{d}\Phi}{\mathrm{d}t} \text{ V.} \tag{5.4}$$

Considering a coil used either as a choke or as the primary to a transformer, and assuming that the flux varies sinusoidally, equation (5.4) becomes

$$e = \frac{N\,\mathrm{d}\Phi_m \sin \omega t}{\mathrm{d}t} = N\,\omega\Phi_m \cos \omega t.$$

The maximum value of e occurs when $\cos \omega t = 1$, when $E_m = N\,\omega\Phi_m$ ($\omega = 2\pi f$). Dividing both sides of the equation by $\sqrt{2}$ gives

$$\frac{E_m}{\sqrt{2}} = \frac{N\,2\pi f\,\Phi_m}{\sqrt{2}}.$$

$$\frac{2\pi}{\sqrt{2}} = 4.44, \qquad \Phi_m = B_m A \text{ (equation 5.1) and } \frac{E_m}{\sqrt{2}} = E \text{ (rms).}$$

$$\text{Therefore, } E = 4.44 B_m A f N \text{ V.} \tag{5.5}$$

It should be noted that B_m is the peak value of flux density under the particular working conditions which will not necessarily be the maximum possible for the material employed.

5.1.3 *Force, current and flux density*

When an electric current passes through a magnetic field in which the lines of force are not parallel with the direction of the current, a force is exerted on that current. The current may be in a conductor or in the form of an electron beam in a vacuum.

For a current in a conductor, with flux density B tesla, current I amperes and length of the conductor in the field ℓ metres, and where the direction of current and flux are at right angles, the value of the force F is given by

$$F = BI\ell \text{ newtons} \tag{5.6}$$

The force is at right angles to both current and flux.

Now consider a single conductor situated in air and carrying a current I_1 amperes. At radius r metres,

$$H = \frac{I_1}{2\pi r} \text{ A/m and } B = \frac{I_1 \mu_0}{2\pi r} \text{ tesla} \qquad \text{(equation 5.3, Fig. 5.2)}$$

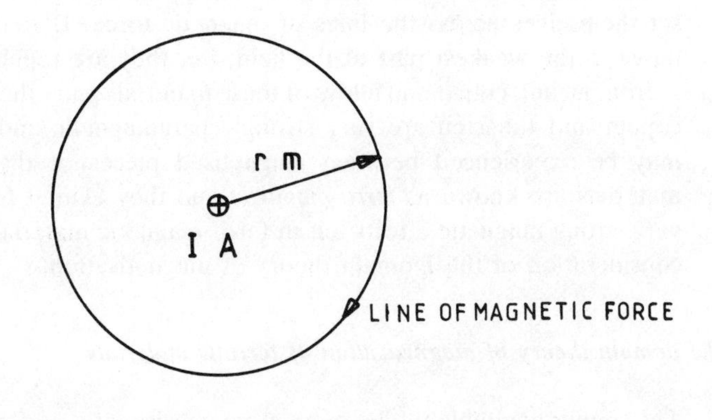

Fig 5.2

Another conductor running parallel with the first and at a distance r metres from it will be situated in this field and will suffer a force $F = BI\ell$ newtons (equation 5.6). If the second conductor is carrying a current of I_2 amperes

$$F = \frac{I_1 \mu_0}{2\pi r} \times I_2 \times \ell \text{ newtons on length } \ell \text{ metres.}$$

Making $\ell = 1$ metre,

$$F = \frac{I_1 I_2 \mu_0}{2\pi r} \text{ N/m length.} \tag{5.7}$$

For I_1 and I_2 in the same direction, the force is one of attraction between the conductors and vice versa.

For the particular case when $I_1 = I_2 = 1$ ampere and $r = 1$ metre

$$F = \frac{\mu_0}{2\pi} = \frac{4\pi \times 10^{-7}}{2\pi} = 2 \times 10^{-7} \text{ N/m length.}$$

This leads to the definition of the ampere which is:

Each of two infinitely long parallel conductors situated one metre apart in vacuum suffer a force of 2×10^{-7} newtons per metre length when the current in each conductor is one ampere.

5.2 Classification of magnetic materials

All materials exhibit some magnetic properties, some reacting strongly to the presence of a magnetic field, others showing little effect. Materials are classified as either *paramagnetic* or *diamagnetic*.

Paramagnetic materials will, if free to move as in the case of a suspended needle of the material, align themselves along the magnetic lines of force of a magnetic field and will be attracted towards the strongest part of the field. They have a relative permeability μ_r greater than unity.

Diamagnetic materials have a relative permeability less than unity and will set themselves across the lines of magnetic force. If free to do so, they will move to the weakest part of the field, i.e. they are repelled by a magnet.

Iron, nickel, cobalt and alloys of these materials and others such as chromium, copper and tungsten are very strongly paramagnetic and considerable forces may be experienced between magnetised pieces of these materials. These materials are known as *ferro-magnetic* and they exhibit ferro-magnetism. The very strong magnetic effects felt in ferro-magnetic materials are explained by a consideration of the Domain theory of magnetisation.

5.2.1 The domain theory of magnetisation of ferrous materials

The simplest suitable model of an atom consists of a positively charged nucleus around which electrons orbit. Since the movement of electrons constitutes an electric current, each orbiting electron sets up a magnetic field at right angles to the plane in which it moves. Now in an atom with many electrons, each in a different orbit, many fields in different directions are produced and the total effect is usually zero. However in certain materials the fields do not cancel and there is a resultant magnetic field in one particular direction. In ferro-magnetic materials groups of about 10^{15} atoms group themselves together in what are termed *domains*. Each domain is in effect a tiny magnet with a north pole on one side and a south pole on the other. In the unmagnetised state the domains form themselves into closed loops as shown in Fig. 5.3(a). Placing the material in an externally produced magnetic field causes the domains to re-orientate themselves progressively as the external field strength is increased.

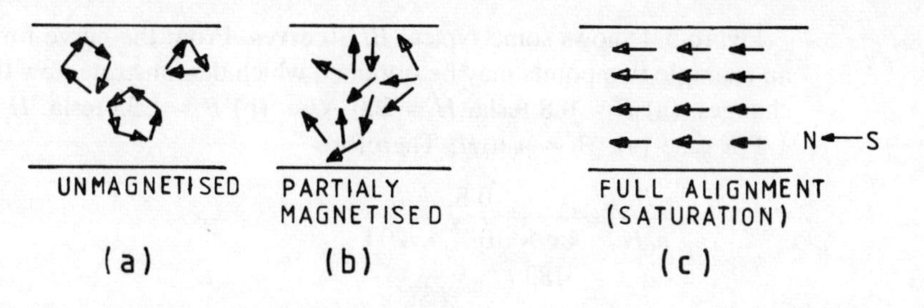

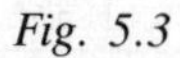

Fig. 5.3

Suppose the current in the coil of N turns in Fig. 5.1 is gradually increased from zero. The torroid is made from a ferrous material. With no current in the coil the domains are arranged in closed loops as in Fig. 5.3(a). As the current is increased, the magnetising force increases and the closed domain loops are broken as they start to align themselves with the external field. There is, as a consequence, a resultant field in one direction (Fig. 5.3(b)). Further increases in current cause the alignment to progress towards that in Fig. 5.3(c). When all the domains are aligned the material is said to be saturated. The flux in the specimen will only rise at a rate equivalent to that in air for additional magnetising force once saturation has occurred.

Now since $B = \mu_0\mu_r H$ (equation 5.2) and B is not proportional to H due to saturation effects, μ_0 being a constant, it follows that μ_r must be variable.

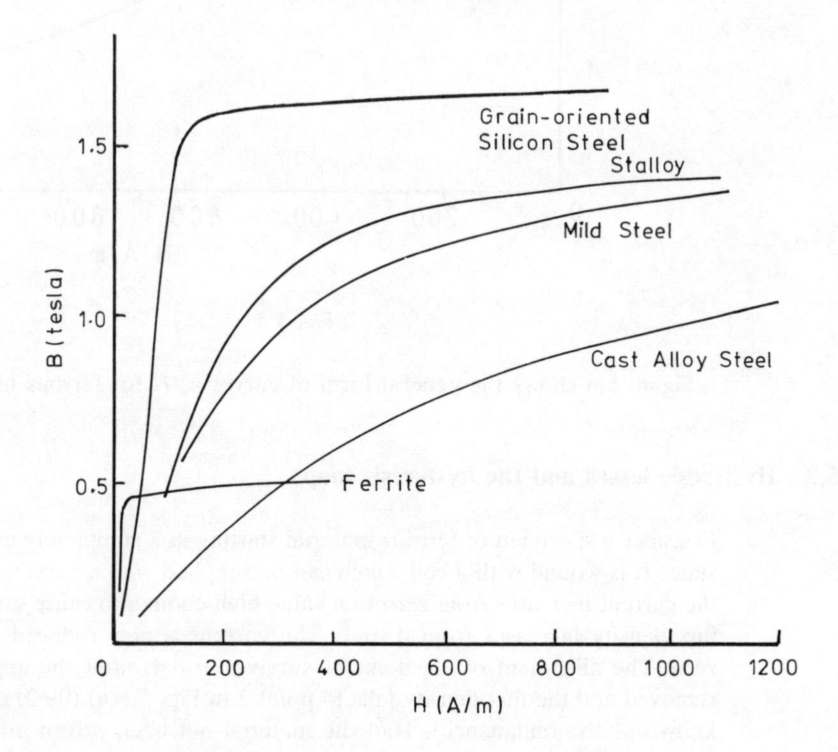

Fig. 5.4

Figure 5.4 shows some typical B/H curves. From the curve for mild steel as an example two points may be obtained which demonstrate how the value of μ_r changes. (a) $B = 0.8$ tesla, $H = 200$ A/m, (b) $B = 1.35$ tesla, $H = 1000$ A/m.

For case (a), $B = \mu_0\mu_r H$. Therefore

$$\mu_r = \frac{B}{\mu_0 H} = \frac{0.8}{4\pi \times 10^{-7} \times 200}$$
$$= 3183$$

For case (b), $\mu_r = \dfrac{1.35}{4\pi \times 10^{-7} \times 1000} = 1074.$

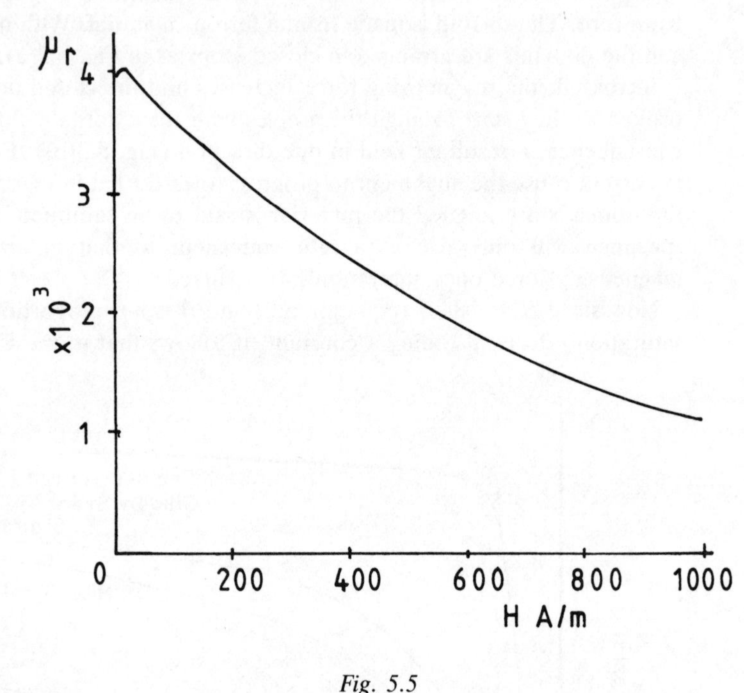

Fig. 5.5

Figure 5.5 shows the general form of curves μ_r/H for ferrous materials.

5.3 Hysteresis losses and the hysteresis loop

Consider a specimen of ferrous material starting in a completely unmagnetised state. It is wound with a coil which can be supplied with a varying current. As the current increases from zero to a value high enough to cause saturation, the flux density increases from 0 to 1. The current is now reduced gradually to zero. The alignment of the domains survives in part after the applied field is removed and the flux density falls to point 2 in Fig. 5.6(a) (0–2, marked R, is known as the remanence). Had the material not been driven into saturation before the reduction of the magnetising force to zero, the general term for the flux remaining would be residual flux or remanent flux.

If the current direction is reversed and increased so as to drive the specimen into saturation with reversed polarity, the curve extends to point 3. The magnitude of reverse magnetising force to cause the flux density to become zero during this procedure is called the coercive force (distance C in Fig. 5.6(a)).

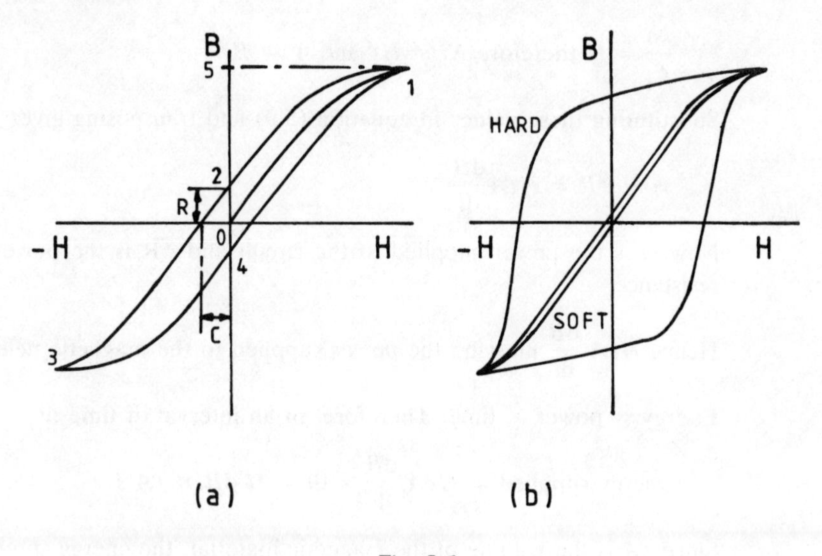

Fig. 5.6

Again reducing the magnetising force to zero brings the curve back to point 4 and then with forward magnetising force to point 1 again.

Supplying the coil with alternating current results in successive excursions through 1, 2, 3, 4. This is known as a hysteresis loop.

Changing the pattern of domains requires the expenditure of energy so that where alternating magnetisation is involved this energy shows up as heat in the material. The energy expended is known as the hysteresis loss. The loss per cycle is a function of the area inside the hysteresis loop so that for minimum loss this area must be small. The softer the material used the narrower is the loop. A permanent magnet material has a very wide loop having high remanence as shown in Fig. 5.6(b).

Let the magnetic circuit shown in Fig. 5.1 be made from a ferrous material. Suppose the current in the coil is increasing. It will be creating a magnetic flux in the circuit.

At a particular instant there will be an induced back e.m.f., e volts. From equation (5.4)

$$e = N\frac{d\Phi}{dt}\ \text{V}.$$

Let the resistance of the coil be $R\ \Omega$.

Applied voltage − back e.m.f. = volt drop in the resistance

$$v - N\frac{d\Phi}{dt} = iR. \tag{5.8}$$

Multiplying equation (5.8) throughout by i gives

$$vi - iN\frac{d\Phi}{dt} = i^2R. \tag{5.9}$$

From the definitions in section 5.1.1

$$\frac{Ni}{\ell} = H \text{ therefore } Ni = H\ell \text{ and } \Phi = BA.$$

Substituting these values in equation (5.9) and transposing gives

$$vi - i^2R = H\ell A\frac{dB}{dt}.$$

Now vi is the power supplied to the circuit and i^2R is the power loss in coil resistance.

Hence $H\ell A \frac{dB}{dt}$ must be the power supplied to the magnetic field.

Energy = power × time. Therefore, in an interval of time dt

$$\text{energy supplied} = H\ell A \frac{dB}{dt} \times dt = H\,dB \times \ell A \text{ J}. \tag{5.10}$$

Since ℓA is the volume of the magnetic material, the energy supplied per unit volume (per m^3) during time dt seconds is H dB joules.

During time dt, B has increased by dB and the value of H dB is the area of the hatched strip in Fig. 5.7.

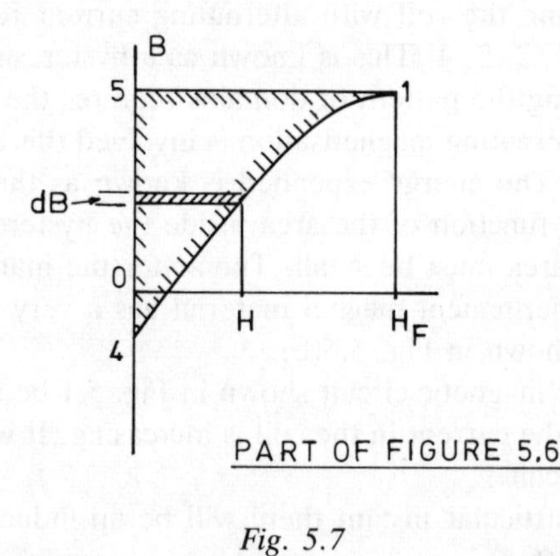

Fig. 5.7

The total energy supplied as H rises from 0 to H_F is therefore the sum of all such strips, ΣH dB. This sum is equal to the total area bounded by the lines 4, 1, 5, 4, the edges of which are cross-hatched.

Reducing the magnetising force to zero so that only the residual flux remains will allow the domains to partially re-orientate themselves so giving energy back to the supply. The corresponding area is that bounded by the lines 2, 1, 5,

2 in Fig. 5.6. The difference between the energy supplied and that returned must be the energy lost and this is represented by the difference between the two areas, i.e. that inside the hysteresis loop. The same loss occurs during the reverse magnetisation of the specimen.

When magnetised by an alternating current at f Hz, the loss occurs f times per second.

Energy loss: (area of loop) $\times f$ J/s = W.

$$\text{The product } H \times B \text{ has units } \frac{\text{ampere-turns}}{\text{metre}} \times \frac{\text{webers}}{(\text{metres})^2} \quad \text{(i)}$$

$$\text{From equation (5.4), } e = N\frac{d\Phi}{dt}\text{ V. Transposing: } d\Phi = \frac{e\,dt}{N} \quad \text{(ii)}$$

$$\text{The units of (ii) are: webers} = \frac{\text{volt seconds}}{\text{turns}}$$

Substituting this in (i)

$$HB \text{ has units } \frac{\text{ampere turns}}{\text{metre}} \times \frac{\text{volt seconds}}{\text{turns (metres)}^2}$$

$$= \frac{\text{volt-ampere seconds}}{(\text{metres})^3} = \text{J/m}^3.$$

For a hysteresis loop plotted to scales (say) 100 A/m = 1 cm and 0.1 tesla = 1 cm, 1 cm^2 corresponds to $100 \times 0.1 = 10$ J/m^3. Knowing the total area of the loop and the actual volume of the core will enable the energy loss per excursion round the loop to be determined.

In general, for H plotted at x A/m per unit length and B plotted at y T per unit length,

power loss = $x \times y \times$ area of the hysteresis loop in $(\text{units})^2 \times$ volume of the core $\times$ frequency watts. (5.11)

5.4 Energy stored in a magnetic system in terms of its self-inductance

A circuit is said to possess a self-inductance of one henry if, when a current is changing at a rate of one ampere per second in it, an e.m.f. of one volt is induced in that circuit.

Written as a formula this is

$$e = L\frac{di}{dt}\text{ V} \quad (5.12)$$

where L is the self-inductance in henrys.

Consider a magnetic circuit as shown in Fig. 5.1. The current in the coil is increasing so creating a magnetic flux in the circuit. At a particular instant there will be a back e.m.f., e volts, induced (see also section 5.3). The relevant equations are

(5.4) $e = N \frac{d\Phi}{dt}$ V (5.12) $e = L \frac{di}{dt}$ V

(5.9) $vi - iN \frac{d\Phi}{dt} = i^2R.$

Equations (5.4) and (5.12) express e in terms of different quantities. Equating these

$$N \frac{d\Phi}{dt} = L \frac{di}{dt}.$$

Hence

$$L = N \frac{d\Phi}{dt} \times \frac{dt}{di} = N \frac{d\Phi}{di}. \qquad (5.13)$$

Transposing equation (5.9) gives

$$vi - i^2R = iN \frac{d\Phi}{dt}.$$

In words: power supplied − power loss = power to the magnetic field.
This is an alternative expression to that in equation (5.10).
During time interval dt, energy supplied to the magnetic field = $iN\,(d\Phi/dt) \times dt$ J.
Transposing equation (5.13): $L\,di = N\,d\Phi$. Substituting in the equation above gives

energy supplied to the magnetic field in time dt = $i \times L\,di$ J.

Total energy supplied to the field as the current increases from zero to a final steady state value of I amperes is found by integration between the limits 0 and I.

$$\text{Energy to the field} = \int_0^I Li\,di = L\left(\frac{i^2}{2}\right)_0^I = \tfrac{1}{2}LI^2 \text{ J}. \qquad (5.14)$$

WORKED EXAMPLE 5.1

A non-magnetic former of a toroid (Fig. 5.1) has a mean circumference of 0.2 m and a uniform cross-sectional area of 3.5 cm^2. The coil consists of 650 turns and carries a steady current of 5 A.

Calculate (a) the m.m.f., (b) the magnetising force, (c) the reluctance, (d) the magnitude of the flux and flux density, (e) the circuit inductance, and (f) the energy stored in the magnetic field.

(a) m.m.f. = $N \times I = 650 \times 5 = 3250$ A.

(b) Magnetising force $= \frac{\text{m.m.f.}}{\text{length of the flux path } (\ell)} = \frac{3250}{0.2} = 16\,250$ A/m.

(c) Reluctance $S = \dfrac{\ell}{\mu_0 A} = \dfrac{0.2}{4\pi \times 10^{-7} \times 3.5 \times 10^{-4}} = 4.547 \times 10^8$ A/Wb.

This means that it would require 4.547×10^8 A to set up a flux of 1 Wb in this toroid.

(d) Flux $\Phi = \dfrac{\text{m.m.f.}}{S} = \dfrac{3250}{4.547 \times 10^8} = 7.148 \times 10^{-6}$ Wb.

Flux density $B = \dfrac{\Phi}{A} = \dfrac{7.148 \times 10^{-6}}{3.5 \times 10^{-4}} = 0.02$ T (tesla = Wb/m^2).

Using equation (5.2) an alternative method is possible. $B = \mu_0 H$ and $BA = \Phi$. Therefore

$$\begin{aligned}\Phi &= \mu_0 HA \\ &= 4\pi \times 10^{-7} \times 16\,250 \times 3.5 \times 10^{-4} \\ &= 7.148 \times 10^{-6} \text{ Wb.}\end{aligned}$$

(e) Inductance $= N(\mathrm{d}\Phi/\mathrm{d}i)$. (equation 5.13)

Since $B = \mu_0 H$, the flux is proportional to current. Any size of change in current produces a proportional change in flux. Hence $\Delta\Phi/\Delta i$ may be written for $\mathrm{d}\Phi/\mathrm{d}i$.

$\Delta i = (0 - 5)$ A and $\Delta\Phi = (0 - 7.148 \times 10^{-6})$ Wb; i.e. the changes in each quantity from zero to final conditions.

$$L = \frac{650 \times 7.148 \times 10^{-6}}{5} = 929.12 \ \mu\text{H}.$$

(f) Energy stored $= \frac{1}{2} LI^2$ J (equation 5.14)

$= \frac{1}{2} \times 929 \times 10^{-6} \times 5^2 = 0.0116$ J.

WORKED EXAMPLE 5.2

A steel toroid has a mean length of 20 cm and cross-sectional area 0.8 cm^2. It has a radial slot 4 mm wide cut through the steel.

Determine the value of current required in a 400-turn coil wound on the toroid to set up a flux of 72 μWb in the steel and air gap. (The relative permeability of the steel at the working flux density = 1800.)

Considering the steel

Flux density $B = \dfrac{\Phi}{A} = \dfrac{72 \times 10^{-6}}{0.8 \times 10^{-4}} = 0.9$ T.

$B = \mu_0 \mu_r H$ T. Transposing, $H = \dfrac{B}{\mu_0 \mu_r}$

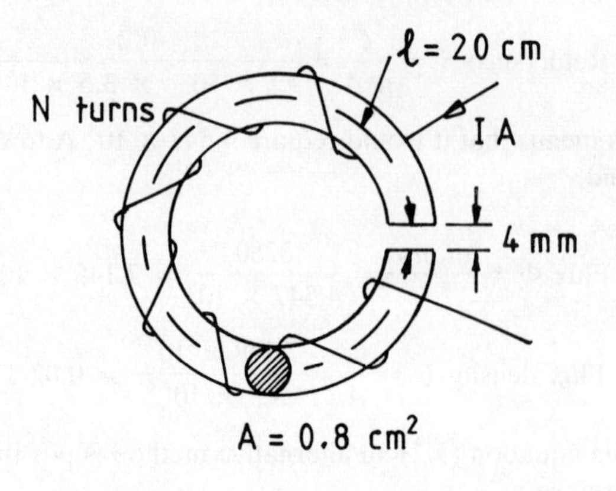

Fig. 5.8

$$= \frac{0.9}{4\pi \times 10^{-7} \times 1800} = 397.9 \text{ A/m.}$$

$H = \dfrac{NI}{\ell}$ A/m. Transposing: $H\ell = NI$ (= m.m.f.)

$$= 397.9 \times 0.2 = 79.6 \text{ A.}$$

Considering the air gap

Assuming no leakage, the flux and flux density in the air gap will be the same as in the steel. In air $B = \mu_0 H$. Therefore

$$H = \frac{0.9}{4\pi \times 10^{-7}} = 716\,197 \text{ A/m.}$$

$$\begin{aligned} NI &= H \times \text{length of the gap} \\ &= 716\,197 \times 4 \times 10^{-3} \\ &= 2864.8 \text{ A.} \end{aligned}$$

$$\begin{aligned} \text{Total m.m.f.} &= (\text{m.m.f. for steel}) + (\text{m.m.f. for air}) \\ &= 79.6 + 2864.8 \\ &= 2944.4 \text{ A.} \end{aligned}$$

Now m.m.f. = $N \times I$. Therefore

$$\begin{aligned} 2944.4 &= 400\,I \\ I &= 7.36 \text{ A.} \end{aligned}$$

SELF-ASSESSMENT EXAMPLE 5.3

A mild steel ring of uniform cross-sectional area 1.5 cm^2 has a mean diameter of 0.15 m. Estimate the current required in a coil of 300 turns wound on the ring to establish a flux of 195 μWb in the steel using the data given in the table.

For mild steel

B (tesla)	1.0	1.15	1.3	1.45
H (A/m)	400	500	800	1650

WORKED EXAMPLE 5.4

The hysteresis loop for a magnetic material used in the core of a transformer is drawn to the following scales

100 A/m ≡ 1 cm on the H scale
0.05 tesla ≡ 1 cm on the B scale.

The area of the hysteresis loop is 158 cm^2. The volume of the transformer core is 0.015 m^3. The operating frequency is 50 Hz.

Determine the power loss due to hysteresis in the core.

Using equation (5.11), $100 \times 0.05 = 5$. This tells us that each square centimetre of the hysteresis loop represents a loss of 5 joules for each cubic metre of the core for each cycle of the input voltage.

Therefore with 0.015 m^3 and a frequency of 50 Hz, the power loss is given by

$$\text{power loss} = (5 \times 0.015 \times 50) \times 158 = 592.5 \text{ W}.$$

SELF-ASSESSMENT EXAMPLE 5.5

(a) An iron toroid with mean length 25 cm and a cross-sectional area 3 cm^3 is wound with a coil of 120 turns. A current in the coil is increased from zero to 10 A when the flux in the toroid increases from zero to 480 μWb. For a coil current of 10 A, determine (i) the value of H, (ii) the value of B, (iii) the mean value of inductance over the range $N(\Delta\Phi/\Delta i)$.

(b) During the increase in current in the coil in part (a), the curve shown in Fig. 5.9 was plotted. The scales chosen are such that $x = y = 16$ cm. The hatched section has an area of 98 cm^2.

Determine the amount of energy involved in orienting the domains during the magnetisation period.

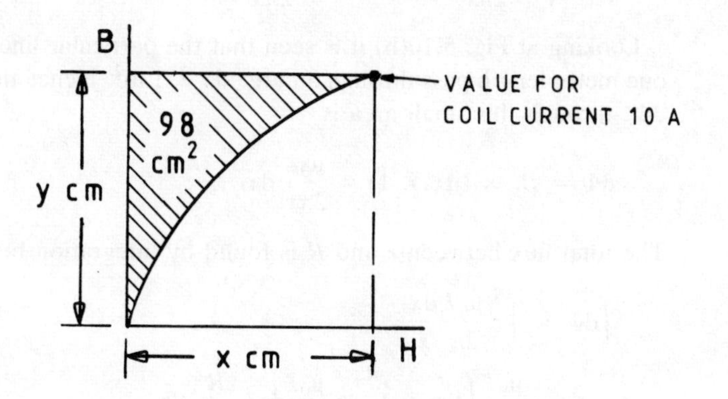

Fig. 5.9

5.5 The inductance of concentric cylinders (co-axial cable)

The core of the cable in Fig. 5.10(a) carries I amperes away from the viewer and to the right while the sheath carries the return current. Since the magnetic fields produced by the go and return currents will be equal and in opposite directions, there will be no magnetic field outside the sheath. A magnetic field exists within the core and between the core and sheath. The magnetic field within the conductor is very small and is ignored at this stage.

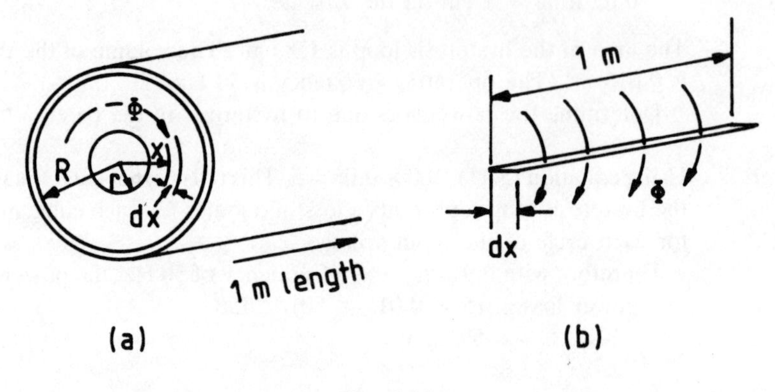

Fig. 5.10

Consider the dielectric, from radius r to radius R. At any radius x a line of magnetic force has length $2\pi x$ metres.
Generally

$$H = \frac{NI}{\ell}. \qquad \text{(section 5.1.1 definitions)}$$

Since a single current loop (go and return) is being considered, $N = 1$. Therefore

$$H = \frac{I}{\ell} = \frac{I}{2\pi x} \quad \text{and } B_x = \frac{\mu_0 I}{2\pi x}\text{T}. \qquad \text{(equation 5.3)}$$

Looking at Fig. 5.10(b) it is seen that the particular lines of force along the one metre length pass through an area $dx \times 1\ \text{m}^2$. Hence the increment of flux, $d\Phi$, through this small area is

$$d\Phi = B_x \times (dx \times 1) = \frac{\mu_0 I}{2\pi x}\,dx.$$

The total flux between r and R is found by integration between these limits.

$$\int d\Phi = \int_r^R \frac{\mu_0 I}{2\pi}\frac{dx}{x}$$

$$\Phi = \frac{\mu_0 I}{2\pi}\left(\log_e x\right)_r^R = \frac{\mu_0 I}{2\pi}\log_e \frac{R}{r}\ \text{Wb}.$$

Inductance = flux linkages per ampere $\left(N\frac{d\Phi}{di}\right)$ (equation 5.13).

For N = 1, in a non-magnetic medium (see worked example 5.1) the following may be written

$$L = \frac{\Delta\Phi}{\Delta I}.$$

In the cable, as i rises from zero to final value I, the flux increases from zero to

$$\frac{\mu_0 I}{2\pi} = \log_e \frac{R}{r} \text{ Wb}.$$

Therefore

$$\frac{\text{change in flux}}{\text{change in current}} = L = \frac{\mu_0}{2\pi} \log_e \frac{R}{r} \text{ H/m length}.$$

5.6 The inductance of a parallel pair of twin conductors remote from earth

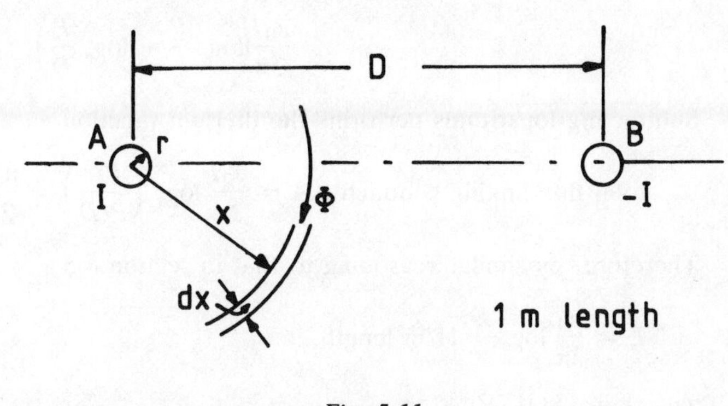

Fig. 5.11

In Fig. 5.11, conductor A carries current away from the viewer, the return current being in conductor B. The distance D between the conductors is much greater than their radii r. A 1 m length of the twin conductor system will be considered.

Conductor A

At any radius x from A, the magnetic flux density is

$$B_x = \frac{\mu_0 I}{2\pi x} \text{ T}. \qquad \text{(equation 5.3)}$$

Theoretically, there is no limit to the distance from the conductor at which a magnetic field will be created. In practice, at only a few metres distance the effect is undetectable. Let the radius at which no effect is felt be P metres. The increment of flux passing through the area $dx \times 1$ m^2 (see Fig. 5.10(b)) is

$$d\Phi = \frac{\mu_0 I}{2\pi x}\,dx.$$

The flux linking with conductor A due to its own current $= \int_r^P \frac{\mu_0 I}{2\pi}\frac{dx}{x}.$

$$\Phi = \frac{\mu_0 I}{2\pi}\log_e \frac{P}{r}\ \text{Wb}.$$

Similarly, the flux which links with (passes round) conductor A due to the current $-I$ in conductor B will be the same integral, but between the limits P and D. Note that at radii less than D from conductor B, lines of force do not pass round conductor A.

The flux linking with conductor A due to the current in conductor B

$$= \int_D^P \frac{\mu_0(-I)}{2\pi}\frac{dx}{x} = \frac{-\mu_0 I}{2\pi}\log_e \frac{P}{D}\ \text{Wb}.$$

Total flux linking conductor A = sum of that due to its own current and that due to current in B

$$= \frac{\mu_0 I}{2\pi}\left(\log_e \frac{P}{r} - \log_e \frac{P}{D}\right).$$

Subtracting logarithms performs the division function

$$\text{total flux linking conductor } A = \frac{\mu_0 I}{2\pi}\log_e\left(\frac{P/r}{P/D}\right) = \frac{\mu_0 I}{2\pi}\log_e \frac{D}{r}.$$

Therefore, by similar reasoning to that in section 5.5

$$L = \frac{\mu_0}{2\pi}\log_e \frac{D}{r}\ \text{H/m length}.$$

This is the inductance of conductor A.

By repeating the process for conductor B an identical expression will be arrived at.

The total inductance of the circuit (both conductors) is therefore

$$L = 2\left(\frac{\mu_0}{2\pi}\log_e \frac{D}{r}\right)\ \text{H/m}$$

$$= \frac{\mu_0}{\pi}\log_e \frac{D}{r}\ \text{H/m length}.$$

5.7 The Steinmetz index

When a magnetic material is not operated up to saturation value but rather to some other lesser value, the area of the hysteresis loop is reduced. From a consideration of Fig. 5.12 it may be seen that reducing the maximum flux density from B_m to $\frac{1}{2}B_m$ reduces the area of the loop by a factor considerably greater than 2.

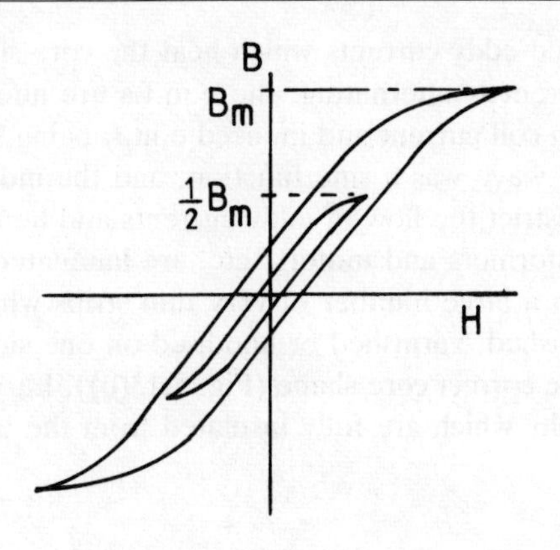

Fig. 5.12

Charles Steinmetz, an American engineer, established an empirical formula connecting the hysteresis losses for the same core to different peak working flux densities. His formula is

$$\text{power loss} = k_h B_m^n f \text{ W.} \tag{5.15}$$

where k_h is a constant for the material, the hysteresis coefficient, and n is called the Steinmetz index. This has values between 1.6 and 3 generally, depending on the material.

As an example, consider a transformer core with Steinmetz index 1.9 and a hysteresis loss of 150 W at 50 Hz when operating at a maximum flux density of 1.4 tesla.

It is required to estimate what the hysteresis loss would be if the peak flux density were reduced to 1.2 tesla.

The original conditions give

$$150 = k_h \times 1.4^{1.9} \times 50. \tag{i}$$

For the new condition

$$P = k_h \times 1.2^{1.9} \times 50. \tag{ii}$$

Dividing equation (i) by equation (ii) gives

$$\frac{150}{P} = \left(\frac{1.4}{1.2}\right)^{1.9} \quad \text{(since } k_h \text{ and frequency cancel)}$$

$$P = \frac{150}{(1.166)^{1.9}} = 111.9 \text{ W.}$$

5.8 Eddy current losses

Consider a coil wound on a conducting core as shown in Fig. 5.13(a). The current in the coil is increasing so that it will produce an increasing flux which will induce e.m.f.s in the coil itself and in the core. The e.m.f. induced in the coil inhibits the rate of rise of that current, while that in the core drives what are called eddy currents which heat the core since it has resistance. When the coil current is alternating, the e.m.f.s are alternating, the phase relationship between coil current and induced e.m.f. being 90°. (See section 5.1.2, in which the flux wave was a sine function, and the induced e.m.f. a cosine function.)

To restrict the flow of eddy currents and hence limit the heat loss, the cores of transformers and motors, etc. are laminated. That is to say they are made up from a large number of very thin strips which are cut from sheet, cleaned and polished, varnished or anodised on one side and then pressed together to form the correct core shape (Fig. 5.13(b)). Large cores are held together using core bolts which are fully insulated from the laminations.

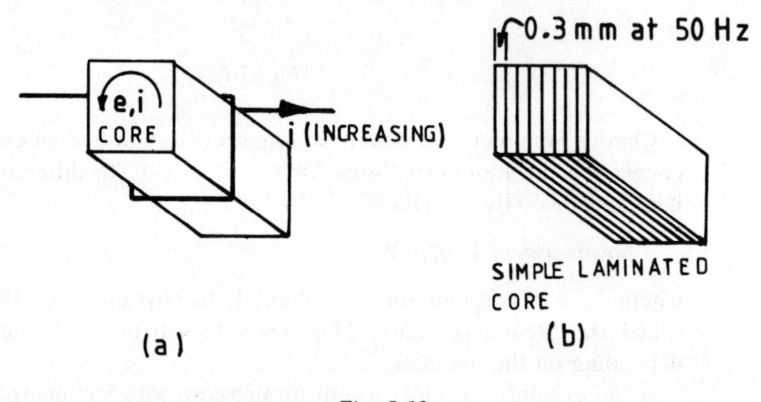

Fig. 5.13

Consider a section of a single lamination situated in a sinusoidally varying magnetic field. The flux enters the narrow side of the lamination parallel to lines XY and PQ, which may be considered to be drawn on the surface of the lamination (Fig. 5.14).

A loop A, B, C, D within the lamination has a width $2x$ metres and is 1 metre in height. Its cross-sectional area is $2x$ m^2. Sinusoidal alternating flux linking with this loop will induce an e.m.f. expressed by equation (5.5)

$$E = 4.44B_m AfN \text{ V}$$
$$= 4.44B_m(2x)f \text{ V (since } N = \text{one loop).}$$

This e.m.f. drives current round the loop A, B, C, D which extends 1 metre into the lamination. Looking down on to the top of the lamination it will be seen that the area through which current must pass is dx thick and 1 metre long. (Distance XY × dx) (i)

The length of the current path = AB + BC + CD + DA metres.

Now since the thickness of the lamination is likely to be a fraction of 1 mm, AB + CD may be ignored compared with BC (= 1 m) and DA (= 1 m).

Length of current path = 2 m (closely). (ii)

The resistance of the path, $R = \dfrac{\ell}{\sigma A}$ (section 5.1.1)

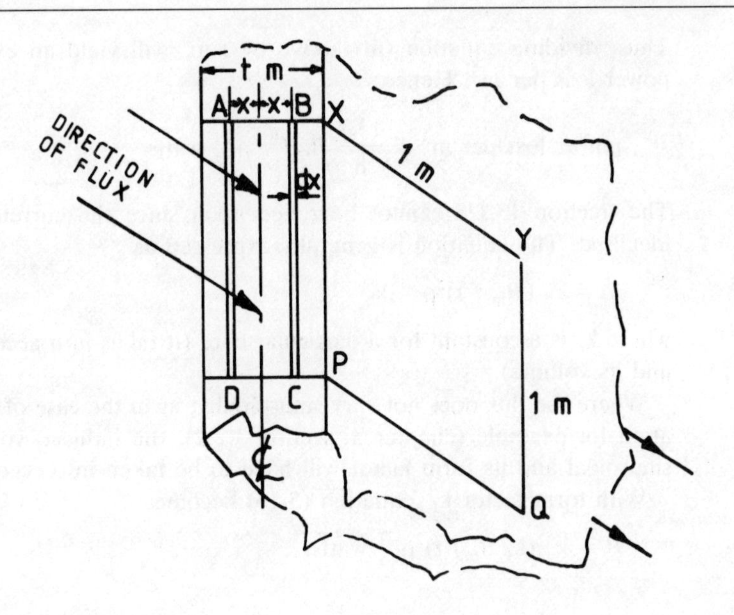

Fig. 5.14

or since $\rho = \dfrac{1}{\sigma}$

$$R = \frac{\rho\ell}{A} \tag{iii}$$

where ρ = resistivity of the lamination material.

Substituting equations (i) and (ii) in equation (iii) gives

$$R = \frac{2\rho}{dx}.$$

Power dissipated in this resistance $= \dfrac{E^2}{R}$ watts

$$= \frac{(4.44B_m\,(2x)f)^2}{2\rho/dx} = \frac{(8.88B_m fx)^2\,dx}{2\rho}.$$

The total loss over the whole thickness is obtained by integrating between the limits $x = 0$ (at the centre line) and $x = t/2$ (at the edge).

$$\text{Total loss} = \frac{(8.88B_m f)^2}{2\rho}\int_0^{t/2} x^2\,dx$$

$$= \frac{(8.88B_m f)^2}{2\rho}\left(\frac{x^3}{3}\right)_0^{t/2} = 13.1\ B_m^2 f^2\rho^{-1}\ (x^3)_0^{t/2}$$

$$= 13.1\ B_m^2 f^2\rho^{-1}\frac{t^3}{8}. \tag{iv}$$

The volume of this section of lamination $= t \times 1 \times 1 = t\ \text{m}^3$.

Thus, dividing equation (iii) above by t m^3 will yield an expression for the power loss per m^3. Hence

$$\text{power loss per m}^3 = \frac{13.1}{8} B_m^2 f^2 t^2 \rho^{-1} \text{ watts.}$$

The fraction 13.1/8 cannot be relied upon since the current path has been idealised. The equation is generally expressed as

$$P = k_e (B_m f t)^2 \rho^{-1} \text{ W} \tag{5.16}$$

where k_e is a constant for a particular core (it takes into account the material and its volume).

Where the flux does not vary sinusoidally, as in the case of forced magnetisation for example (chapter 3, section 3.2.2), the induced voltage will not be sinusoidal and its form factor will have to be taken into account.

With form factor k_f, equation (5.16) becomes

$$P = k_e (k_f B_m f t)^2 \rho^{-1} \text{ watts.}$$

WORKED EXAMPLE 5.6

A core of a choke has a hysteresis loss of 25 W and an eddy current loss of 15.6 W when operated at 50 Hz.

Assuming that the flux density remains unchanged, determine the separate losses if (a) the frequency is increased to 60 Hz, (b) a new core is built using the same material but using laminations one half the thickness of those in (a). (The operating frequency is 50 Hz.)

(a) The total core loss = hysteresis loss + eddy current loss

$$= k_h B_m^n f + k_e B_m^2 f^2 t^2 \rho^{-1}. \text{ (equations 5.15 and 5.16)}$$

Since the flux density is to remain constant, B_m may be combined with the existing constants

$$P = k_1 f + k_2 f^2 t^2 \text{ W } (k_1 = k_h B_m^n \text{ and } k_2 = k_e B_m^2 \rho^{-1}).$$

Hysteresis loss Original conditions, $k_1 \times 50 = 25$ W

$$k_1 = \frac{25}{50}.$$

At 60 Hz, $k_1 \times 60 = P$ (hysteresis).

Substituting for k_1

$$P_H = \frac{25}{50} \times 60 = 30 \text{ W.}$$

Eddy current loss Original conditions, $k_2 \times 50^2 t^2 = 15.6$

$$k_2 t^2 = \frac{15.6}{50^2}.$$

At 60 Hz, $k_2 t^2 \times 60^2 = P$ (eddy currents).

Substituting for $k_2 t^2$

$$P_E = \frac{15.6}{50^2} \times 60^2 = 22.46 \text{ W}.$$

(b) The hysteresis loss is independent of lamination thickness so that at one half of the lamination thickness the hysteresis loss is 25 W.

Eddy current loss $= k_2 f^2 t^2$ W.

Original conditions, $15.6 = k_2 \times 50^2 \times t^2$. (i)

New conditions, $P_E = k_2 \times 50^2 \times (\frac{1}{2}t)^2$. (ii)

Dividing equation (i) by equation (ii) gives

$$\frac{15.6}{P_E} = \frac{t^2}{(\frac{1}{2}t)^2} = 2^2$$

$$P_E = \frac{15.6}{4} = 3.9 \text{ W}.$$

SELF-ASSESSMENT EXAMPLE 5.7

A transformer core has an eddy current loss of 125 W at 50 Hz. The laminations are 0.34 mm thick. The design is to be modified so that it will operate at a different voltage and at 400 Hz with the same eddy current loss. At the new voltage the flux density will be one half the original value. What is the necessary thickness of the laminations assuming all other parameters remain unchanged?

5.9 Determination of the separate hysteresis and eddy current losses in the ferrous core of a coil or transformer

Consider the condition of free magnetisation of an iron core (flux varies sinusoidally). The total core loss

$$P_C = k_h B_m^n f + k_e B_m^2 f^2 t^2 \rho^{-1} \text{ W} \quad \text{(equations 5.15 and 5.16)}$$

Dividing both sides of the equation by the frequency f gives

$$\frac{P_C}{f} = k_h B_m^n + k_e B_m^2 f t^2 \rho^{-1} \quad (5.17)$$

If the coil is fed from a variable-voltage variable-frequency supply, it is possible to measure the total core loss for a number of different frequencies while maintaining the flux density constant. The conditions for constant flux density can be realised from the transformer e.m.f. equation (5.5).

$$E = 4.44 B_m A f N \text{ V}$$

Dividing both sides of this equation by f yields

$$\frac{E}{f} = 4.44B_mAN.$$

Since 4.44, A and N are constant for a particular device,

$$\frac{E}{f} = k_1B_m \ (k_1 = 4.44AN).$$

Hence, if the frequency is varied while keeping the ratio E/f constant, B_m will be constant.

For a constant flux density B_m, and realising that k_h, k_e, t and ρ are all constants for a particular device, equation (5.17) reduces to

$$\frac{P_C}{f} = k_2 + k_3f \tag{5.18}$$

where $k_2 = k_hB_m^n$, $k_3 = k_eB_m^2t^2\ \rho^{-1}$.

This is of the form $y = c + mx$, which is the equation of a straight line.

To separate the hysteresis and eddy current losses, firstly the d.c. resistance of the coil under test is determined. The coil is then connected as shown in Fig. 5.15. The variable supply may be an oscillator with sufficient range and power or an alternator set, the speed and output voltage of which can be varied as required.

For a particular ratio of E/f, a number of readings of input power are taken. Ignoring any wattmeter errors, the core loss P_C in each case is the wattmeter indication less the coil power loss $IR_{d.c.}$.

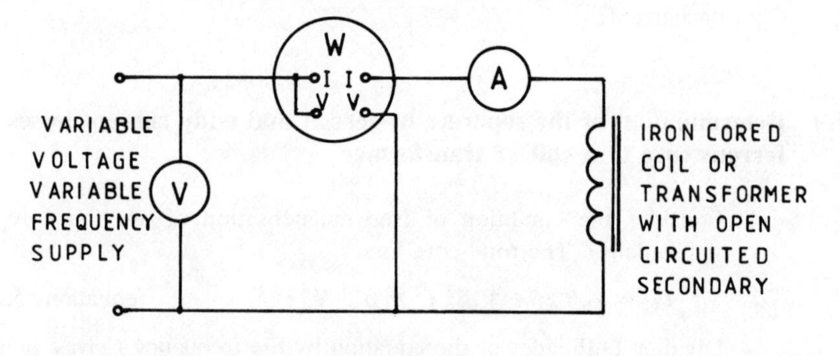

Fig. 5.15

Plotting P_C/f against a base of the frequency f produces a straight line as shown in Fig. 5.16.

Now, in equation (5.18)

$$\frac{P_C}{f} = k_2 + k_3f$$

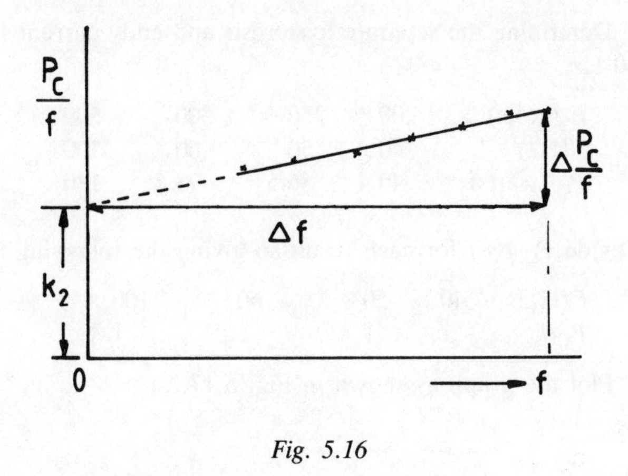

Fig. 5.16

k_2 is the intercept on the P_C/f axis at $f = 0$. But $k_2 = k_h B_m^n$, so that this value is now known.

$$k_3 \text{ is the slope of the line } = \frac{\Delta P_C/f}{\Delta f}$$
$$= k_e B_m^2 t^2 \rho^{-1}$$

so that a value can now be obtained for the combined product of all the constants involved in the eddy current loss calculations.

The separate losses at any frequency at the particular flux density at which the tests were done is therefore arrived at by re-transposing equation (5.18) into

$$P_C = k_2 f \text{ (the hysteresis loss)} + k_3 f^2 \text{ (the eddy current loss).} \qquad (5.19)$$

The assumptions made were:

(a) constant flux density
(b) flux variation having sinusoidal form or if not then constant in each of the tests (k_f ignored or considered constant)
(c) the Steinmetz index constant.

Assumptions (a) and (b) are justified by keeping E/f constant and employing the conditions for free magnetisation of the core (chapter 3, section 3.2.1). Assumption (c) is not really justified since, in fact, the Steinmetz index is not a constant but tends to increase with frequency. The method is therefore only approximate and is suitable for establishing approximately the relative sizes of the hysteresis and eddy current losses in a specimen of iron, the core of a transformer or a choke.

WORKED EXAMPLE 5.8

The following test results refer to a sample of steel used as the core of a coil. The connections were as in Fig. 5.15.

Determine the separate hysteresis and eddy current losses at (a) 50 Hz, (b) 60 Hz.

E (volts)	200	250	300	500
f (Hz)	40	50	60	100
P_C (watts)	40	56.5	73.8	180

Divide P_C by f for each result so giving the following table

f (Hz)	40	50	60	100
P_C/f	1	1.132	1.23	1.8

Plot the graph as shown in Fig. 5.17.

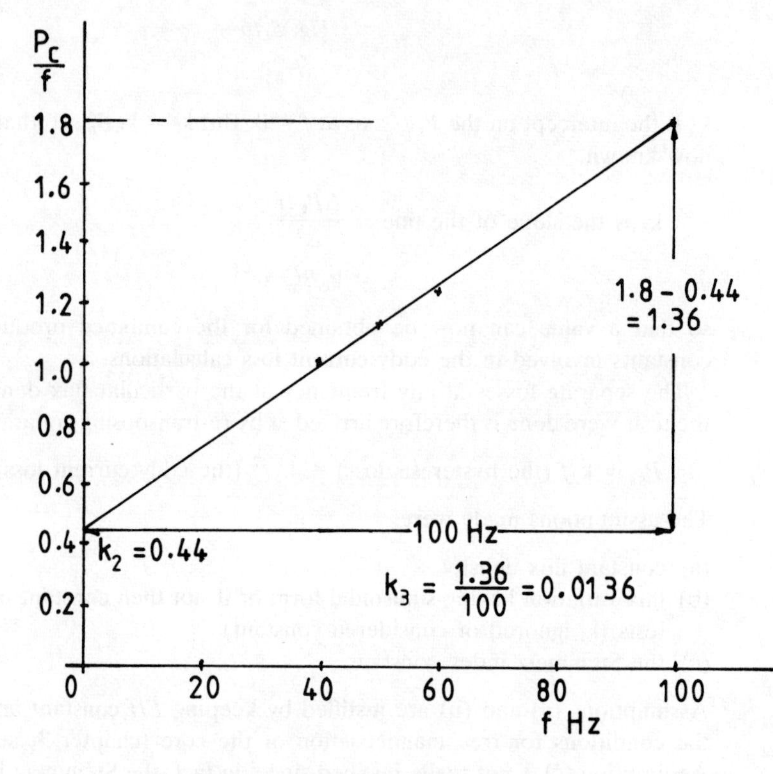

Fig. 5.17

The equation of the line is

$$P_C/f = 0.44 + 0.0136f.$$

Transposing to the form of equation 5.19 gives

$$P_C = 0.44f + 0.0136f^2.$$

The separate losses at 50 Hz $= 0.44 \times 50 + 0.0136 \times 50^2$
$= 22 + 34.$

The hysteresis loss = 22 W. The eddy current loss = 34 W. Total loss = 56 W. At 60 Hz, the losses are

hysteresis, $0.44 \times 60 = 26.4$ W
eddy current, $0.0136 \times 60^2 = 48.96$ W
total loss = 75.36 W.

Comparing these values of total losses with those in the original table it is seen that they are slightly in error. This is frequently the case due to an optimum line being drawn through experimental points with a little scatter and due to the variation in the Steinmetz index as already discussed.

SELF-ASSESSMENT EXAMPLE 5.9

A test carried out on a transformer primary to separate the hysteresis and eddy current losses in the core yielded the following results with E/f held constant

P_C (watts)	92.5	110	171.5	220
f (Hz)	50	55	70	80

Determine the values of the separate hysteresis and eddy current losses at 50 Hz.

5.10 Determination of the value of the Steinmetz index

If the value of the Steinmetz index is required it may be obtained by carrying out two sets of tests as outlined previously (section 5.7).

One test is carried out at normal flux density and the other at a fraction of normal. For ease of calculation the second test is often carried out at one half of the normal value. The values of P_C/f are plotted against f for each test.

Let the intercept on the P_C/f axis for normal flux density (B_m) have a value $k_{2(1)}$ and that for the test at one half normal flux density ($\frac{1}{2}B_m$) have a value $k_{2(2)}$. From equation (5.18)

$$k_{2(1)} = k_h B_m^n \quad \text{(i)}$$

$$\text{and } k_{2(2)} = k_h (\tfrac{1}{2}B_m)^n \quad \text{(ii)}$$

Dividing equation (i) by equation (ii) gives

$$\frac{k_{2(1)}}{k_{2(2)}} = \frac{B_m^n}{(\frac{1}{2}B_m)^n} = 2^n. \quad (5.20)$$

The Steinmetz index may therefore be calculated.

WORKED EXAMPLE 5.10

Two tests were carried out on a transformer primary to determine the Steinmetz index for the core material. The results were

Test 1 With $E/f = 5$, which gives normal flux density,

P_C (watts)	62.5	204
f (Hz)	50	100

Test 2 With $E/f = 2.5$, which gives one half normal flux density,

P_C (watts)	15.15	50
f (Hz)	50	100

Determine the value of the Steinmetz index.

Firstly it is worth observing that from equation (5.5), $E = 4.44B_m AfN$ volts. Dividing both sides by f gives

$$E/f = 4.44B_m AN = k_1 B_m \qquad \text{(as in section 5.9).}$$

Thus, reducing E/f from 5 to 2.5 reduces B_m in the same proportion.

From the results the following table is produced, as in worked example 5.8.

	P_C/f	
f (Hz)	Test 1	Test 2
50	1.25	0.303
100	2.04	0.5

A graph drawn to scale may be used to determine the values of the two intercepts. Alternatively it may be argued that since, in test 1, the value of P_C/f falls by $2.04 - 1.25 = 0.79$ as the frequency falls from 100 Hz to 50 Hz, it will fall by the same amount again as the frequency falls from 50 Hz to zero. Hence

$$k_{2(1)} = 1.25 - 0.79 = 0.46.$$

By similar reasoning

$$k_{2(2)} = 0.106.$$

The graph is shown in Fig. 5.18.

From equation (5.20)

$$\frac{k_{2(1)}}{k_{2(2)}} = 2^n$$

$$\frac{0.46}{0.106} = 2^n.$$

Taking logarithms to base 10

$$\log_{10} 4.3396 = n \times \log_{10} 2$$

$$0.637 = n \times 0.3010$$

$$n = 2.12.$$

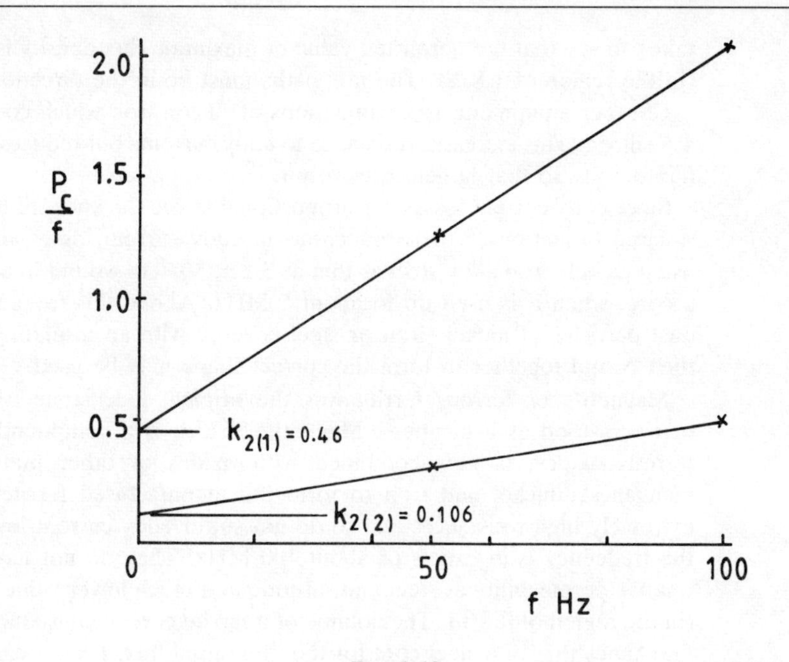

Fig. 5.18

SELF-ASSESSMENT EXAMPLE 5.11

The results of two tests on the core of a choke are as follows

Test 1 With $E/f = 4$: at 30 Hz, $P_C/f = 0.96$, at 60 Hz, $P_C/f = 1.32$.
Test 2 With $E/f = 2$: at 35 Hz, $P_C/f = 0.265$, at 70 Hz, $P_C/f = 0.37$.

Determine the values of (a) the hysteresis loss for the flux density in test 1 at 50 Hz, (b) the eddy current loss for the flux density in test 2 at 40 Hz, (c) the Steinmetz index for the core material.

5.11 Low retentivity ferro-magnetic materials

In electrical machinery, transformers, motors, etc., a high flux density is required to minimise the cross-sectional area and hence the weight of the core or yoke of the apparatus. This high flux density should be produced with the minimum magnetising force or the current and hence power loss (I^2R) will be large. Where the flux reverses periodically, the hysteresis and eddy current losses must be small. The hysteresis losses are determined by the softness of the material while the eddy current losses are determined by the resistance of the core material which is increased by the use of laminations.

Almost pure iron is used and for very high quality equipment it is cold-rolled to the correct thickness and then annealed, which has the effect of orienting the grain of the steel in the direction of rolling. It is easy to magnetise in this direction so that high flux density is achieved with low magnetising force. It does however saturate more rapidly than normal mild steel and care must be taken to see that the permitted value of maximum flux density is not exceeded (in the region of 1.8 T). The flux paths must lie in the direction of rolling.

Cheaper equipment uses laminations of silicon iron which contains 0.5% to 4% silicon. This increases resistance to eddy currents but reduces the saturation flux density so that larger cores result.

Since eddy current losses are proportional to the thickness of the laminations squared (equation (5.16)), reductions in eddy current losses are achieved by using nickel–iron alloy strip as thin as 3.5×10^{-6} m wound in a spiral to form a core, when it is used up to about 5 MHz. Above this frequency individual dust particles of nickel–iron or steel covered with an insulating material and then bound together to form the correct shape may be used.

Magnetite or ferrous ferrite was the original lode stone which the early mariners used as a compass. Magnetite, which is a compound of ferric and ferrous oxides, is now combined with oxides of other materials such as manganese, nickel and iron to form the manufactured ferrites. These have extremely high resistances and so do not suffer eddy current losses at all until the frequency is in excess of about 100 MHz. They do not have such a high relative permeability as steel and saturate at a much lower value of flux density (in the region of 0.4 T). The volume of a ferrite core in consequence is three to four times that of a steel core for the same total flux. Ferrite cores are used in chokes and small transformers in audio work.

FURTHER PROBLEMS

5.12 A hysteresis loop is plotted with scales: horizontally, 1000 A/m = 1 cm, vertically, 0.25 T = 1 cm. The area of the loop is 3.1 cm^2. The maximum vertical height (from zero) is 6.35 cm.

Calculate (a) the maximum flux density in the material, (b) the energy lost in the material in joules per m^3 once round the loop, (c) the power loss in a core for a choke made from this material which has a mass of 2.87 kg. The density of the material is 7800 kg/m^3. The operating frequency is 100 Hz.

5.13 A transformer core has a mass of 6.9 kg. The hysteresis loss at 50 Hz is 15.75 W. The density of the material is 7800 kg/m^3. The hysteresis loop for the material has an area of 7.12 cm^2 when the horizontal axis is scaled at 250 A/m per cm. The maximum height of the loop (from zero) is 7.25 cm.

Determine (a) the value of the maximum flux density, (b) the induced e.m.f. in the primary coil which has 540 turns if the core cross-sectional area is 25 cm^2.

5.14 A coil consisting of 500 turns is wound on a toroidal former with mean length 0.26 m and a cross-sectional area 3 cm^2. The former is non-magnetic and the coil carries a current of 2 A.

Determine (a) the value of the magnetising force, H, (b) the flux density within the toroid, (c) the inductance of the coil, (d) the energy stored in the magnetic field, (e) the value of the induced e.m.f. in the coil if the current is reversed, taking 0.06 s to complete the reversal.

5.15 An iron ring 18 cm in diameter and 5 cm^2 in cross-sectional area is wound with 250 turns of wire. It is to operate at a flux density of 1.1 T when the relative permeability of the material is 2175.

Determine (a) the steady coil current required, (b) the inductance of the coil, (c) the energy stored.

5.16 A ring made of magnetic material is wound with 200 turns. When the current is increased from 1.5 A to 2.5 A the flux in the ring increases from 150 μWb to 300 μWb. Calculate the inductance of the coil.

If an alternating current flows in the coil with peak value 1.5 A, what is the mean value of inductance? (Hint: what is the total flux change as the current changes from +1.5 A to −1.5 A?)

5.17 Derive the relationship which exists between the supply frequency, the maximum flux density in the core, the core cross-section, the number of winding turns and the induced e.m.f., in a single-phase transformer. Hence or otherwise explain the effect upon the core flux density of supplying a transformer at its rated voltage but at twice the normal frequency.

A single-phase transformer is to operate at 240 V:110 V. The core flux density is to be close to, but not exceeding, 1.5 T when the cross-sectional area is 2500 mm^2 and the frequency is 50 Hz. Determine to the nearest whole number the number of turns required on both primary and secondary windings.

5.18 A concentric cable has a core radius of 0.5 cm. It is to have an inductance of 2×10^{-7} H/m length. Determine the required inner radius of the sheath.

5.19 A single-phase power line comprises two conductors each with radius 0.75 cm, spaced 1.5 m apart in air. Determine the inductance of the line per metre length.

What change in conductor radius with spacing 1.5 m would have the same effect on the inductance as using the original conductors at 1.25 m spacing?

5.20 (a) Describe a method of separating the eddy current loss and the hysteresis loss from the total core loss in a transformer core.

(b) The following results were noted during a test on a small transformer

Applied e.m.f. (volts)	200	180	160	140	120	100	80
Frequency (Hz)	50	45	40	35	30	25	20
Power input (W)	70	62	51	44	36	29	21

(The I^2R loss in the coil is small enough to be ignored.)

Draw a suitable graph from these results and hence determine the separate hysteresis and eddy current losses at (i) 25 Hz, and (ii) 50 Hz.

5.21 The total iron losses in the laminated core of an alternator are 240 W at 50 Hz and 300 W at 60 Hz, the ratio E/f being constant for the two tests. Find the separate hysteresis and eddy current losses at 60 Hz given that the Steinmetz index is 1.6.

Determine the corresponding values with a 30% increase in flux density and a 50% increase in the thickness of the laminations.

5.22 What are the requirements for magnetic materials for the following applications?

(a) the core of a high-quality power transformer operating at 50 Hz,
(b) the core of a radio tuning coil operating at 100 MHz.

Specify in each case a suitable material.

5.23 In a 440 V, 50 Hz, single-phase transformer the total iron loss is 2500 W. When the applied voltage is reduced to 220 V at a frequency of 25 Hz the corresponding loss is 850 W. Calculate the value of the eddy current loss at normal voltage and frequency.

5.24 Two sets of tests were carried out on a transformer in order to determine the Steinmetz index for the core material. The results were

Test 1 With $E/f = 5$.

Core loss (W)	17	37.5	65	120
Frequency (Hz)	30	50	70	100

Test 2 With $E/f = 2.5$.

Core loss (W)	3.9	8.8	15.4	28.8
Frequency (Hz)	30	50	70	100

Determine the separate hysteresis and eddy current losses at 50 Hz for each flux density and the value of the Steinmetz index.

5.25 What losses occur in a transformer core? How is each of these minimised? A single-phase, 50 Hz transformer is to be connected to a 440 V supply. The output from the transformer is to be 110 V on no load. The 110 V winding is to have 800 turns while the maximum core flux density is to be 1.3 T.

Calculate the cross-sectional area of the core and the number of turns on the 440 V winding. Prove any formulae used.

Answers

5.3 $B = 1.3$ T; $H = 800$ A/m; $I = 1.257$ A
5.5 (a) (i) 4800 A/m (ii) 1.6 T (iii) 5.76 mH (b) Scales: 300 A/m = 1 cm; 0.1 T = 1 cm; volume of core = 75×10^{-6} m^3; energy = 0.2205 J
5.7 0.085 mm
5.9 17.5 W; 75 W
5.11 (a) 30 W (b) 4.8 W (c) 1.907
5.12 (a) 1.588 T (b) 1587.5 J/m^3 (c) 58.4 W
5.13 (a) 1.45 T (b) 435 V
5.14 (a) 3846 A/m (b) 4.833 mWb/m^2 (c) 362.5 μH (d) 725 μJ (e) 0.024 V
5.15 (a) 0.91 A (b) 0.151 H (c) 0.063 J
5.16 0.03 H; 0.02 H on a.c.
5.17 Halves B_m; ratio 289 : 133
5.18 1.36 cm
5.19 2.12 μH/m; new conditions: 2.046 μH/m; radius 0.9 cm

5.20 (b) (i) H 22.95 W; E 6.82 W (ii) H 45.8 W; E 27.2 W
5.21 H 228 W; E 72 W; new: H 346.9 W; E 273.8 W
5.23 1600 W
5.24 E/f = 5: H 15 W; E 22.5 W; E/f = 2.5: H 3.15 W; E 5.6 W; n = 2.25
5.25 A = 0.000476 m^2; primary 3200 turns

Chapter 6

Two-port Networks

A two-port network is a network with two input terminals and two output terminals. The network may contain an active device such as a transistor or consist of passive components only. In this chapter the two-port network will be treated as a transmission element containing passive devices only; our concern being with input and output impedance of the network and the proportion of the input which appears at the output.

There are several conventions for use with two-port networks, the two most common being shown in Fig. 6.1. This book uses that in Fig. 6.1(a). If that in Fig. 6.1(b) is preferred, all that is required is a sign change for I_2, i.e. I_2 in Fig. 6.1(b) $= -I_2$ in Fig. 6.1(a).

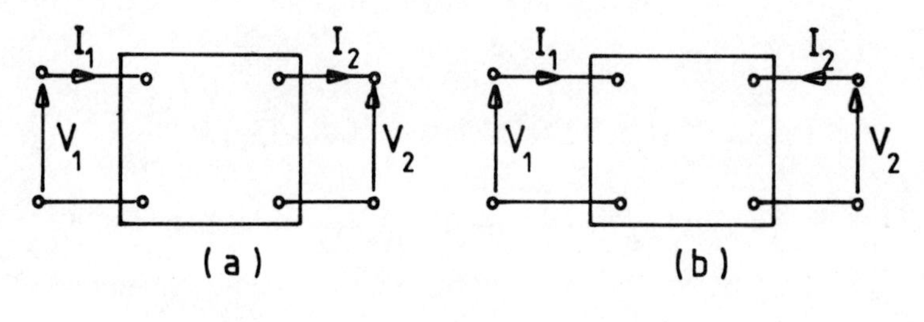

Fig. 6.1

6.1 General equations for V_1 and I_1 in terms of V_2 and I_2

For any four-terminal network consisting of linear components there will be a linear relationship between input voltage and input current (V_1 and I_1) and output voltage and output current (V_2 and I_2). The relationships are expressed as follows:

$$V_1 = AV_2 + BI_2 \tag{6.1}$$

$$I_1 = CV_2 + DI_2 \tag{6.2}$$

where A and D are dimensionless constants, C is a constant with the dimensions of admittance, and B is a constant with the dimensions of impedance. These constants are known as the 'four-terminal' constants or parameters and may be determined using the open-circuit and short-circuit tests.

6.2 Four-terminal network parameters from open-circuit and short-circuit tests

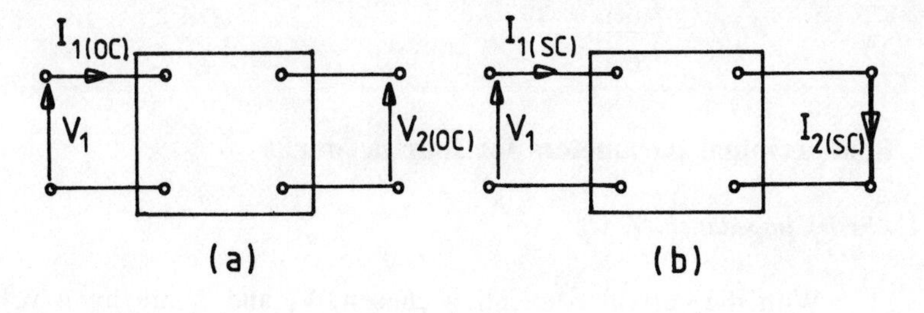

Fig. 6.2

The general equations are

$$V_1 = AV_2 + BI_2 \qquad \text{(equation 6.1)}$$
$$I_1 = CV_2 + DI_2 \qquad \text{(equation 6.2)}$$

6.2.1 The open-circuit test

With the output terminals open-circuited, as in Fig. 6.2(a), a voltage V_1 is applied to the input terminals when a current $I_{1(oc)}$ enters the network. The output current is zero ($I_2 = 0$). Hence from equation (6.1)

$$V_1 = AV_{2(oc)} + 0$$
$$A = \frac{V_1}{V_{2(oc)}}.$$

From equation (6.2)

$$I_1 = CV_{2(oc)} + 0$$
$$C = \frac{I_1}{V_{2(oc)}}.$$

6.2.2 The short-circuit test (Fig. 6.2(b))

The output terminals are short-circuited through a low-impedance ammeter. The network is supplied with a voltage V_1 and a current $I_{1(sc)}$ enters the network. The potential difference across the output terminals is zero ($V_2 = 0$). Hence, from equation (6.1)

$$V_1 = 0 + BI_{2(sc)}$$
$$B = \frac{V_1}{I_{2(sc)}}.$$

From equation (6.2)

$$I_{1(sc)} = 0 + DI_{2(sc)}$$

$$D = \frac{I_{1(sc)}}{I_{2(sc)}}.$$

6.3 Four-terminal parameters for four networks

6.3.1 *Series impedance (Z Ω)*

With the current convention chosen, V_1 and I_1 are input voltage and input current respectively and V_2 and I_2 the corresponding outputs.
Using Kirchhoff's first law, $I_1 = I_2$.
Using Kirchhoff's second law, $V_1 - I_1Z = V_2$, or since $I_1 = I_2$,
$V_1 - I_2Z = V_2$.
Transposing gives $V_1 = V_2 + ZI_2$.
Compare this with equation (6.1) $V_1 = AV_2 + BI_2$.
Hence $A = 1$ and $B = Z$. (6.3)

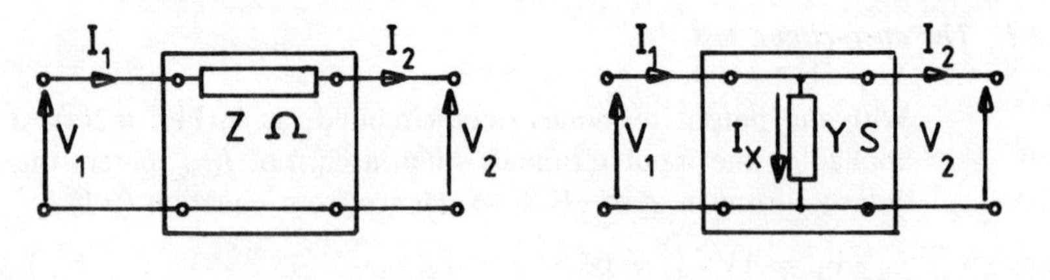

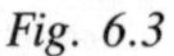

Fig. 6.3 *Fig. 6.4*

From above $I_1 = I_2$.
Compare with equation (6.2) $I_1 = CV_2 + DI_2$.

Hence $C = 0$ and $D = 1$. (6.4)

6.3.2 *Shunt admittance (Y S)*

In Fig. 6.4, since there is no series impedance $V_1 = V_2$.
Comparing this with equation (6.1) $V_1 = AV_2 + BI_2$.
Hence $A = 1$ and $B = 0$. (6.5)
By Kirchhoff's first law, $I_2 = I_1 - I_x$. (i)
Now $I_x = YV_1 = YV_2$. Substituting for I_x in (i) and transposing gives

$$I_1 = YV_2 + I_2.$$

Comparing this with equation (6.2) $I_1 = CV_2 + DI_2$.
Hence $C = Y$ and $D = 1$. (6.6)

6.3.3 *The T-network*

In Fig. 6.5(a)

Current I_2 flows in Z_2, hence $V_x = V_2 + I_2Z_2$. (i)

$$I_x = V_xY. \quad \text{(ii)}$$

$$I_1 = I_2 + I_x. \qquad \text{(iii)}$$

Substituting equation (ii) in equation (iii) $I_1 = I_2 + V_xY$. (iv)

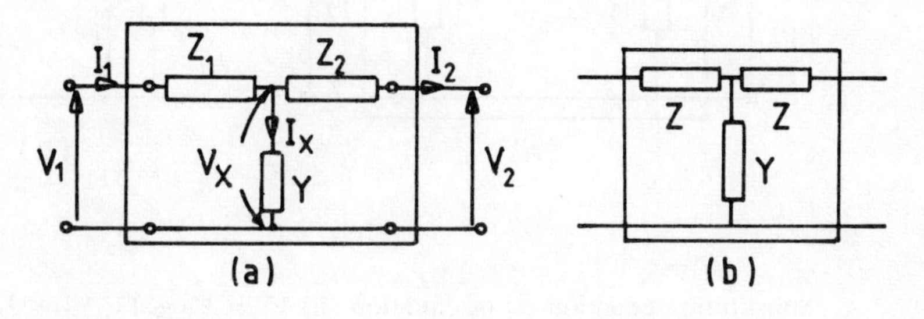

Fig. 6.5

Substituting equation (i) in equation (iv) $I_1 = I_2 + (V_2 + I_2Z_2)Y$
$= I_2 + V_2Y + I_2Z_2Y$.

Regrouping $I_1 = YV_2 + (Z_2Y + 1)\, I_2$.
Compare with equation (6.2) $I_1 = CV_2 + DI_2$.

Hence $C = Y$ and $D = (1 + Z_2Y)$. (6.7)

Referring back to Fig. 6.5(a) $V_1 = V_x + I_1Z_1$. (v)

Substituting equation (iv) in equation (v) $V_1 = V_x + (I_2 + V_xY)Z_1$. (vi)

Substituting equation (i) in equation (vi) $V_1 = V_2 + I_2Z_2 + (I_2 + (V_2 + I_2Z_2)Y)Z_1$.

$V_1 = V_2 + I_2Z_2 + I_2Z_1 + V_2YZ_1 + I_2Z_2YZ_1$.

Regrouping $V_1 = (1 + YZ_1)V_2 + (Z_2 + Z_1 + Z_2Z_1Y)I_2$.

Compare with equation (6.1) $V_1 = A\, V_2 + B\, I_2$.

Hence $A = (1 + YZ_1)$ and $B = (Z_2 + Z_1 + Z_2Z_1Y)$. (6.8)

If the network is symmetrical, as shown in Fig. 6.5(b), $Z_1 = Z_2 = Z$ and the four-terminal parameters reduce to

$$A = (1 + YZ) \qquad B = (2Z + Z^2Y) \qquad C = Y \qquad D = (1 + YZ). \qquad (6.9)$$

Making the network symmetrical has made the two constants A and D equal. This is true generally. In any symmetrical, four-terminal network which does not contain sources of e.m.f., i.e. a passive network, $A = D$.

6.3.4 *The π-network*

Current in Y_2 is V_2Y_2, and I_x = current in $Y_2 + I_2$.
Hence $I_x = V_2Y_2 + I_2$. (i)

Also $V_1 = V_2 + I_xZ$. (ii)

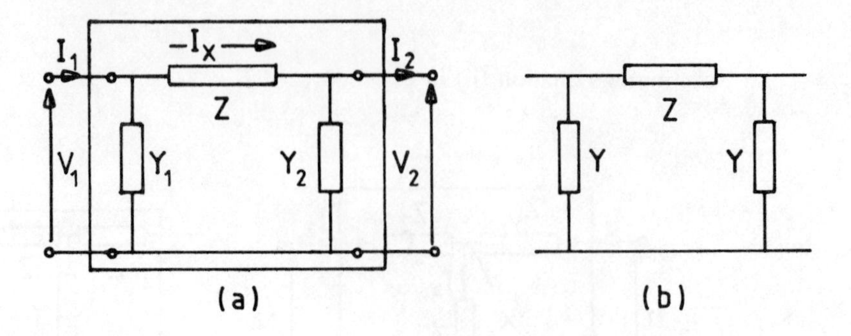

Fig. 6.6

Substituting equation (i) in equation (ii) $V_1 = V_2 + (V_2Y_2 + I_2)Z = V_2 + V_2Y_2Z + I_2Z$.

Rearranging $\quad V_1 = (1 + Y_2Z)V_2 + ZI_2$. (iii)

Compare with equation (6.1) $\quad V_1 = A\ V_2 + BI_2$.

Hence $A = (1 + Y_2Z)$ and $B = Z$. (6.10)

Again, in Fig. 6.6(a): current in $Y_1 = V_1Y_1$ (iv)

and I_1 = current in $Y_1 + I_x$. (v)

Substituting equation (i) and equation (iv) in equation (v) $\quad I_1 = V_1Y_1\ (V_2Y_2 + I_2)$. (vi)

Substituting equation (iii) for V_1 in equation (vi)

$$\begin{aligned} I_1 &= [(1 + Y_2Z)V_2 + ZI_2]Y_1 + (V_2Y_2 + I_2) \\ &= V_2Y_1 + Y_2Y_1ZV_2 + Y_1ZI_2 + V_2Y_2 + I_2 \\ &= (Y_1 + Y_2 + Y_1Y_2Z)V_2 + (1 + Y_1Z)I_2 \end{aligned}$$

Compare with equation (6.2) $\quad I_1 = C\ V_2 + D\ I_2$.

Hence $C = (Y_1 + Y_2 + Y_1Y_2Z)$ and $D = (1 + Y_1Z)$. (6.11)

For a symmetrical π-network, as shown in Fig. 6.6(b), $Y_1 = Y_2 = Y$ and the four-terminal parameters reduce to

$$A = (1 + YZ) \quad B = Z \quad C = (2Y + Y^2Z) \quad D = (1 + YZ). \qquad (6.12)$$

Symmetry results in the two constants A and D being equal as in the case of the symmetrical T-network.

6.4 $AD - BC = 1$

A general property of a four-terminal passive network, whether symmetrical or not, is that $AD - BC = 1$. This will be examined for the asymmetrical π-network shown in Fig. 6.6(a).

Using equations (6.10) and (6.11)

$$AD = (1 + Y_2Z)\ (1 + Y_1Z) = 1 + Y_1Z + Y_2Z + Y_1Y_2Z^2$$

$$BC = Z\,(Y_1 + Y_2 + Y_1Y_2Z) = \quad Y_1Z + Y_2Z + Y_1Y_2Z^2$$

Hence, $AD - BC = 1$.

SELF-ASSESSMENT EXAMPLE 6.1

Using the $ABCD$ paraemeters for the asymmetric T-network in Fig. 6.5(a), deduce that $AD - BC = 1$.

WORKED EXAMPLE 6.2

A four-terminal symmetrical π-network has a series impedance of $100\angle 65°\ \Omega$. Each of the shunt admittances is $2 \times 10^{-3}\angle 85°$ S, all at a particular frequency.

Calculate (a) the values of the constants A, B, C and D, (b) the values of the input voltage and current (V_1 and I_1) if, when a 50 Ω resistor is connected across the output terminals, a current of 2.5 mA flows in it.

(a) Using equation (6.12)

$$\begin{aligned} A = D = (1 + YZ) &= 1 + 2 \times 10^{-3}\angle 85° \times 100\angle 65° \\ &= 1 + 0.2\angle 150° \\ &= 1 - 0.1732 + \mathrm{j}0.1 \\ &= 0.827 + \mathrm{j}0.1 = 0.833\angle 6.9°. \end{aligned}$$

$$B = Z = 100\angle 65°.$$

$$\begin{aligned} C = 2Y + Y^2Z &= 2 \times 2 \times 10^{-3}\angle 85° + (2 \times 10^{-3})^2 \times 100\angle 65° \\ &= 4 \times 10^{-3}\angle 85° + 4 \times 10^{-4}\angle(85° + 85° + 65°) \\ &= 3.486 \times 10^{-4} + \mathrm{j}3.985 \times 10^{-3} + (-2.29 \times 10^{-4} - \\ &\quad \mathrm{j}3.28 \times 10^{-4}) \\ &= 1.19 \times 10^{-4} + \mathrm{j}3.66 \times 10^{-3} = 3.66 \times 10^{-3}\angle 88.13°. \end{aligned}$$

(b) The potential difference across the output terminals is I_2R_L.
Hence

$$V_2 = 2.5 \times 10^{-3} \times 50 = 0.125 \text{ V}.$$

Using V_2 as reference, i.e. writing $V_2 = 0.125\angle 0°$ V, then $I_2 = 2.5 \times 10^{-3}\angle 0°$ A, since in a resistor voltage and current are in phase.

Using equation (6.1)

$$\begin{aligned} V_1 &= AV_2 + BI_2 \\ &= 0.833\angle 6.9° \times 0.125\angle 0° + 100\angle 65° \times 2.5 \times 10^{-3}\angle 0° \\ &= 0.103 + \mathrm{j}0.0125 + 0.106 + \mathrm{j}0.227 \\ &= 0.209 + \mathrm{j}0.239 = 0.318\angle 48.9° \text{ V}. \end{aligned}$$

Using equation (6.2)

$$\begin{aligned} I_1 &= CV_2 + DI_2 \\ &= 3.66 \times 10^{-3}\ \angle 88.13° \times 0.125\angle 0° + 0.833\angle 6.9° \times 2.5 \times 10^{-3}\angle 0° \end{aligned}$$

$$= 14.93 \times 10^{-6} + j457.3 \times 10^{-6} + 2.067 \times 10^{-3} + j250.2 \times 10^{-6}$$
$$= 2.082 \times 10^{-3} + j707.4 \times 10^{-6} = 2.2\angle 18.76° \text{ mA}.$$

SELF-ASSESSMENT EXAMPLE 6.3

A single-phase power transmission line may be represented by the symmetrical T-network shown in Fig. 6.7. $V_2 = 158\,770\angle 0°$ V and $I_2 = 126\angle -36.87°$ A.
Determine (a) the values of the four-terminal network parameters, (b) the magnitude and phase of V_1, (c) the magnitude and phase of I_1.

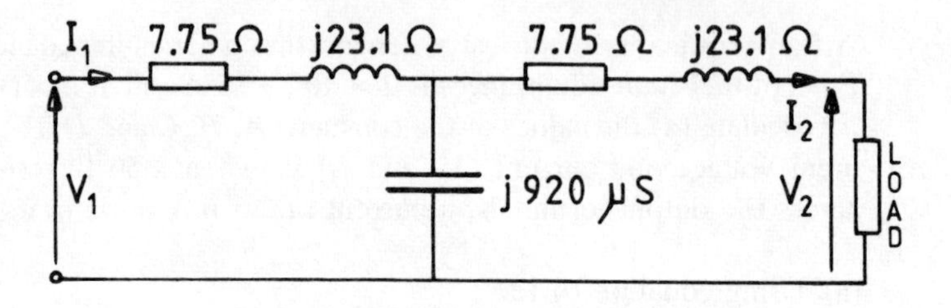

Fig. 6.7

6.5 General equations for V_2 and I_2 in terms of V_1 and I_1

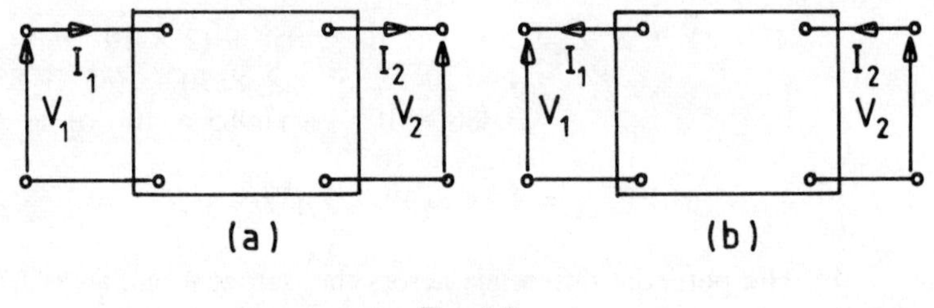

Fig. 6.8

For the network as shown in Fig. 6.8(a) (identical to Fig. 6.1(a)), equation (6.1) is $V_1 = AV_2 + BI_2$, equation (6.2) is $I_1 = CV_2 + DI_2$.
Multiplying equation (6.1) by D and equation (6.2) by B gives

$$DV_1 = ADV_2 + BDI_2 \qquad \text{(i)}$$
$$BI_1 = BCV_2 + BDI_2. \qquad \text{(ii)}$$

Subtracting equation (ii) from equation (i) gives

$$DV_1 - BI_1 = (AD - BC)V_2.$$

From section 6.4, $AD - BC = 1$, therefore

$$V_2 = DV_1 - BI_2. \qquad (6.13)$$

Now multiplying equation (6.1) by C and equation (6.2) by A gives

$$CV_1 = CAV_2 + BCI_2 \tag{iii}$$
$$AI_1 = CAV_2 + ADI_2. \tag{iv}$$

Subtracting equation (iv) from equation (iii) gives

$$CV_1 - AI_1 = (BC - AD)I_2.$$

Since $AD - BC = 1$, $BC - AD = -1$, therefore

$$-I_2 = CV_1 - AI_1. \tag{6.14}$$

These equations are relevant to the current directions in Fig. 6.8(a).

It is often more useful to consider the currents reversed in direction, and hence in sign, as shown in Fig. 6.8(b). The network is now being fed with V_2 volts and the load is considered to be at the front of the network. The voltage polarities are unchanged.

Reversing the signs for both currents yields

$$V_2 = DV_1 + BI_1 \tag{6.15}$$

$$I_2 = CV_1 + AI_1. \tag{6.16}$$

For symmetrical networks $A = D$, and feeding the network from either end results in the same relationships between the sending and receiving conditions. For asymmetric networks, reversing the network will change its performance.

6.6 Image impedance

From the maximum power transfer theorem (chapter 2, section 2.7), it is known that for maximum power transfer from a generator with internal impedance Z_I into a load with impedance Z_L, the condition $|Z_I| = |Z_L|$ must be satisfied.

Where Z_L is not equal to Z_I, the load may be matched to the generator using either a transformer or a four-terminal network. A network which is designed to satisfy this condition is said to be image-matched.

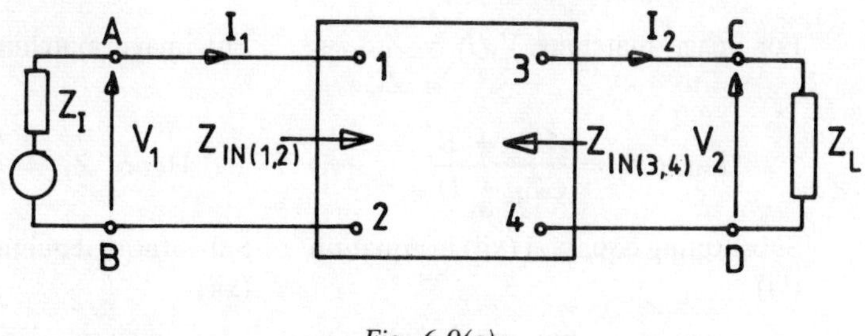

Fig. 6.9(a)

For image matching, the impedance viewed from terminals AB in Fig. 6.9(a) in either direction will be the same. Looking to the left the impedance is Z_I and to the right it is $Z_{IN(1,2)}$.

$$Z_{IN(1,2)} = Z_I.$$

The same is true for the load end; the impedance viewed either way from terminals CD will be the same.

$$Z_{IN(3,4)} = Z_L.$$

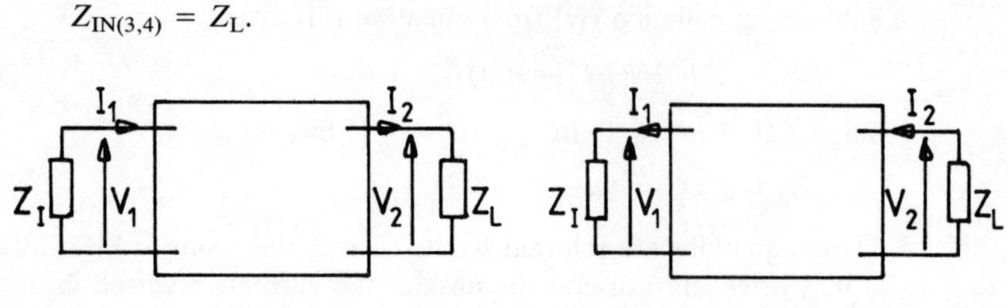

Fig. 6.9(b)

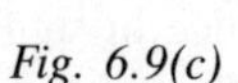

Fig. 6.9(c)

Equation (6.1) $V_1 = AV_2 + BI_2$.
Equation (6.2) $I_1 = CV_2 + DI_2$.
Dividing throughout by I_2

$$\frac{V_1}{I_2} = \frac{AV_2}{I_2} + \frac{BI_2}{I_2} \qquad \text{(i)}$$

$$\frac{I_1}{I_2} = \frac{CV_2}{I_2} + \frac{DI_2}{I_2}. \qquad \text{(ii)}$$

Dividing equation (i) by equation (ii)

$$\frac{I_2}{I_2}\frac{V_1}{I_1} = \frac{A(V_2/I_2) + B}{C(V_2/I_2) + D} \qquad \text{(iii)}$$

$$V_2/I_2 = Z_L. \qquad \text{(iv)}$$

Substituting equation (iv) in equation (iii)

$$\frac{V_1}{I_1} = \frac{AZ_L + B}{CZ_L + D}. \qquad \text{(v)}$$

For image matching $V_1/I_1 = Z_{IN(1,2)}$
$= Z_I$.

$$\text{Hence } Z_I = \frac{AZ_L + B}{CZ_L + D}. \qquad \text{(vi)}$$

Substituting equation (xii) in equation (vi)

$$Z_I = \frac{A(DZ_I + B)/(CZ_I + A) + B}{C(DZ_I + B)/(CZ_I + A) + D}. \qquad (6.17)$$

Equation (6.15) $V_2 = DV_1 + BI_1$.
Equation (6.16) $I_2 = CV_1 + AI_1$.
Dividing throughout by I_1

$$\frac{V_2}{I_1} = \frac{DV_1}{I_1} + \frac{BI_1}{I_1} \qquad \text{(vii)}$$

$$\frac{I_2}{I_1} = \frac{CV_1}{I_1} + \frac{AI_1}{I_1}. \qquad \text{(viii)}$$

Dividing equation (vii) by equation (viii)

$$\frac{I_1}{I_1}\frac{V_2}{I_2} = \frac{D(V_1/I_1) + B}{C(V_1 + I_1) + A} \qquad \text{(ix)}$$

$$V_1/I_1 = Z_I. \qquad \text{(x)}$$

Substituting equation (x) in equation (ix)

$$\frac{V_2}{I_2} = \frac{DZ_I + B}{CZ_I + A}. \qquad \text{(xi)}$$

For image matching $V_2/I_2 = Z_{IN(3,4)}$
$= Z_L$.

$$\text{Hence } Z_L = \frac{DZ_I + B}{CZ_I + A}. \qquad \text{(xii)}$$

Substituting equation (vi) in equation (xii)

$$Z_L = \frac{D(AZ_L + B)/CZ_L + D) + B}{C(AZ_L + B)/CZ_L + D) + A}. \qquad (6.18)$$

Either of the two final equations may be solved. The solution to equation (6.17) will now be examined.

In the numerator, to bring B over the common denominator $(CZ_I + A)$, the two quantities are multiplied together. Similarly in the denominator, multiplying D by $(CZ_I + A)$ brings it over the common denominator.

Multiplying out the parentheses yields

$$Z_I = \frac{ADZ_I + AB + BCZ_I + AB}{CDZ_I + CB + CDZ_I + AD} \times \frac{CZ_I + A}{CZ_I + A}.$$

Cross-multiplying

$$CDZ_I^2 + CBZ_I + CDZ_I^2 + ADZ_I = ADZ_I + AB + BCZ_I + AB$$

$$2CDZ_I^2 = 2AB$$

$$Z_I = \sqrt{\left(\frac{AB}{CD}\right)}. \tag{6.19}$$

Solving equation (6.18) in the same manner yields the result

$$Z_L = \sqrt{\left(\frac{DB}{CA}\right)}. \tag{6.20}$$

The result may be deduced by a consideration of section 6.5 in which it is seen that viewing the network from the V_2I_2-end results in changing the positions of D and A in the general equations.

As an alternative to using the four-terminal network parameters the values of the image impedances may be determined using the open-circuit and short-circuit tests. (See section 6.2 and Fig. 6.2.)

In Fig. 6.9(a), supplying terminals 1 and 2, and with terminals 3 and 4 open-circuited, then $I_2 = 0$.

Equation (6.1) reduces to $\quad V_1 = AV_2. \quad$ (i)

Equation (6.2) reduces to $\quad I_1 = CV_2. \quad$ (ii)

Dividing equation (i) by equation (ii) $\dfrac{V_1}{I_1} = \dfrac{AV_2}{CV_2} = \dfrac{A}{C}$

$$\frac{V_1}{I_1} = Z_{IN(1,2)(oc)}$$

(where subscript 1,2 = at terminals 1 and 2, and subscript oc = output terminals open-circuited).

Next supplying terminals 1 and 2, with terminals 3 and 4 short-circuited

equation (6.1) reduces to $\quad V_1 = BI_2. \quad$ (iii)

equation (6.2) reduces to $\quad I_1 = DI_2. \quad$ (iv)

Dividing equation (iii) by equation (iv) $\dfrac{V_1}{I_1} = \dfrac{B}{D}$

$$\frac{V_1}{I_1} = Z_{IN(1,2)(sc)}.$$

(sc = output terminals short-circuited).

Therefore, $Z_{IN(1,2)(oc)} \times Z_{IN(1,2)(sc)} = \frac{A}{C} \times \frac{B}{D}$.

Now from equation (6.19) $Z_I = \sqrt{\left(\frac{AB}{CD}\right)}$

hence, $Z_I = \sqrt{(Z_{IN(1,2)(oc)} \times Z_{IN(1,2)(sc)})}$. (6.21)

Similarly, viewed from terminals 3 and 4, open-circuiting and short-circuiting terminals 1 and 2 yield a similar result.

$Z_L = \sqrt{(Z_{IN(3,4)(oc)} \times Z_{IN(3,4)(sc)})}$. (6.22)

WORKED EXAMPLE 6.4

Determine the values of Z_A and Z_B which, when connected to terminals 1, 2 and 3, 4 respectively in Fig. 6.10, give image matching.

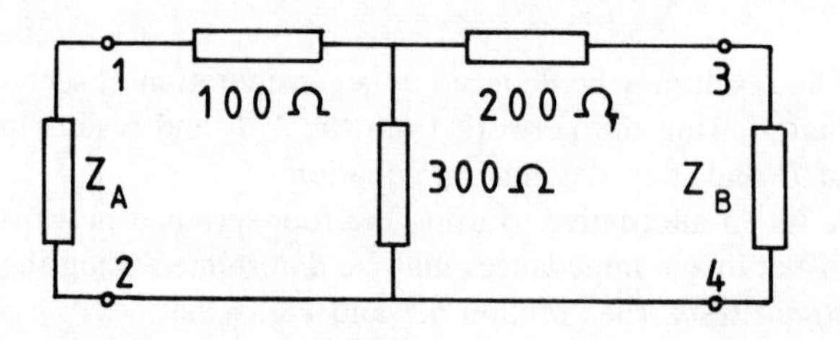

Fig. 6.10

The two methods available will be examined: (a) using the *ABCD* parameters, (b) using the open-circuit and short-circuit tests.

(a) Using equations (6.7) and (6.8)

$$A = 1 + YZ_1 = 1 + \frac{100}{300} = 1.333$$

$$B = Z_1 + Z_2 + Z_1Z_2Y = 100 + 200 + \frac{100 \times 200}{300} = 366.66$$

$$C = Y = \frac{1}{300}$$

$$D = 1 + YZ_2 = 1 + \frac{200}{300} = 1.66.$$

From equation (6.19) $Z_I = Z_A = \sqrt{\left(\frac{AB}{CD}\right)} = \sqrt{\left(\frac{1.33 \times 366.66 \times 300}{1.66}\right)}$

$= 296.6\ \Omega$.

From equation (6.20) $Z_L = Z_B = \sqrt{\left(\frac{DB}{CA}\right)} = \sqrt{\left(\frac{1.66 \times 366.66 \times 300}{1.33}\right)}$

$= 370.8\ \Omega$.

(b)

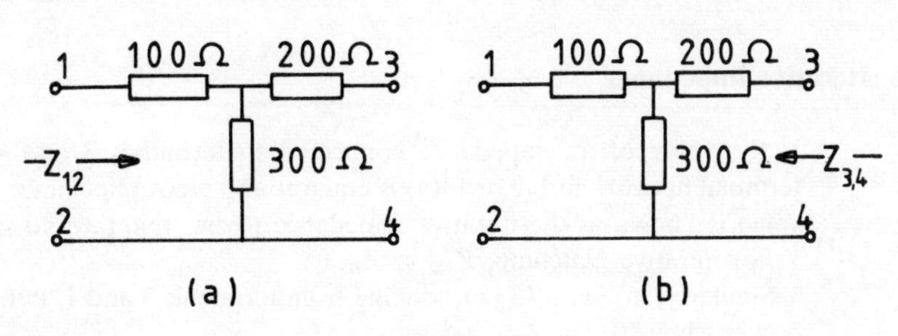

Fig. 6.11

In Fig. 6.11 (a), with an open circuit at terminals 3 and 4

$$Z_{1,2} = 100 + 300$$
$$= 400\ \Omega.$$

A short circuit across terminals 3 and 4 places the 200 Ω resistor in parallel with the 300 Ω resistor

$$Z_{1,2} = 100 + \frac{200 \times 300}{200 + 300} = 220\ \Omega.$$

Using equation (6.21) $Z_A = \sqrt{(400 \times 220)} = 296.6\ \Omega$.

In Fig. 6.11 (b), with an open circuit at terminals 1 and 2

$$Z_{3,4} = 200 + 300$$
$$= 500\ \Omega.$$

A short circuit across terminals 1 and 2 places the 100 Ω resistor in parallel with the 300 Ω resistor

$$Z_{3,4} = 200 + \frac{100 \times 300}{100 + 300} = 275\ \Omega.$$

Using equation (6.22) $Z_B = \sqrt{(500 \times 275)} = 370.8\ \Omega$.

SELF-ASSESSMENT EXAMPLE 6.5

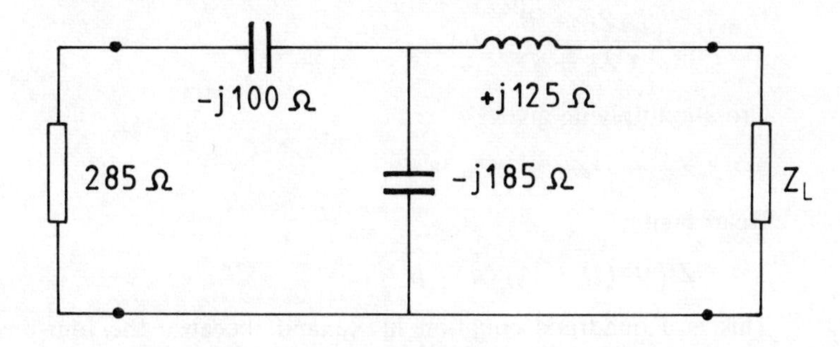

Fig. 6.12

Determine the value of Z_L in Fig. 6.12 which will give image matching.

6.7 Iterative impedance

If the value of an impedance connected to terminals 3 and 4 of the four-terminal network in Fig. 6.13(a) is equal to the input impedance $Z_{1,2}$, then this value is known as the iterative impedance (from 'reiterate' to state again).

For iterative matching, $Z_{1,2} = Z_B$.

Similarly, in Fig. 6.13 (b), looking from terminals 3 and 4, iterative matching occurs when $Z_{3,4} = Z_A$.

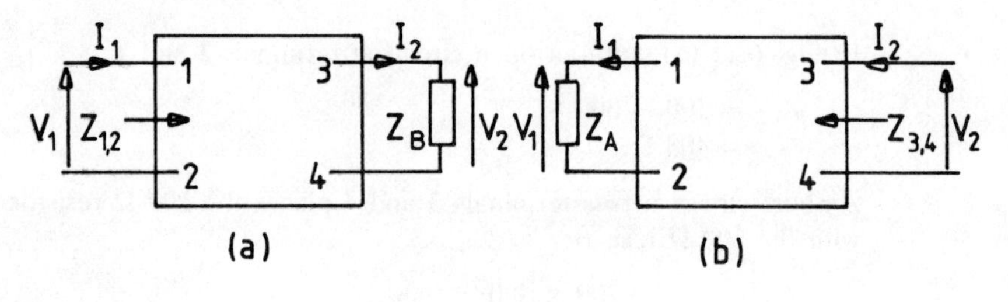

Fig. 6.13

Using equations (6.1) and (6.2) which apply to Fig. 6.13(a), dividing both equations by I_2 gives

$$\frac{V_1}{I_2} = A\frac{V_2}{I_2} + B\frac{I_2}{I_2} \qquad \text{(i)}$$

$$\frac{I_1}{I_2} = C\frac{V_2}{I_2} + D\frac{I_2}{I_2}. \qquad \text{(ii)}$$

Dividing equation (i) by equation (ii) gives

$$\left(\frac{I_2}{I_2}\right)\frac{V_1}{I_1} = \frac{A(V_2/I_2) + B}{C(V_2/I_2) + D}.$$

But $V_1/I_1 = Z_{1,2} = Z_B$

and V_2/I_2 is also equal to Z_B. Therefore

$$Z_B = \frac{AZ_B + B}{CZ_B + D}.$$

Cross-multiplying gives

$$CZ_B^2 + DZ_B = AZ_B + B.$$

Rearranging

$$CZ_B^2 + (D - A)Z_B = B. \qquad (6.23)$$

This is a quadratic equation in Z_B and, because the four-terminal network parameters are known, Z_B may be determined.

With reference to Fig. 6.13(b), dividing equations (6.15) and (6.16) by I_1 yields

$$\frac{V_2}{I_1} = D\frac{V_1}{I_1} + B\frac{I_1}{I_1} \quad \text{(iii)}$$

$$\frac{I_2}{I_1} = C\frac{V_1}{I_1} + A\frac{I_1}{I_1}. \quad \text{(iv)}$$

Dividing equation (iii) by equation (iv) and substituting $V_1/I_1 = Z_A$, $V_2/I_2 = Z_{3,4} = Z_A$ gives

$$\frac{I_1}{I_1}\frac{V_2}{I_2} = \frac{DZ_A + B}{CZ_A + A}.$$

Cross-multiplying and rearranging: $CZ_A^2 + (A - D)Z_A = B$. (6.24)

Again, knowing the four-terminal network parameters, the quadratic may be solved. Where the values of the circuit components are known it is often more direct to work from first principles and this will be demonstrated in worked example 6.6.

WORKED EXAMPLE 6.6

Determine the value of Z_A in Fig. 6.14 to give iterative matching.

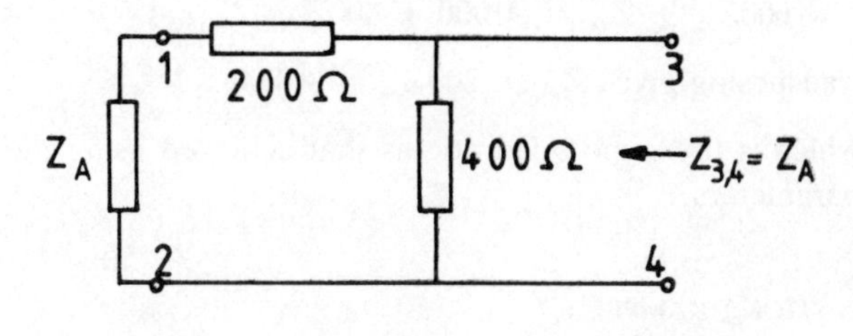

Fig. 6.14

The 200 Ω resistor and the 400 Ω resistor form Z_1 and Y in a T-network with Z_2 equal to zero (Fig. 6.5(a)).

Using equations (6.7) and (6.8)

$$A = 1 + YZ_1 = 1 + \frac{200}{400} = 1.5$$

$$B = Z_2 + Z_1 + Z_2Z_1Y = 0 + 200 + 0 \times \frac{200}{400} = 200$$

$$C = Y = \frac{1}{400}$$

$$D = 1 + Z_2Y = 1 + 0 \times \frac{1}{400} = 1.$$

Using equation (6.24) $CZ_A^2 + (A - D)Z_A = B$

$$\frac{1}{400} Z_A^2 + (1.5 - 1)Z_A = 200.$$

Multiplying by 400 gives

$$Z_A^2 + 200Z_A = 80\,000.$$

The quadratic may be solved by any known method. One way is to complete the square

$$Z_A^2 + 200Z_A + 100^2 = 80\,000 + 100^2 = 90\,000$$

$$Z_A + 100 = \pm \sqrt{90\,000}$$

$$Z_A = \pm\, 300 - 100 = 200\ \Omega\ (\text{or } -400\ \Omega).$$

The value ($-400\ \Omega$) is a mathematical solution but not achievable using passive components. Therefore $Z_A = 200\ \Omega$.

The same result is achievable working from first principles. The impedance, looking in at terminals 3 and 4 in Fig. 6.14, has to be equal to Z_A.

The impedance $Z_{3,4}$ consists of a resistor of value 400 Ω in parallel with $(200 + Z_A)\ \Omega$, hence

$$Z_{3,4} = Z_A = \frac{400\ (200 + Z_A)}{400 + 200 + Z_A}$$

cross-multiplying

$$600Z_A + Z_A^2 = 80\,000 + 400Z_A$$

transposing gives $Z_A^2 + 200Z_A = 80\,000$

which is the same quadratic as that achieved using the four-terminal network parameters.

SELF-EXAMINATION EXAMPLE 6.7

Determine the value of an impedance Z_B which when connected to terminals 3 and 4 in Fig. 6.14 will give iterative matching.

6.8 Characteristic impedance

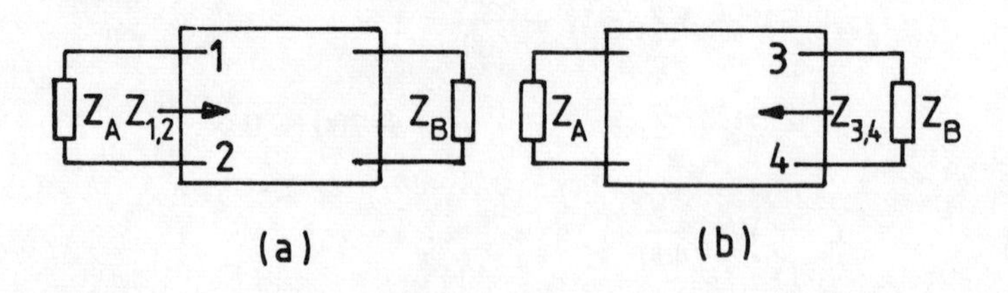

Fig. 6.15

In Fig. 6.15(a), for image impedance

$$Z_{1,2} = Z_A = \sqrt{\left(\frac{AB}{CD}\right)}. \qquad \text{(equation 6.19)}$$

For a symmetrical network $A = D$, hence

$$Z_A = \sqrt{\left(\frac{B}{C}\right)}.$$

In Fig. 6.15(b), for iterative impedance

$$Z_{3,4} = Z_A.$$

From equation (6.24)

$$CZ_A^2 + (A - D)Z_A = B.$$

For a symmetrical network $A = D$, and equation (6.24) reduces to

$$CZ_A^2 = B.$$

Transposing $Z_A = \sqrt{\left(\frac{B}{C}\right)}$.

Hence for a symmetrical network, image and iterative impedances Z_A are equal. Looking the other way, this is found to be true for Z_B, and because of the symmetry, $Z_A = Z_B$.

This unique impedance, where image and iterative impedances have the same value looking in either direction, is called the characteristic impedance, symbol Z_0.

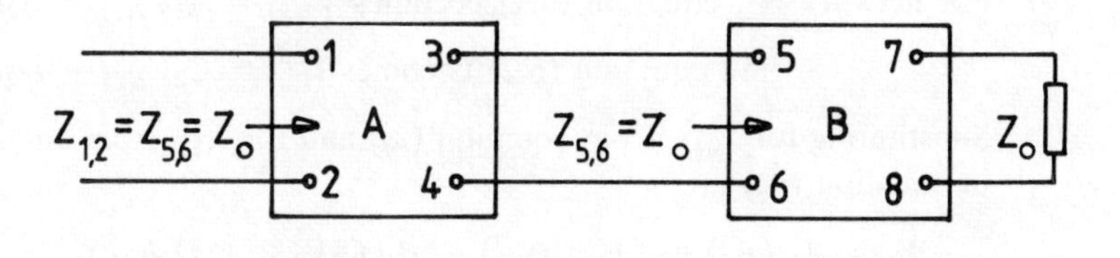

Fig. 6.16

In Fig. 6.15(a), let $Z_B = Z_0$. Now $Z_{1,2} = Z_0$. Instead of an actual impedance Z_0, an identical network terminated in Z_0 could be used. In Fig. 6.16, network B is terminated in Z_0 so that $Z_{5,6} = Z_0$. Connecting this to network A results in $Z_{1,2}$ being equal to $Z_{5,6}$, which is in turn equal to Z_0.

The process could be continued until an infinite number of identical networks were connected in series, the final one being terminated in Z_0. In fact the more networks there are in series, the smaller is the effect of the termination on the input impedance.

Following this reasoning, characteristic impedance is sometimes defined as the impedance of an infinite number of identical symmetrical two-port networks connected in series.

Terminating a symmetrical network or a number of identical such networks in series with an impedance Z_0 ensures maximum power transfer and no reflection of power from the termination towards the source. The system is said to be correctly terminated. (See further work on transmission lines, chapter 9.)

6.9 Cascade connection of networks

Networks are said to be connected in cascade when the output from the first network becomes the input to the second, and so on, as shown in Fig. 6.17.

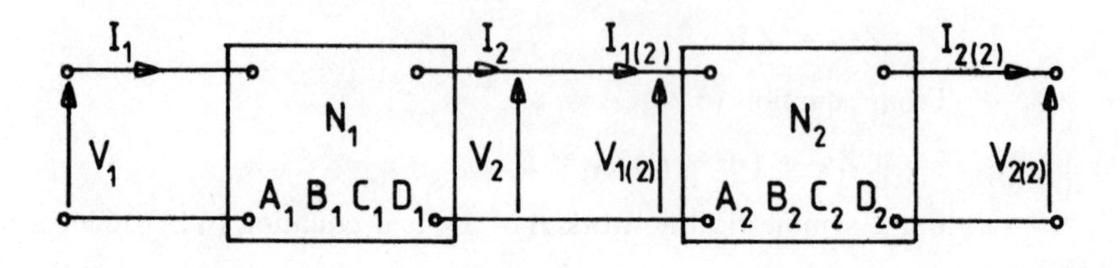

Fig. 6.17

Two, two-port networks, N_1 and N_2 with parameters A_1, B_1, C_1, D_1 and A_2, B_2, C_2, D_2 respectively are connected so that the output from N_1 becomes the input to N_2.

$$V_2 = V_{1(2)} \quad \text{and } I_2 = I_{1(2)}. \tag{i}$$

Using equation (6.1) in network N_1 $\quad V_1 = A_1V_2 + B_1I_2$.

Substituting for V_2 and I_2 from equation (i) $V_1 = A_1V_{1(2)} + B_1I_{1(2)}$. (ii)

For network N_2, equation (6.1) becomes $V_{1(2)} = A_2V_{2(2)} + B_2I_{2(2)}$ (iii)

and equation (6.2) becomes $I_{1(2)} = C_2V_{2(2)} + D_2I_{2(2)}$. (iv)

Substituting for $V_{1(2)}$ from equation (iii) and for $I_{1(2)}$ from equation (iv) (both in equation (ii)) gives

$$\begin{aligned} V_1 &= A_1\,(A_2V_{2(2)} + B_2I_{2(2)}) + B_1(C_2V_{2(2)} + D_2I_{2(2)}) \\ &= A_1A_2V_{2(2)} + A_1B_2I_{2(2)} + B_1C_2V_{2(2)} + B_1D_2I_{2(2)} \\ &= (A_1A_2 + B_1C_2)V_{2(2)} + (A_1B_2 + B_1D_2)I_{2(2)}. \end{aligned} \tag{6.25}$$

This formula relates the input voltage V_1 of the first network to $V_{2(2)}$ and $I_{2(2)}$, the output quantities from the second network.

Now using equation (6.2) in network N_1 $\quad I_1 = C_1V_2 + D_1I_2$.
Substituting values for V_2 and I_2 from equation (i) $I_1 = C_1V_{1(2)} + D_1I_{1(2)}$. (v)

Substituting equation (iii) for $V_{1(2)}$ and equation (iv) for $I_{1(2)}$ in equation (v) gives

$$\begin{aligned} I_1 &= C_1(A_2V_{2(2)} + B_2I_{2(2)}) + D_1(C_2V_{2(2)} + D_2I_{2(2)}) \\ &= C_1A_2V_{2(2)} + C_1B_2I_{2(2)} + D_1C_2V_{2(2)} + D_1D_2I_{2(2)} \\ &= (C_1A_2 + D_1C_2)V_{2(2)} + (C_1B_2 + D_1D_2)I_{2(2)}. \end{aligned} \tag{6.26}$$

This formula relates the input current I_1 of the first network to $V_{2(2)}$ and $I_{2(2)}$, the output quantities from the second network.

The effect of the two cascaded networks is the same as that of a single network with four-terminal network parameters A, B, C and D

$$\text{when } V_1 = AV_{2(2)} + BI_{2(2)} \quad \text{(vi)}$$
$$\text{and } I_1 = CV_{2(2)} + DI_{2(2)}. \quad \text{(vii)}$$

Comparing equation (vi) with equation (6.25) $A = A_1A_2 + B_1C_2$ (6.27)

and $B = A_1B_2 + B_1D_2$. (6.28)

Comparing equation (vii) with equation (6.26) $C = C_1A_2 + D_1C_2$ (6.29)

and $D = C_1B_2 + D_1D_2$. (6.30)

(*Note*: if the reader is familiar with matrix algebra, it will be realised that these results come from the product of the two individual matrices for the networks N_1 and N_2.)

WORKED EXAMPLE 6.8

Two networks N_1 and N_2 are to be connected in cascade by joining terminals 3 and 5, and 4 and 6, as shown in Fig. 6.18. Determine the overall values of the four-terminal network parameters for the combined network and hence determine the required input conditions for a potential difference of 2 V to be developed across a 1.6 kΩ resistor connected across the output terminals 7 and 8.

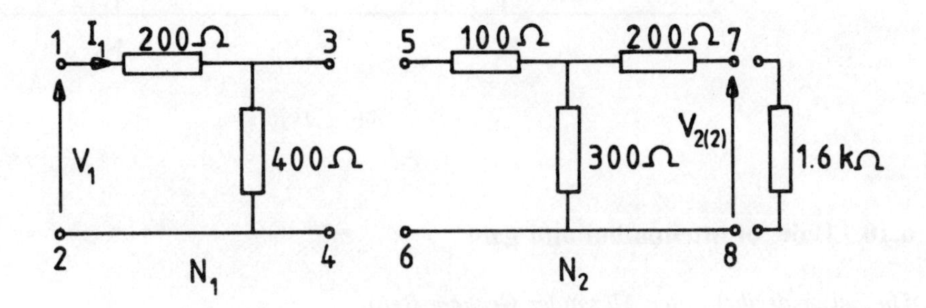

Fig. 6.18

From worked example 6.6, for network N_1 $A_1 = 1.5$, $B_1 = 200$,

$$C_1 = \frac{1}{400},\ D_1 = 1.$$

From worked example 6.4, for network N_2 $A_2 = 1.33$, $B_2 = 366.66$,

$$C_2 = \frac{1}{300},\ D_2 = 1.66.$$

Equation (6.27) $A = A_1A_2 + B_1C_2 = 1.5 \times 1.33 + 200 \times \dfrac{1}{300} = 2.666$

Equation (6.28) $B = A_1B_2 + B_1D_2 = 1.5 \times 366.66 + 200 \times 1.66 = 883.33$

Equation (6.29) $C = C_1A_2 + D_1C_2 = \dfrac{1}{400} \times 1.33 + 1 \times \dfrac{1}{300} = 0.006667$

Equation (6.30) $D = C_1B_2 + D_1D_2 = \dfrac{1}{400} \times 366.66 + 1 \times 1.66 = 2.583$

$$V_1 = AV_{2(2)} + BI_{2(2)} \quad \text{and } I_{2(2)} = \frac{2}{1600} = 1.25 \text{ mA}$$
$$= 2.666 \times 2 + 883.33 \times 1.25 \times 10^{-3}$$
$$= 6.437 \text{ V.}$$

$$I_1 = CV_{2(2)} + DI_{2(2)}$$
$$= 0.006667 \times 2 + 2.583 \times 1.25 \times 10^{-3}$$
$$= 16.56 \text{ mA.}$$

SELF-ASSESSMENT EXAMPLE 6.9

Determine the values of the combined *ABCD* parameters for the networks N_1 and N_2 connected in cascade as shown in Fig. 6.19.

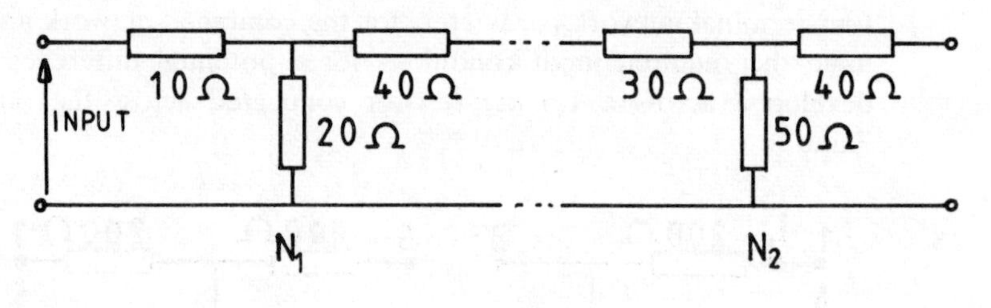

Fig. 6.19

6.10 Units of attenuation and gain

The bel or decibel (after Alexander Graham Bell)

Gain or attenuation in this system is given in terms of power. If a network is considered with passive components only, the power in R_L (Fig. 6.20) will be less than the input power.

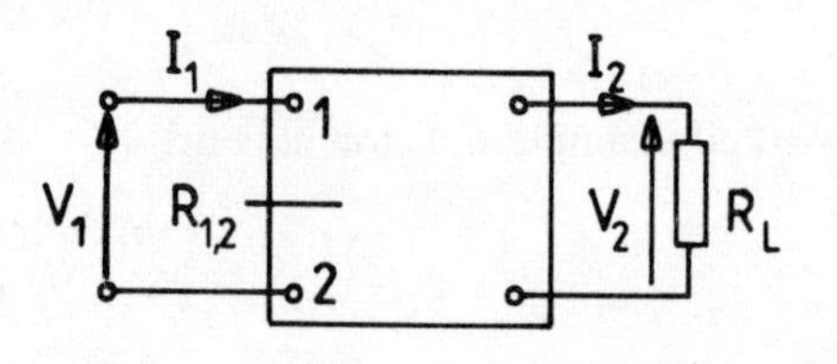

Fig. 6.20

The quantity $\log_{10}$ (power out/power in) is defined as the attenuation in bels. Multiplying by 10 gives the sub-unit, the decibel (dB).
Now, power output $= I_2^2 R_L$ W and power input $= I_1^2 R_{1,2}$ W.

$$\text{Therefore, attenuation} = 10 \log_{10} \left(\frac{I_2^2 R_L}{I_1^2 R_{1,2}} \right) \text{dB}$$

$$= 20 \log_{10} \frac{I_2}{I_1} + 10 \log_{10} \frac{R_L}{R_{1,2}} \text{ dB.}$$

Now, if R_L is equal in value to $R_{1,2}$, as in the case of iterative matching, then the attenuation expression reduces to

$$\text{attenuation} = 20 \log_{10} \frac{I_2}{I_1} \text{ dB.} \tag{6.31}$$

By similar reasoning, the attenuation may be expressed in terms of output and input voltages. For $R_L = R_{1,2}$

$$\text{attenuation} = 20 \log_{10} \frac{V_2}{V_1} \text{ dB.} \tag{6.32}$$

The neper

In the neper system the attenuation is expressed in terms of voltage or current using natural logarithms.

$$\text{Attenuation} = \log_e \frac{I_2}{I_1} \text{ neper (N)}$$

$$\text{or attenuation} = \log_e \frac{V_2}{V_1} \text{ neper.}$$

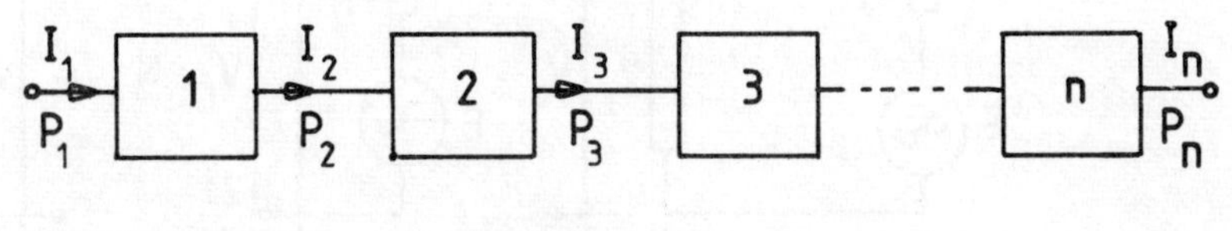

Fig. 6.21

For several networks connected in series as shown in Fig. 6.21, the overall attenuation may be expressed using either system. For power, the attenuation over the first network is expressed as

$$\text{attenuation} = 10 \log_{10} \frac{P_2}{P_1} \text{ dB}$$

and for the second network as

$$\text{attenuation} = 10 \log_{10} \frac{P_3}{P_2} \text{ dB, etc.}$$

Hence the overall attenuation of n networks in cascade is

$$10\left(\log_{10}\frac{P_2}{P_1}+\log_{10}\frac{P_3}{P_2}+\log_{10}\frac{P_4}{P_3}+\ldots+\log_{10}\frac{P_n}{P_{(n-1)}}\right)\text{dB}$$

$$=10\left(\log_{10}\frac{P_2}{P_1}\times\frac{P_3}{P_2}\times\frac{P_4}{P_3}\times\ldots\times\frac{P_n}{P_{(n-1)}}\right)$$

$$=10\log_{10}\frac{P_n}{P_1}\ \text{dB.} \tag{6.33}$$

If all the networks are iteratively matched, this may be expressed in terms of output and input current (or voltage)

$$\text{attenuation}=20\log_{10}\frac{I_n}{I_1}\ \text{dB.} \tag{6.34}$$

In the neper system, the overall current attenuation will be

$$\text{attenuation}=\log_e\frac{I_n}{I_1}\ \text{neper.}$$

The overall attenuation in both systems is the sum of the individual attenuations in logarithmic form.

6.11 Insertion loss ratio

The insertion loss ratio of a four-terminal network is defined as the ratio of voltage or current at the load without the network in circuit to the corresponding voltage or current with the network in circuit.

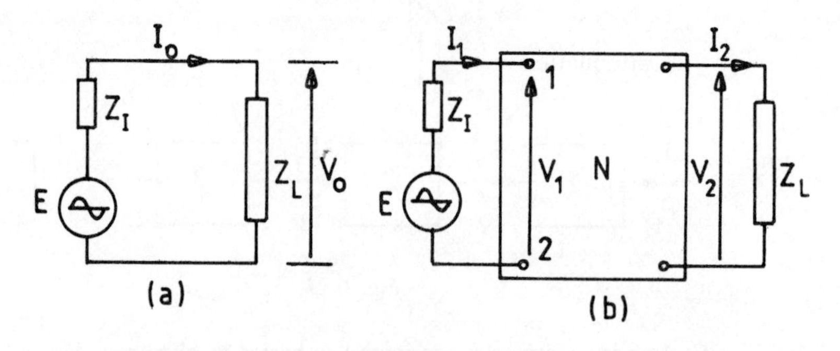

Fig. 6.22

In Fig. 6.22(a), the load voltage without the network is V_0 volts. In Fig. 6.22(b), the load voltage after the insertion of the network is V_2 volts.

$$\text{Insertion loss ratio}=\log_e\frac{V_0}{V_2}=\log_e\frac{I_0}{I_2}.$$

Generally V_0/V_2 or I_0/I_2 will be a complex quantity and there will be an insertion phase shift of ($\arg V_0-\arg V_2$).

If the network is terminated in its iterative impedance, the input impedance to the network is equal to Z_L, $V_0=V_1$ and $I_0=I_1$ (Fig. 6.22(b)). Hence

$$\text{insertion loss ratio} = \log_e \frac{V_1}{V_2} = \log_e \frac{I_1}{I_2}. \tag{6.35}$$

Using decibels, for the same condition

$$\text{insertion loss ratio} = 20 \log_{10} \frac{V_1}{V_2} = 20 \log_{10} \frac{I_1}{I_2} \text{ dB}. \tag{6.36}$$

For iterative matching the loss is therefore a function of the network alone being concerned with $V_1(I_1)$ and $V_2(I_2)$, which leads to the definitions of the transfer coefficient γ (gamma).

6.12 Transfer coefficient γ

6.12.1 Iterative transfer coefficient

When a network is terminated in its iterative impedance, $Z_{1,2} = Z_L$ (Fig. 6.22(b)). Hence

$$\frac{V_1}{I_1} = \frac{V_2}{I_2} \quad \text{or transposing } \frac{V_1}{V_2} = \frac{I_1}{I_2}.$$

The iterative transfer coefficient is defined as the natural logarithm of the ratio of input to output voltage or current. Thus

$$\text{iterative transfer coefficient } \gamma = \log_e \frac{V_1}{V_2} = \log_e \frac{I_1}{I_2} \text{ N}. \tag{6.37}$$

Equation (6.37) is identical to that for insertion loss ratio for the iteratively terminated network.

6.12.2 Propagation coefficient

Where a network is symmetrical, both iterative impedances are the same and are known as the characteristic impedance of the network (section 6.8). In this case the constant is known as the propagation constant or coefficient. It has particular significance in work on transmission lines. The propagation coefficient is defined by equation (6.37). Using the current relationship, since

$$\gamma = \log_e \frac{I_1}{I_2}, \text{ then } e^{\gamma} = \frac{I_1}{I_2}.$$

Generally I_2 differs from I_1 in both magnitude and phase so the transfer coefficient is complex.

$$\gamma = \alpha + j\beta, \qquad \text{therefore } e^{\gamma} = e^{\alpha + j\beta} = e^{\alpha} \times e^{j\beta}.$$

But $e^{j\beta} = (\cos\beta + j\sin\beta)$ (Euler's formula).

Therefore, $e^{\gamma} = e^{\alpha}(\cos\beta + j\sin\beta)$.

In polar form, $e^{\gamma} = e^{\alpha}\angle\beta$. (6.38)

α is the attenuation coefficient and it determines the relationship between the moduli of input and output currents.

$$e^{\alpha} = \frac{|I_1|}{|I_2|}.$$

β is the phase-change coefficient and is the phase angle between I_1 and I_2. For example,

$$I_1 = 5\angle 0° \text{ and } I_2 = 2.5\angle -45° \text{ A}$$

$$\frac{I_1}{I_2} = \frac{5}{2.5\angle -45°} = 2\angle 45°.$$

$$e^{\gamma} = 2\angle 45°; \qquad e^{\alpha} = 2,\ \beta = 45°.$$

WORKED EXAMPLE 6.10

(a) An oscillator with internal resistance 600 Ω and e.m.f. 5 V feeds a 600 Ω load directly. Determine the power developed in the load. (b) It is required to attenuate the load power by 3 dB using an iterative network of the form shown in Fig. 6.23. Determine (i) the values of R_1 and R_2, (ii) the value of the load power with the attenuator in circuit.

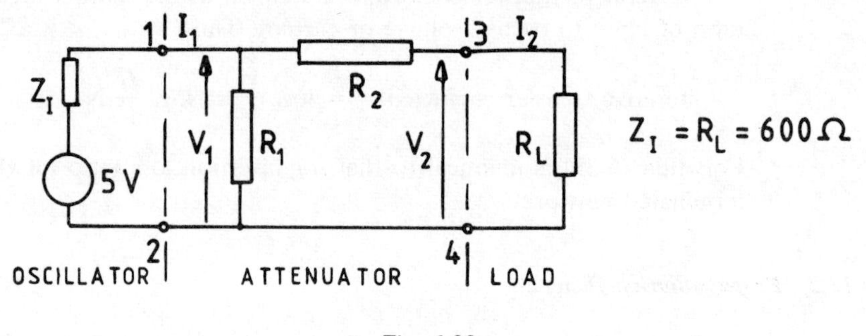

Fig. 6.23

(a) Without attenuator, load current $= \dfrac{E}{R_L + R_I} = \dfrac{5}{600 + 600} = 4.166$ mA.

$$\text{Load power} = I^2R_L = (4.166 \times 10^{-3})^2 \times 600 = 0.0104 \text{ W}.$$

$$\text{Load voltage} = IR_L = 2.5 \text{ V}.$$

(b) (i) Since the network is to be iterative, the resistance looking to the right from terminals 1 and 2 is 600 Ω.

From terminals 1 and 2 the resistance consists of R_1 in parallel with (R_2 + 600) and this is itself to be 600 Ω. Hence

$$600 = \frac{R_1\,(R_2 + 600)}{R_1 + R_2 + 600}. \qquad \text{(i)}$$

For 3 dB attenuation: $20 \log_{10} \dfrac{V_1}{V_2} = 3$ (equation 6.32)

$$\text{Transposing:} \quad \log_{10} \frac{V_1}{V_2} = \frac{3}{20}$$

$$10^{3/20} = \frac{V_1}{V_2}$$

$$\therefore \quad \frac{V_1}{V_2} = 1.4126. \qquad \text{(ii)}$$

To determine V_2, consider R_L and R_2 as a potential divider, the voltage across which is the voltage V_1. Therefore

$$V_2 = \left(\frac{R_L}{R_L + R_2}\right) V_1$$

$$\text{Transposing} \quad \frac{V_1}{V_2} = \frac{R_L + R_2}{R_L}. \qquad \text{(iii)}$$

But $V_1/V_2 = 1.4126$ (equation ii). Substituting this value and R_L in equation (iii) gives

$$1.4126 = \frac{600 + R_2}{600}$$

$$847.56 = 600 + R_2$$

$$R_2 = 247.56\ \Omega.$$

Substituting for R_2 in equation (i) above gives

$$600 = \frac{R_1\ (247.56 + 600)}{R_1 + 247.56 + 600}$$

$$847.56R_1 = 600R_1 + 508\,536$$

$$R_1 = 2054.2\ \Omega.$$

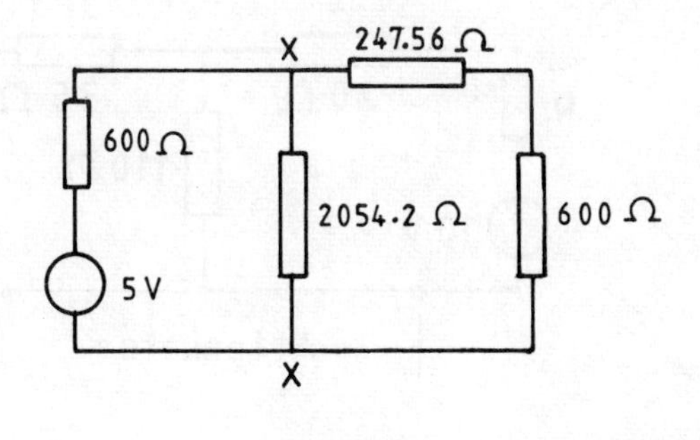

Fig. 6.24

(ii) The circuit is now as shown in Fig. 6.24. The resistance of the network and load resistor to the right of points XX is 600 Ω. (The reader might care to check whether 2054.2 Ω in parallel with (247.56 + 600) Ω is indeed 600 Ω.)

$$\text{The current leaving the oscillator} = \frac{5}{600 + 600} = 4.166 \text{ mA as in (a).}$$

$$\text{The voltage } V_{XX} = V_1 \text{ (in Fig. 6.23)} = IR_{XX} = 4.166 \times 10^{-3} \times 600$$
$$= 2.5 \text{ V (again, as in (a)).}$$

$$\text{Current in } R_L = \frac{V_{XX}}{247.56 + 600}$$

$$= \frac{2.5}{847.56} = 2.95 \text{ mA.}$$

$$\text{Load power} = I^2R_L = (2.95 \times 10^{-3})^2 \times 600 = 5.22 \text{ mW.}$$

This is exactly one half of the value in (a) which is the requirement for 3 dB attenuation since

$$10 \log_{10}\left(\frac{P_{(a)}}{\frac{1}{2}P_{(a)}}\right) = 10 \log_{10}2 = 3 \text{ dB } (P_{(a)} = \text{power in part (a)).}$$

It should be observed that although the load power is reduced, the oscillator is still delivering the original power into the same resistance as it did when the 600 Ω resistor was connected directly to its output terminals. It is the proportion of that power which reaches the load which is changed by the attenuator.

SELF-ASSESSMENT EXAMPLE 6.11

Figure 6.25 shows a source of e.m.f. with internal resistance R_I driving a load R_L through an attenuating network. Calculate (a) the values of R_I and R_L for iterative matching, (b) the power which would be supplied by the source when connected directly to R_L (using the values from (a)), (c) the power delivered to R_L with the attenuating network in circuit, (d) the insertion loss ratio in nepers and decibels.

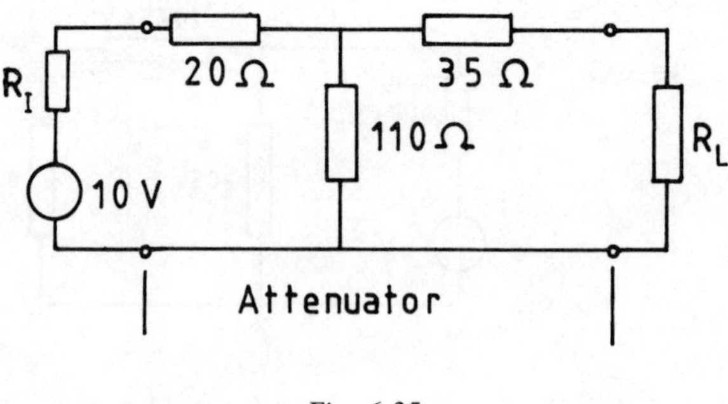

Fig. 6.25

WORKED EXAMPLE 6.12

Determine the values of the attenuation coefficient α and phase-change coefficient β for the network shown in Fig. 6.26.

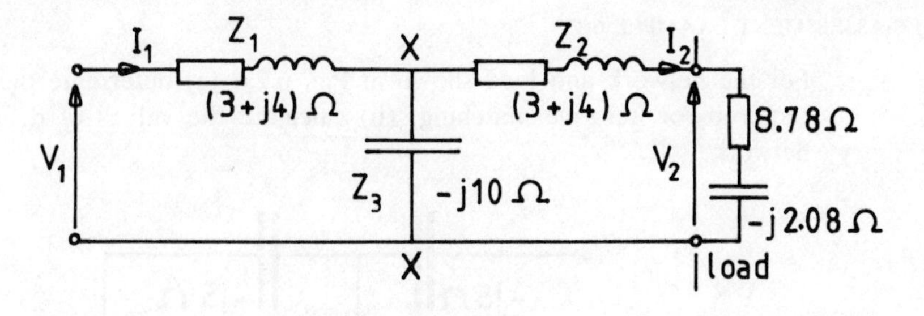

Fig. 6.26

Impedance to the right of terminals XX $= Z_{XX} = \dfrac{Z_3(Z_2 + Z_L)}{Z_3 + Z_2 + Z_L}$. (i)

$$V_{XX} = I_1 Z_{XX} \quad \text{(ii)}$$

and $I_2 = \dfrac{V_{XX}}{Z_2 + Z_L}$.

Substituting equation (ii) for V_{XX} gives

$$I_2 = \frac{I_1 Z_{XX}}{Z_2 + Z_L}.$$

Transposing $\dfrac{I_1}{I_2} = \dfrac{Z_2 + Z_L}{Z_{XX}}$. (iii)

Substituting equation (i) for Z_{XX} in equation (iii) gives

$$\frac{I_1}{I_2} = \frac{Z_2 + Z_L}{Z_3(Z_2 + Z_L)/Z_3 + Z_2 + Z_L} = \frac{Z_3 + Z_2 + Z_L}{Z_3}.$$

Introducing the actual values gives

$$\frac{I_1}{I_2} = \frac{-\text{j}10 + (3 + \text{j}4) + (8.78 - \text{j}2.08)}{-\text{j}10}$$

$$= \frac{11.78 - \text{j}8.08}{-\text{j}10} = \frac{14.28\angle -34.4°}{10\angle -90°}$$

$$= 1.428\angle 55.6°.$$

This is in the form of equation (6.38), $e^{\alpha}\angle\beta = I_1/I_2$. Hence

$$e^{\alpha} = 1.428, \qquad \text{therefore, } \alpha = \log_e 1.428 = 0.356$$

and $\beta = 55.6°$.

Now since $I_1/I_2 = 1.428\angle 55.6°$, transposing gives

$$I_2 = \frac{I_1}{1.428\angle 55.6°}$$

$$= 0.7 I_1 \angle -55.6° \text{ A}.$$

I_2 has a modulus 0.7 times that of I_1 and lags I_1 by 55.6°.

Further work would show that the same relationship exists between V_1 and V_2.

SELF-ASSESSMENT EXAMPLE 6.13

For the network and load shown in Fig. 6.27, (a) determine the value of Z_L required for iterative matching, (b) calculate the values of α and β for the network.

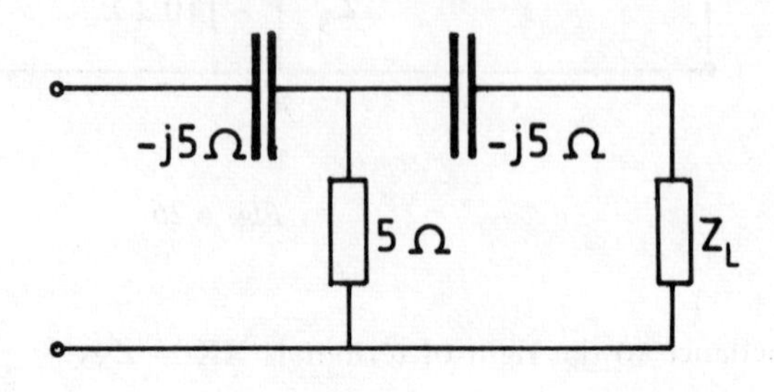

Fig. 6.27

6.13 Passive filters

A filter is a four-terminal network which, ideally, is conducting between input and output terminals only over a specific frequency range.

A low-pass filter should transmit a signal between its input and output terminals from d.c. up to a specific frequency at which it would suddenly give zero output. A high-pass filter would give zero transmission up to a specific frequency at which it would become conducting, giving an output.

Band-pass filters may be formed by using two filters, a high-pass filter turning on at a particular frequency, in series with a low-pass filter which would shut off at the required higher frequency.

Passive filters are formed from inductors and capacitors only. Active filters use capacitors and resistors disposed around an amplifying element such as an operational amplifier.

No filter is perfect in that the commencement of conduction and the shutting off takes place over a frequency range and not suddenly.

We will consider π and T passive configurations which are known as *constant k* filters. Such filters have a ratio of total series reactance to shunt susceptance of:

$$\frac{j\omega L}{j\omega C} = \frac{L}{C} = k$$

The value of k is seen to be independent of the operating frequency.

6.13.1 *The π-section low-pass filter*

A low-pass filter is a network which is designed to have zero attenuation up to a given frequency above which frequency, ideally, it should have infinite attenuation (zero transmission).

This can never be realised completely in practice but some filters come closer to the ideal than others.

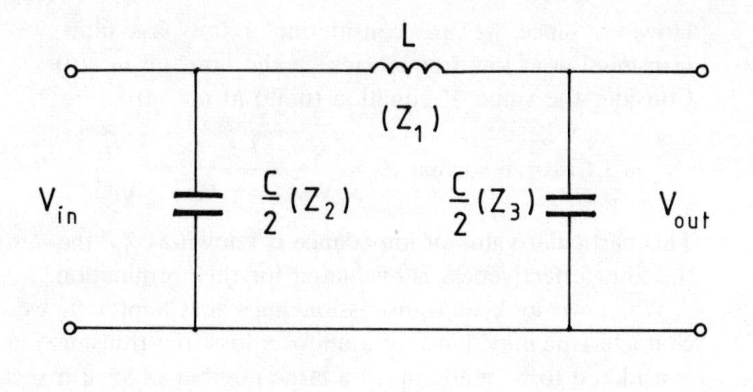

Fig. 6.28

First, using equations (6.10) and (6.11) we will find the A,B,C,D parameters on the π filter shown in Fig. 6.28.

For a π network $A = 1 + \dfrac{Z_1}{Z_3} \qquad B = Z_1$

$$C = \frac{1}{Z_2} + \frac{1}{Z_3} + \frac{Z_1}{Z_2 Z_3} \qquad D = 1 + \frac{Z_1}{Z_2}$$

$$\text{A} = 1 + \frac{j\omega L}{\dfrac{1}{j\omega C/2}} = 1 + \frac{j^2\omega^2 LC}{2}$$

$$= 1 - \frac{\omega^2 LC}{2} \qquad (= D \text{ since the network is symmetrical})$$

$$B = j\omega L$$

$$C = \frac{1}{\dfrac{1}{j\omega C/2}} + \frac{1}{\dfrac{1}{j\omega C/2}} + \frac{j\omega L}{\dfrac{1}{j\omega C/2} \times \dfrac{1}{j\omega C/2}}$$

$$C = \frac{j\omega C}{2} + \frac{j\omega C}{2} + \frac{j\omega L\ (j^2\omega^2 C^2)}{4}$$

$$= j\omega C - \frac{j\omega^3 LC^2}{4}.$$

What would be the value of Z_{load} for correct termination? For a symmetrical network image impedance = iterative impedance. Taking the image approach:

$$\text{equation (6.19) } Z = \sqrt{\frac{AB}{CD}} \qquad \text{and } A = D$$

$$\therefore\ Z = \sqrt{\frac{B}{C}} = \sqrt{\frac{j\omega L}{j\omega C - j\omega^3 LC^2/4}} = \sqrt{\frac{j\omega L}{j\omega C[1 - \omega^2 LC/4]}}$$

$$= \sqrt{\frac{L}{C[1 - \omega^2 LC/4]}} \qquad (6.39)$$

Z could then have any value dependent upon frequency.

However since we are considering a low-pass filter we require maximum transmission at low frequencies, in the limit, at $\omega = 0$.
Consider the value of equation (6.39) at $\omega = 0$

$$\omega^2 LC/4 = 0 \text{ so that } Z = \sqrt{\frac{L}{C\,(1-0)}} = \sqrt{\frac{L}{C}}$$

This particular value of impedance is known as Z_0, the *design impedance*, and the filter effectiveness is evaluated for this termination.

When we look at transmission lines in Chapter 9, we shall find that the characteristic impedance of a uniform loss-free transmission line which may be considered to be made up of a large number of such π sections is $\sqrt{(L/C)}\ \Omega$. (Loss-free implies either very low resistance, which is the case with the filter, or that the operating frequency is high enough to make the reactive components predominate.)
Using equation (6.1)

$$V_{in} = AV_0 + BI_0 \quad \text{and for } Z_0 = \sqrt{\frac{L}{C}};\ \ I_0 = \frac{V_0}{Z_0} = \frac{V_0}{\sqrt{L/C}} = V_0\sqrt{\frac{C}{L}}.$$

Hence $V_{in} = AV_0 + BV_0\sqrt{\dfrac{C}{L}}$.

Substitute for A and B. Divide both sides by V_0

$$\frac{V_{in}}{V_0} = 1 - \frac{\omega^2 LC}{2} + j\omega L\sqrt{\frac{C}{L}}$$

$$= 1 - \frac{\omega^2 LC}{2} + j\omega\sqrt{(LC)} \qquad (6.40)$$

Equation (6.40) can be evaluated for a range of values of ω until an appreciable reduction in V_0 is apparent, i.e. V_{in}/V_{out} becomes, say, 2 or more. (In electronics, when considering the roll-off of amplifiers we use f_2 or f_β, the half-power points, as corner frequencies at which the reduction in gain become appreciable.)

It is found that when $\omega = 2/\sqrt{(LC)}$ rad/s such a reduction in gain occurs. This particular angular velocity is denoted by ω_0
Transposing gives

$$\sqrt{(LC)} = \frac{2}{\omega_0} \quad \text{or} \quad LC = \frac{4}{\omega_0^2}.$$

Substituting for $\sqrt{LC}$ and LC in equation (6.40)

$$\frac{V_{in}}{V_{out}} = 1 - \frac{\omega^2}{2}\frac{4}{\omega_0^2} + j\omega\frac{2}{\omega_0}$$

$$= 1 - \frac{2\omega^2}{\omega_0^2} + j2\frac{\omega}{\omega_0} \qquad (6.41)$$

Evaluating equation (6.41) for $\omega = \omega_0$

$$\frac{\omega^2}{\omega_0^2} = 1 \qquad \frac{\omega}{\omega_0} = 1$$

$$\therefore \frac{V_{in}}{V_{out}} = 1 - 2 + j2 = \sqrt{5}\angle 116.3° = 2.236\angle 116.3°.$$

For $\omega = 0.5\omega_0$ equation (6.41) gives $V_{in}/V_{out} = 1 - 0.5 + j1 = 0.5 + j1 = 1.118\angle 63.4°$.

Thus, attenuation has only just commenced at an angular velocity of $\frac{1}{2}\omega_0$ but at ω_0 considerable attenuation has occurred.

Table of values from equation (6.41)

Angular velocity ω	Expression	V_{in}/V_{out}	Angle
$0.5\omega_0$	$1 - 2 \times \frac{1}{4} + j2 \times \frac{1}{2}$		
	$1 - 0.5 + j1$	1.118	63.4°
ω_0	$1 - 2 + j2$		
	$-1 + j2$	2.236	116.3°
$2\omega_0$	$1 - 2\ (2^2) + j2 \times 2$		
	$-7 + j4$	8.06	150.25°
$5\omega_0$	$1 - 2\ (5^2) + j2 \times 5$	50.01	168.46°
$10\omega_0$	$1 - 2\ (10^2) + j2 \times 10$		
	$-199 + j20$	200	174.26°

As shut off occurs, the phase angle between the output and input voltages approaches 180° (lagging).

ω_0 is generally referred to as the *shut-off frequency* but this is far from a clear cut case, shut off being gradual, and is not complete even at $\omega = 10\omega_0$.

WORKED EXAMPLE 6.14

Suggest values of components suitable for use in the π filter as shown in Fig. 6.28 to feed a load of 60 Ω with a nominal shut-off frequency of 1 kHz.

The design impedance $Z_0 = 60\ \Omega$

$$\therefore \sqrt{L/C} = 60 \qquad \text{and squaring } L/C = 3600.$$

Transposing $L = 3600C$

$\omega_0 = 2\pi \times 1000$

$$\text{and } \omega_0 = \frac{2}{\sqrt{(LC)}} \qquad \therefore 2\pi \times 1000 = \frac{2}{\sqrt{3600C \times C}} = \frac{2}{60\ C}.$$

$$\text{Transposing } C = \frac{2}{60 \times 2\pi \times 1000} = 5.305\ \mu\text{F}.$$

Now since the filter employs one half of the capacitance at the input end and the other half across the output terminals, each capacitor will have a value of 2.65 μF.

$$L = 3600C = 19 \text{ mH}.$$

Using two or more such sections in series gives faster shut off but the cumulative phase changes must not be forgotten.

6.13.2 The T-section, low-pass filter

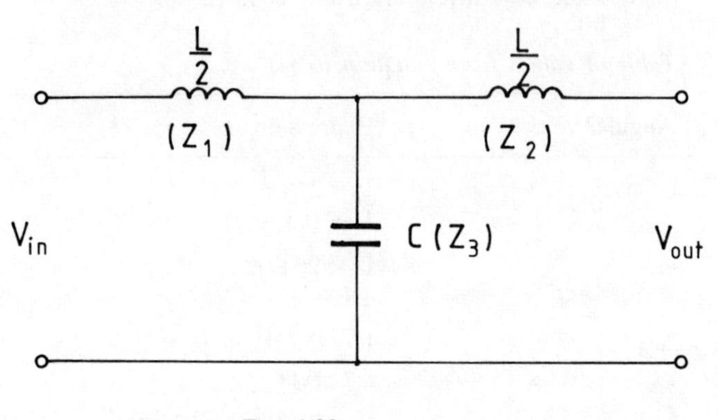

Fig. 6.29

Using equations (6.7) and (6.8), the A, B, C, D parameters of the T-network are found.

$$A = 1 + \frac{Z_1}{Z_3} = 1 + \frac{j\omega L/2}{1/j\omega C} = 1 + \frac{j^2\omega^2 LC}{2}$$

$$= 1 - \frac{\omega^2 LC}{2} \qquad (= D)$$

$$B = Z_1 + Z_2 + \frac{Z_1\,Z_2}{Z_3} = \frac{j\omega L}{2} + \frac{j\omega L}{2} + \frac{\dfrac{(j\omega L)^2}{4}}{1/j\omega C}$$

$$= j\omega L + \frac{j^3\omega^3 L^2 C}{4} = j\omega L - \frac{j\omega^3 L^2 C}{4}$$

$$C = \frac{1}{Z_3} = j\omega C.$$

Now, as with the π section

$$Z_0 = \sqrt{\frac{B}{C}} = \sqrt{\frac{j\omega L - \dfrac{j\omega^3 L^2 C}{4}}{j\omega C}} = \sqrt{\frac{L - \dfrac{\omega^2 L^2 C}{4}}{C}}$$

$$= \sqrt{\frac{L}{C}\left(1 - \frac{\omega^2 LC}{4}\right)} \quad \text{and as } \omega \to 0$$

$$Z_0 \to \sqrt{\frac{L}{C}} \quad \text{as in section 6.13.1.}$$

From equation (6.1)
$V_{in} = AV_0 + BI_0$ Using the design impedance

$$I_0 = \frac{V_0}{Z_0} = \frac{V_0}{\sqrt{\frac{L}{C}}} = \sqrt{\frac{C}{L}}\, V_0.$$

$$V_{in} = \left[1 - \frac{\omega^2 LC}{2}\right] V_0 + \left[j\omega L - \frac{j\omega^3 L^2 C}{4}\right] V_0 \sqrt{\frac{C}{L}}.$$

$$\frac{V_{in}}{V_0} = 1 - \frac{\omega^2 LC}{2} + j\omega L\sqrt{\frac{C}{L}} - \frac{j\omega^3 L^2 C}{4}\sqrt{\frac{C}{L}}.$$

As with the π section, making $\sqrt{(LC)} = 2/\omega_0$ $\quad LC = 4/\omega_0^2$.

$$\frac{V_{in}}{V_0} = 1 - \frac{\omega^2}{2}\frac{4}{\omega_0^2} + j\omega\sqrt{(LC)} - \frac{j\omega^3 LC}{4}\sqrt{(CL)}$$

$$= 1 - \frac{\omega^2}{\omega_0^2} \times 2 + \frac{j\omega}{\omega_0} \times 2 - \frac{j\omega^3\, 4}{\omega_0^2} \times \frac{2}{\omega_0}$$

$$= 1 - 2\frac{\omega^2}{\omega_0^2} + j2\frac{\omega}{\omega_0} - j\frac{2\omega^3}{\omega_0^3}. \tag{6.42}$$

Evaluating equation (6.42) using values of ω is shown in the table below.

Angular velocity ω	Expression	V_{in}/V_{out}	Angle
0	$1 - 0 + j0 - j0$	1	0°
$\frac{1}{2}\omega_0$	$1 - 2\,(\frac{1}{2})^2 + j2\,(\frac{1}{2}) - j2\,(\frac{1}{2})^3$	0.901	56.3°
$0.707\omega_0$	$1 - 2\,(0.707)^2 + j2\,(0.707) - j2\,(0.707)^3$	0.707	90°
ω_0	$1 - 2 + j2 - j2$	1	180°
$2\omega_0$	$1 - 2\,(2)^2 + j2\,(2) - j2\,(2)^3$	13.89	−120.2°
$10\omega_0$	$1 - 2\,(10)^2 + j2\,(10) - j2\,(10)^3$	1989	−95.7°

Note that for $\omega = \frac{1}{2}\omega_0$ and $\omega = 0.707\omega_0$ $V_{in}/V_{out} < 1$, i.e. V_{out} is greater than V_{in}. At ω_0 there is no attenuation but a 180° phase shift. At higher frequencies the filter rapidly shuts off.

6.13.3 The T-section high-pass filter

Observe that in Fig. 6.30 there are two capacitors in series. In order to maintain the constant k relationship each of the capacitors requires to have a capacitance of $2C$ giving a total series reactance equivalent to a single capacitor of magnitude C.

Using equations (6.7) and (6.8) the A, B, C, D parameters of the network are found.

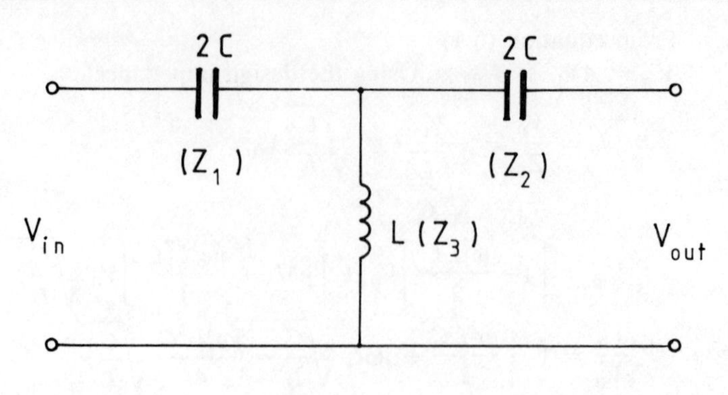

Fig. 6.30

$$A = 1 + \frac{Z_1}{Z_3} = 1 + \frac{1/\text{j}\omega 2C}{\text{j}\omega L} = 1 + \frac{1}{\text{j}^2\omega^2 \, CL \times 2}$$

$$= 1 - \frac{1}{\omega^2 \, 2 \, CL} \qquad (= D)$$

$$B = Z_1 + Z_2 + \frac{Z_1 \, Z_2}{Z_3} = \frac{1}{2\text{j}\omega C} + \frac{1}{2\text{j}\omega C} + \frac{1/2\text{j}\omega C}{\text{j}\omega L} \times 1/2\text{j}\omega C$$

$$= \frac{1}{\text{j}\omega C} + \frac{1}{4\text{j}^3\omega^3 C^2 L} = \frac{1}{\text{j}\omega C} - \frac{1}{\text{j}4\omega^3 C^2 L}$$

$$C = \frac{1}{Z_3} = \frac{1}{\text{j}\omega L}.$$

Using the image approach as in sections 6.13.1 and 6.13.2

$$Z_0 = \sqrt{\frac{B}{C}} = \sqrt{\frac{\frac{1}{\text{j}\omega C} - \frac{1}{\text{j}4\omega^3 C^2 L}}{\frac{1}{j\omega L}}}$$

$$= \sqrt{\frac{\text{j}\omega L}{\text{j}\omega C} - \frac{\text{j}\omega L}{\text{j}4\omega^3 C^2 L}} = \sqrt{\frac{L}{C}\left[1 - \frac{1}{4\omega^2 CL}\right]}. \qquad (6.43)$$

This is a high-pass filter requiring maximum transmission at high frequencies, in the limit, when $\omega = \infty$. Equation (6.43) with $\omega = \infty$ becomes

$$Z_0 = \sqrt{\frac{L}{C}(1 - 0)} = \sqrt{\frac{L}{C}} \quad \text{as for low-pass filters.}$$

Using equation (6.1), and correctly terminating the filter we get

$$V_{\text{in}} = AV_0 + BI_0 = AV_0 + BV_0/\sqrt{(L/C)}$$

$$= \left(1 - \frac{1}{\omega^2 \, 2CL}\right) V_0 + \left(\frac{1}{\text{j}\omega C} - \frac{1}{\text{j}4\omega^3 C^2 L}\right) V_0 \sqrt{\frac{C}{L}}$$

$$\frac{V_{\text{in}}}{V_{\text{out}}} = 1 - \frac{1}{\omega^2 2CL} + \left(\frac{1}{\text{j}\omega C} - \frac{1}{\text{j}4\omega^3 C^2 L}\right) \sqrt{\frac{C}{L}}$$

$$= 1 - \frac{1}{\omega^2 2CL} + \frac{1}{j\omega C}\sqrt{\frac{C}{L}} - \frac{\sqrt{C}}{j4\omega^3 C^2 L\sqrt{L}}.$$

Again, as before, let $\sqrt{(LC)} = 2/\omega_0$: $LC = 4/\omega_0^2$

$$\frac{V_{in}}{V_{out}} = 1 - \frac{1}{\omega^2}\frac{\omega_0^2}{2 \times 4} + \frac{1}{j\omega}\frac{1}{\sqrt{(CL)}} - \frac{\sqrt{C}}{j\omega^3 4CL\ C\sqrt{L}}$$

$$= 1 - \frac{1}{\omega^2}\frac{\omega_0^2}{2 \times 4} + \frac{1}{j\omega}\frac{\omega_0}{2} - \frac{\sqrt{C}\ \omega_0^2\ \omega_0}{j\omega^3\ 4 \times 4\ C\sqrt{L}\ 2}$$

$$= 1 - \frac{\omega_0^2}{8\omega^2} - \frac{j\omega_0}{2\omega} + \frac{j\omega_0^3}{32\omega^3}.$$

At $\omega = \omega_0$ $V_{in}/V_{out} = (1 - \frac{1}{8}) - j\frac{1}{2} + j1/32$
$= 0.875 - j0.46875$
$= 1\angle{-28°}$.

At $\omega = 0$

$$V_{in}/V_{out} = 1 - \infty = \text{shut off}.$$

At $\omega = 2\omega_0$, $\omega_0/\omega = 0.5$
$= 1 - 0.5^2/8 - j0.5/2 + j0.5^3/32$
$= 1 - 0.03125 - j0.246 = 1\angle{-14°}$.

We see that at $\omega = 0$, the filter is shut off. At frequencies of ω_0 and above the output is equal in magnitude to the input but there is a phase shift. The reader might care to examine the relationship V_{in}/V_{out} at $\omega = \frac{1}{2}\omega_0$ and draw a conclusion about the rate at which this filter turns on.

SELF-ASSESSMENT EXAMPLE 6.15

Suggest values for the capacitors and inductor to make a high-pass filter feeding a 40 Ω load given $\omega_0 = 2000 \times 2\pi$ rad/s.

FURTHER PROBLEMS

6.16 (a) What relationships exist between the A, B, C and D parameters of a linear two-port network (i) if it is passive, and (ii) if it is symmetrical.
(b) For Fig. 6.31, calculate the two values of impedance necessary for image matching at input and output terminals.

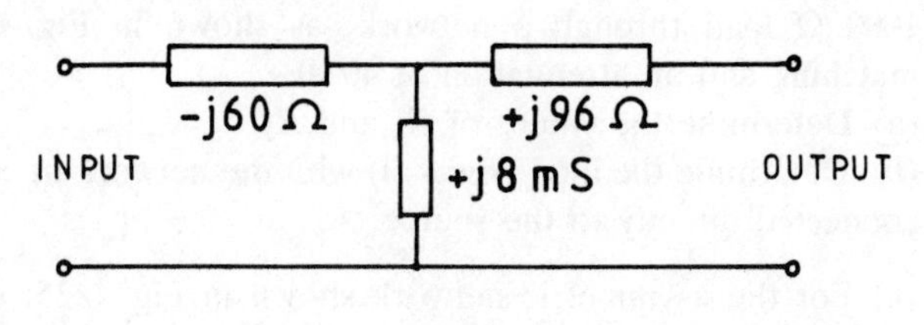

Fig. 6.31

6.17 Tests on a linear two-port network gave the following results:

Open-circuit test: input 10 V and 0.125 A; output 6.25 V.
Short-circuit test: input 10 V and 0.2 A; current in short-circuit 0.12 A.

Determine the values of the four-terminal network parameters *A*, *B*, *C* and *D*. Confirm that the network is passive.

6.18 Tests on a linear two-port network gave the following results:

Open-circuit test: input (6 + j0) V and (4.8 + j2.4) mA; output (1.2 − j2.4) V.
Short-circuit test: input (6 + j0) V and (4.62 + j0.924) mA; output (2.77 − j1.848) mA (in short circuit).

Determine the values of the four-terminal network parameters and hence find the output voltage and current when the input current is $7.5\angle{+30°}$ mA from a source of e.m.f. with negligible internal impedance giving $10\angle 0°$ V.
What is the value of the component connected to the output terminals?

6.19 (a) Define the terms image impedance and iterative impedance as applied to a four-terminal network.
(b) Calculate the insertion loss in decibels or nepers when the LC attenuator shown in Fig. 6.32 is placed between the two resistors R_I and R_L ($R_I = R_L =$ 141.42 Ω. ω = 1414.2 rad/s).

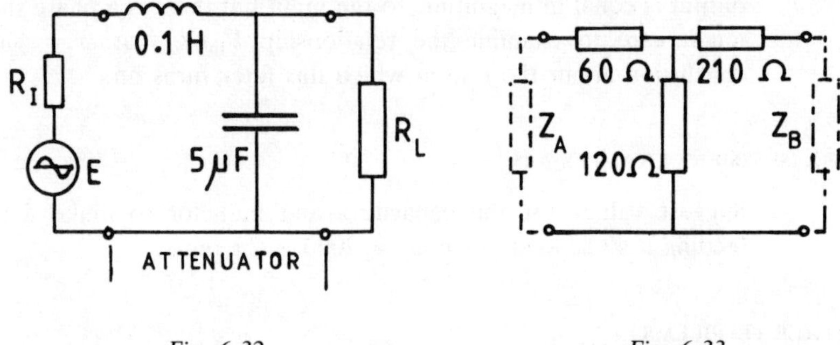

Fig. 6.32 *Fig. 6.33*

6.20 Calculate, for the network shown in Fig. 6.33, (a) the two iterative impedances, and (b) the insertion loss when the network is inserted between these iterative impedances.

6.21 A voltage source with internal resistance 1000 Ω and e.m.f. 10 V feeds a 1000 Ω load through a network, as shown in Fig. 6.34. It gives iterative matching and an attenuation of 40 dB.
(a) Determine the values of R_1 and R_2.
(b) Determine the load power (i) with the network in circuit, (ii) with the load connected directly to the source.

6.22 (a) For the asymmetric network shown in Fig. 6.35, derive the *A*, *B*, *C*, *D* parameters in terms of Z_1, Z_2 and Z_3.

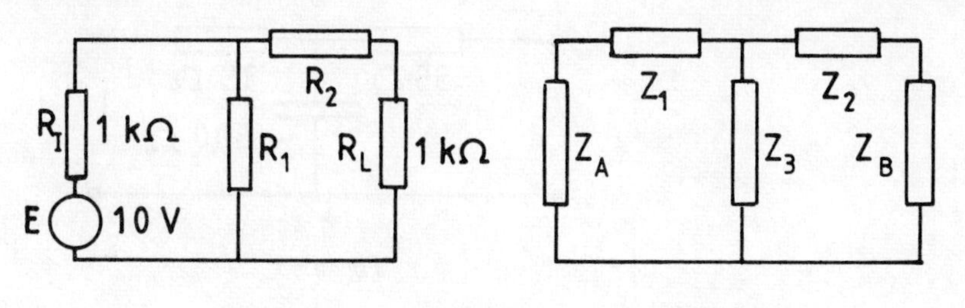

Fig. 6.34 Fig. 6.35

(b) Given that the values of the components are $Z_1 = -j40\ \Omega$, $Z_2 = j60\ \Omega$ and $Z_3 = -j200\ \Omega$, (i) calculate the values of the four-terminal network parameters, (ii) using these parameters, determine the values of the two impedances Z_A and Z_B which will give image matching.

6.23 Two identical passive two-port networks N_1 and N_2 have the following parameters

$$A = (1 + j1),\ B = 50,\ C = j0.02 \text{ and } D = 1.$$

They are connected in cascade. A voltage of $5\angle 0°$ V is applied to the input terminals and a resistive load of 20 Ω is connected to the output terminals.
Calculate (a) the output voltage, (b) the input current.

6.24 The filter shown in Fig. 6.36 is terminated in its characteristic impedance Z_0. Determine (a) the value of Z_0, (b) the value of the attenuation coefficient for the network when terminated in Z_0.

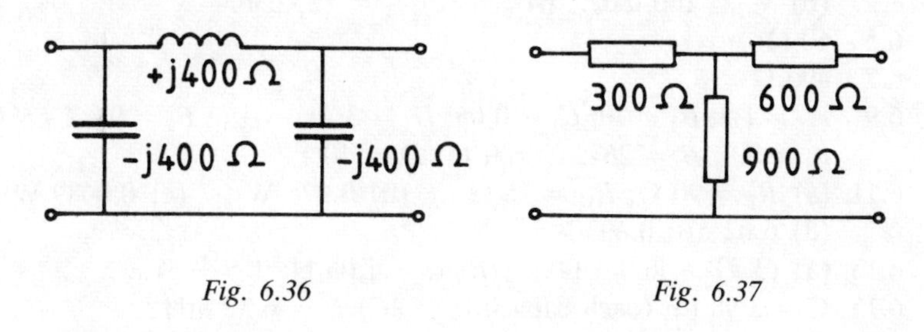

Fig. 6.36 Fig. 6.37

6.25 Calculate the values of the two image impedances for the network shown in Fig. 6.37. What is the insertion loss brought about when this network is inserted between the two image impedances?

6.26 A four-terminal resistive network has input terminals A and B and output terminals C and D. The resistance measured across AB with CD short-circuited is 720 Ω. The resistance measured across AB with CD open-circuited is 900 Ω. The resistance measured across CD with AB open-circuited is 500 Ω. Determine the values of the resistors, assuming a T-configuration.

6.27 (a) Determine the value of the characteristic impedance of the network shown in Fig. 6.38.
(b) Calculate the values of α and β for the network correctly terminated.

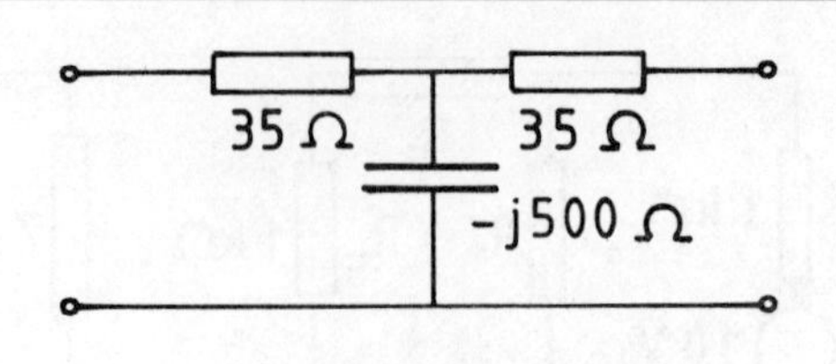

Fig. 6.38

6.28 (a) Deduce the values of components to form a π-section low-pass filter to feed a 75 Ω load. The shut-off frequency is to be 63 Hz. (b) Given that the input voltage to the filter is 5 V at a frequency of 315 Hz, what is the value of the output voltage?

6.29 A T-section low-pass filter is formed using two inductors, each of which has inductance 19.1 mH. The capacitor has a capacitance of 100 nF. Working from first principles deduce: (a) the value of the design impedance, (b) the shut off frequency.

6.30 (a) Deduce the values of components necessary to form a T-section high-pass filter with shut-off frequency 1.5 kHz. It is to have a design impedance of 150 Ω. (b) What relationship exists between V_{out} and V_{in} at frequencies corresponding to (i) $0.1\omega_0$ and (ii) $0.2\omega_0$?

Answers

6.3 (a) $A = D = 0.979\angle 0.417°$; $B = 48.2\angle 71.7°$; $C = j920 \times 10^{-6}$
(b) $V_1 = 160\,480\angle 1.64°$ (c) $I_1 = 123\angle 36°$

6.5 60 Ω

6.7 400 Ω

6.9 $A_1 = 1.5$; $B_1 = 70$; $C_1 = 0.05$; $D_1 = 3$; $A_2 = 1.6$; $B_2 = 94$; $C_2 = 0.02$; $D_2 = 1.8$; $A = 3.8$; $B = 267$; $C = 0.14$; $D = 10.1$

6.11 (a) $R_I = 90$ Ω; $R_L = 75$ Ω (b) 0.275 W (c) 0.0689 W
(d) 6.02 dB; 0.693 N

6.13 (a) $(3.93 - j6.36)$ Ω (b) $\alpha = 1.06$ N; $\beta = -51.8°$

6.15 $C = 3.98$ μF (each capacitor $= 2C$) $L = 6.37$ mH

6.16 (b) $Z_{in} = 256$ Ω; $Z_{out} = 40$ Ω

6.17 $A = 1.6$: B 83.33; $C = 0.02$; $D = 1.66$; $AD - BC = 1$

6.18 $A = (1 + j2)$; $B = (1.5 + j1)\,10^3$; $C = j2 \times 10^{-3}$; $D = (1 + j1)$; $V_{out} = 4.53\angle -27.9°$ V; $I_{out} = 3.41 \times 10^{-3}\angle 72.94°$ A; $Z_{load} = 1328.4\angle 79.16°$ Ω

6.19 (b) 0.97 dB; 0.11 N

6.20 (a) $Z_A = 300$ Ω; $Z_B = 150$ Ω (b) 1.39 N; 12 dB

6.21 (a) $R_2 = 99$ kΩ; $R_1 = 1010$ Ω (b) (i) 2.5 μW (ii) 0.025 W

6.22 (b) (i) $A = 1.2$; $B = j32$; $C = j5 \times 10^{-3}$; $D = 0.7$
(ii) $Z_A = 104.7$ Ω; $Z_B = 61.1$ Ω

6.23 $A = j3$; $B = 100 + j50$; $C = -0.02 + j0.04$; $D = (1 + j1)$
(a) $V_2 = 0.673\angle -47.72°$ V (b) $I_1 = 63\angle 23.8°$ mA

6.24 (a) $Z_0 = 400$ Ω (b) $\alpha = 0$ (low-pass filter designed for zero attenuation)

6.25 890 Ω; 1120 Ω; 8.2 dB

6.26 $Z_1 = 600$ Ω; $Z_2 = 200$ Ω; $Z_3 = 300$ Ω

6.27 (a) $Z_0 = 187.3\angle -44°\ \Omega$ (b) $\alpha = 0.264$ N (2.29 dB); $\beta = 15°$
6.28 (a) Each capacitor = 33.7 μF, $L = 0.379$ H; (b) 0.1 V
6.29 (a) 618 Ω (b) 5150 Hz
6.30 (a) Each capacitor = 2.83 μF; $L = 31.83$ mH
(b) (i) $V_{out} = 0.0349V_{in}$ (phase shift −114°);
(ii) $V_{out} = 0.39V_{in}$ (phase shift −146°)

Chapter 7

Resonance, *Q*-factor and Coupled a.c. Circuits

7.1 Series resonance

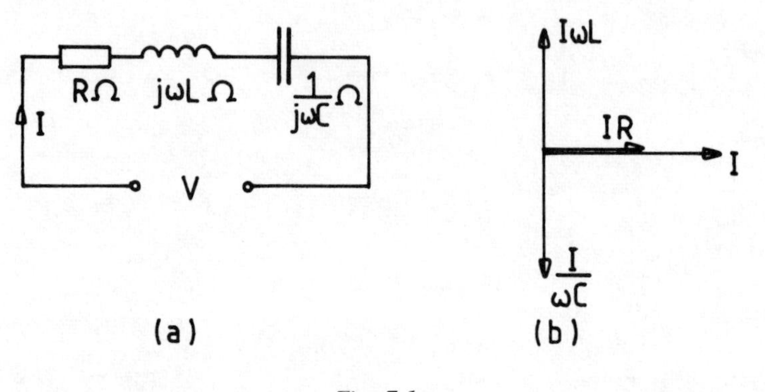

Fig. 7.1

Consider a circuit consisting of a resistive coil in series with a capacitor as shown in Fig. 7.1(a). With angular velocity ω rad/s, the impedance Z of the circuit is expressed as

$$Z = \left(R + j\omega L + \frac{1}{j\omega C}\right)$$

$$= \left(R + j\left(\omega L - \frac{1}{\omega C}\right)\right) \; \Omega.$$

When $\omega L = 1/\omega C$, the reactive terms disappear and the impedance of the circuit is purely resistive. This condition is termed 'series resonance'. $Z = R\ \Omega$, and the angular velocity is written as ω_0 rad/s and the current as I_0 amperes. Hence

$$\omega_0 L = \frac{1}{\omega_0 C}$$

$$\omega_0^2 = \frac{1}{LC} \quad \text{or } \omega_0 = \sqrt{\left(\frac{1}{LC}\right)} \text{ rad/s.} \tag{7.1}$$

At resonance, the circuit current I_0 is V/R A. This is the greatest value the current can achieve, since at angular velocities above and below ω_0 the impedance increases, so reducing the current. For this reason the circuit is sometimes called an 'acceptor' circuit.

At resonance, the voltage developed across the capacitor is $I_0 X_C$ volts.

$$V_C = I_0 \times \frac{1}{\omega_0 C} = \frac{V}{R\omega_0 C}\ \text{V}.$$

The ratio V_C/V is the circuit voltage magnification factor, symbol Q. Hence

$$Q = \frac{\dfrac{V}{R\omega_0 C}}{V} = \frac{1}{R\omega_0 C}. \qquad (7.2)$$

The voltage developed across the inductive part of the coil is $I_0 X_L$ volts.

$$V_L = \frac{V}{R}\omega_0\, L\ \text{V}.$$

Again, $$Q = \frac{V_L}{V} = \frac{\dfrac{V}{R}\omega_0\, L}{V} = \frac{\omega_0\, L}{R}. \qquad (7.3)$$

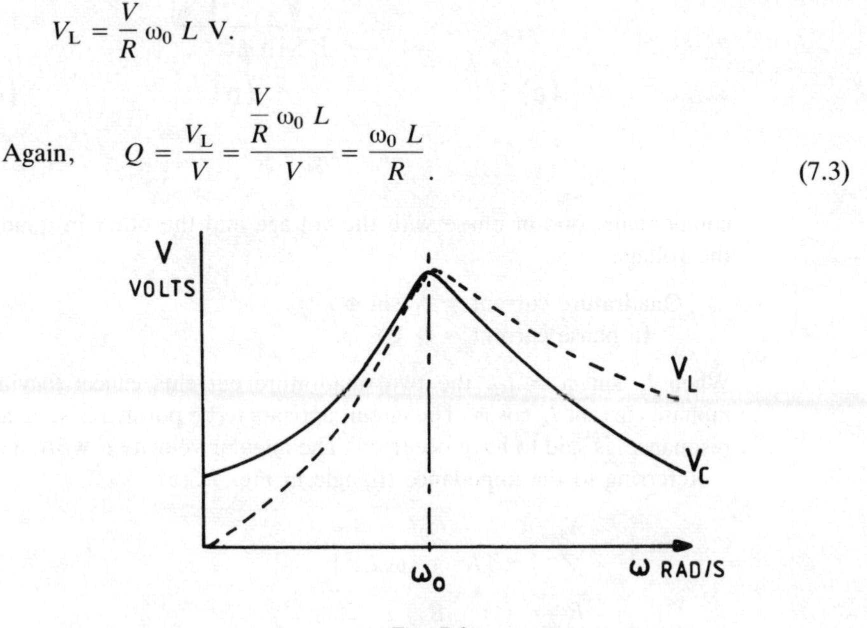

Fig. 7.2

If the resistance of the coil is small, the voltage developed across the coil terminals will not be greatly different from the value of $I_0 X_L$. Figure 7.2 shows the curves of V_C and V_L plotted against angular velocity for the circuit in Fig. 7.1(a). The voltage V_C is at a maximum at ω_0 rad/s, while the maximum value of coil voltage occurs at a slightly higher frequency because of the coil resistance. The two maxima would coincide at ω_0 if the resistance could be eliminated.

7.2 Parallel resonance

The resistive coil is now connected in parallel with a perfect capacitor as shown in Fig. 7.3(a). The current I_C leads the applied voltage by exactly 90°. The current I_L in the coil lags the voltage by an angle ϕ which is determined by the ratio of R to X_L in the impedance triangle for the coil shown in Fig. 7.3(c). The complete phasor diagram (Fig. 7.3(b)) shows I_L being resolved into two

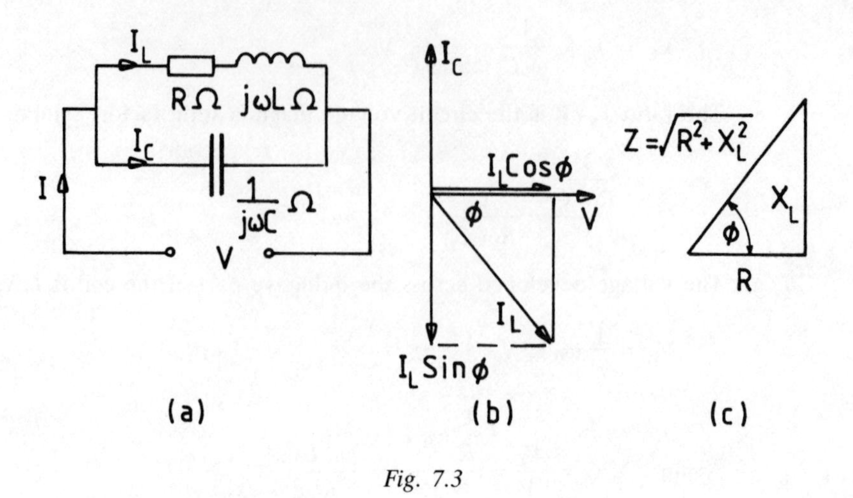

Fig. 7.3

components, one in phase with the voltage and the other in quadrature with the voltage.

Quadrature current = $I_L \sin \phi$
In-phase current = $I_L \cos \phi$.

When $I_L \sin \phi = I_C$, the two quadrature currents cancel leaving only the inphase current $I_L \cos \phi$. The circuit appears to be purely resistive and 'parallel resonance' is said to have occurred. The angular velocity is written as ω_0 rad/s.

Referring to the impedance triangle in Fig. 7.3(c)

$$\sin \phi = \frac{X_L}{Z} = \frac{\omega_0 L}{\sqrt{[R^2 + (\omega_0 L)^2]}} \quad \text{(i)}$$

$$\cos \phi = \frac{R}{Z} = \frac{R}{\sqrt{[R^2 + (\omega_0 L)^2]}} \quad \text{(ii)}$$

$$I_C = \frac{V}{X_C} = \frac{V}{\dfrac{1}{\omega_0 C}} = V\omega_0\, C\ A \quad \text{(iii)}$$

$$I_L = \frac{V}{Z_{coil}} = \frac{V}{\sqrt{[R^2 + (\omega_0\, L)^2]}} \quad \text{(iv)}$$

At resonance, $I_C = I_L \sin \phi$. Substituting for I_C, I_L and $\sin \phi$ from equations (iii), (iv) and (i) above gives

$$V\omega_0\, C = \frac{V}{\sqrt{[R^2 + (\omega_0 L)^2]}} \times \frac{\omega_0 L}{\sqrt{[R^2 + (\omega_0 L)^2]}} = \frac{V\omega_0 L}{R^2 + (\omega_0 L)^2}.$$

Cancelling $V\omega_0$ on both sides of the equation and then transposing gives

$$C = \frac{L}{R^2 + (\omega_0 L)^2}.$$

Transposing

$$R^2 + (\omega_0 L)^2 = \frac{L}{C}. \tag{v}$$

Further transpositions give the result

$$\omega_0 = \sqrt{\left(\frac{1}{LC} - \frac{R^2}{L^2}\right)}. \tag{7.4}$$

Notice that parallel resonance does *not* occur when $\omega L = 1/\omega C$ as in the series case. However, when R is small enough for R^2/L^2 to become insignificant compared with $1/LC$, equation (7.4) reduces to that for series resonance (equation 7.1).

The impedance at resonance is called the 'dynamic impedance' Z_D.

$$Z_D = \frac{V}{\text{circuit current}} = \frac{V}{I_L \cos\phi}.$$

Substituting for I_L and $\cos\phi$ from equations (iv) and (ii) gives

$$Z_D = \frac{V}{V/\sqrt{[R^2 + (\omega_0 L)^2]} \times R/\sqrt{[R^2 + (\omega_0 L)^2]}}$$

$$= \frac{[R^2 + (\omega_0 L)^2]}{R}.$$

From equation (v), $R^2 + (\omega_0 L)^2 = L/C$. Hence

$$Z_D = \frac{L/C}{R} = \frac{L}{CR}\ \Omega. \tag{7.5}$$

At resonance the circuit has its highest possible impedance so that the current I (Fig. 7.3(a)) is a minimum . This is called the 'make-up' current. If the resistance could be made zero then the dynamic impedance would be infinitely great and no make-up current would be required to maintain the branch currents I_C and I_L.

The parallel circuit exhibits current magnification

$$Q = I_C/I_{\text{make-up}}.$$

From equation (iii) $I_C = V\omega_0 C$ and $I_{\text{make-up}} = \dfrac{V}{Z_D}$.

Therefore, $Q = \dfrac{V\omega_0 C}{V/(L/CR)} = \dfrac{V\omega_0 C}{V}\dfrac{L}{CR} = \dfrac{\omega_0 L}{R}$ as for the series circuit.

Also, for the inductive branch $Q = I_L/I_{\text{make-up}}$.

For small values of resistance $I_L = \dfrac{V}{\omega_0 L}$ (closely).

Hence Q is approximately equal to $\dfrac{V/\omega_0 L}{V/(L/CR)} = \dfrac{VL/CR}{V\omega_0 L} = \dfrac{1}{\omega_0 CR}$.

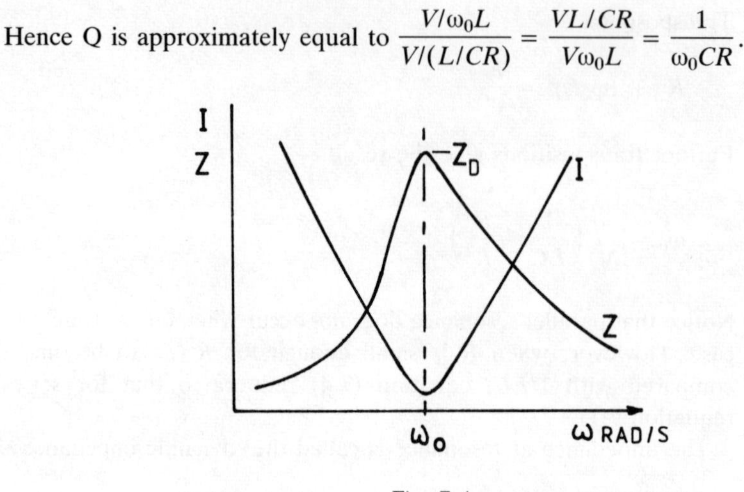

Fig. 7.4

Figure 7.4 shows the variations in impedance and make-up current with angular velocity for the circuit in Fig. 7.3(a).

7.3 *Q*-factor and bandwidth

The bandwidth of a circuit is defined arbitrarily as the range of frequencies within which the circuit power remains in excess of 50% of its peak value. Considering the series circuit (Fig. 7.1(a)), the power developed in the resistance at resonance is given by

$$\text{power} = I_0^2 R \text{ watts, where } I_0 = \frac{V}{R} \text{ amperes.}$$

In order to fall to one half of this value, i.e. to $I_0^2R/2$ watts, the current must fall to I amperes, when $I^2R = I_0^2R/2$. Transposing gives

$$I = \sqrt{\left(\frac{I_0^2R}{2R}\right)} = \frac{I_0}{\sqrt{2}}.$$

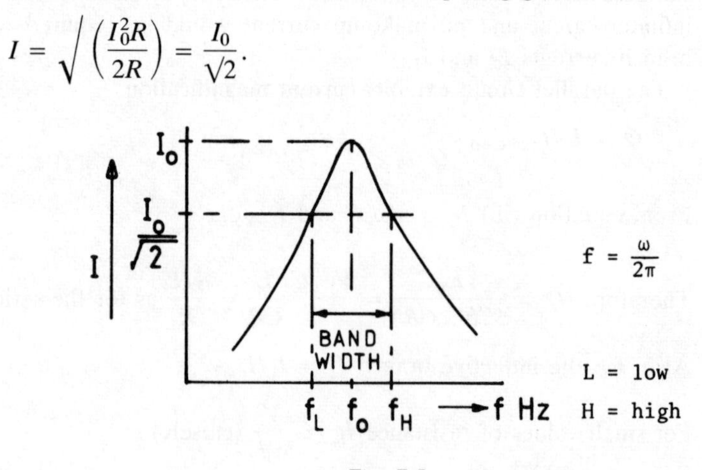

Fig. 7.5

Figure 7.5 shows the two half-power points and the bandwidth.

The ratio of powers is often expressed in decibels (dB) (see section 6.10). In the case envisaged, the power at the half-power points is I^2R watts and the peak power is I_0^2R watts, hence

$$\text{ratio of powers (in dB)} = 10\log_{10}\frac{I^2R}{I_0^2R} = 10\log_{10}\frac{(I_0/\sqrt{2})^2}{I_0^2}$$

$$= 10\log_{10}\left(\frac{1}{\sqrt{2}}\right)^2 = 20\log_{10}\frac{1}{\sqrt{2}}$$

$$= -3\text{ dB}.$$

There is an attenuation of 3 dB between the peak power and the half-power points, which are often referred to as the 3 dB points.

For the current to fall to $I_0/\sqrt{2}$, the impedance must rise to $\sqrt{2}R$. This occurs at two frequencies.

(a) f_H, when $(\omega L - 1/\omega C) = R$. The circuit is predominantly inductive since with the frequency greater than f_0, ωL is greater than $1/\omega C$.
(b) f_L, when $(1/\omega C - \omega L) = $ R. The circuit is predominantly capacitive.

The impedance of the series circuit $= R + \text{j}(\omega L - 1/\omega C)\ \Omega$ (i)

From equation (i), considering Fig. 7.6, it should be clear how the two impedances for half power are realised.

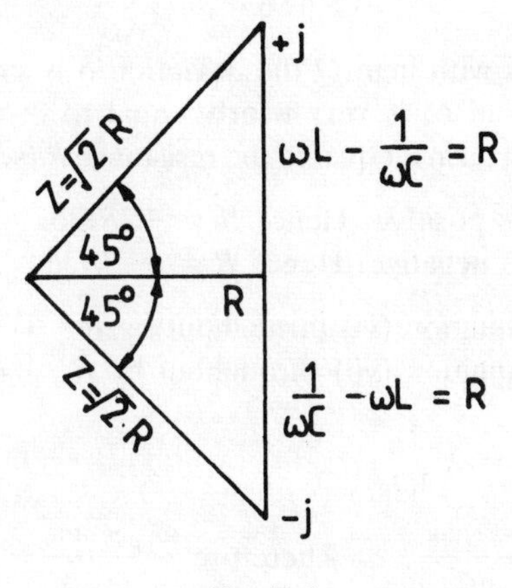

Fig. 7.6

Let the circuit be de-tuned by a fractional amount δ.

$$\delta = \frac{\omega - \omega_0}{\omega_0}.$$

Transposing gives

$$\omega_0\delta = \omega - \omega_0$$

$$\omega = \omega_0(1 + \delta). \qquad \text{(ii)}$$

For δ positive, ω is higher than ω_0, the resonant angular velocity, and vice versa.

Two values of ω corresponding to f_L and f_H in Fig. 7.5 are now going to be determined in terms of the circuit Q-factor.

Substituting for equation (ii) in equation (i) gives

$$Z = R + j\left(\omega_0(1+\delta)L - \frac{1}{\omega_0(1+\delta)C}\right). \qquad \text{(iii)}$$

From equation (7.3), $\omega_0 L/R = Q$. Transposing gives $\omega_0 L = RQ$. (iv)

From equation (7.2), $1/\omega_0 CR = Q$. Transposing gives $1/\omega_0 C = RQ$. (v)

Substituting equations (iv) and (v) in equation (iii) gives

$$Z = R + j\left(RQ(1+\delta) - \frac{RQ}{(1+\delta)}\right).$$

Bringing the terms inside the parentheses to a common denominator $(1 + \delta)$ and the common RQ outside the parentheses gives

$$Z = R + jRQ\left(\frac{(1+\delta)(1+\delta) - 1}{(1+\delta)}\right)$$

$$= R + jRQ\left(\frac{(1 + 2\delta + \delta^2 - 1)}{(1+\delta)}\right)$$

$$= R + jRQ\delta\left(\frac{2+\delta}{1+\delta}\right).$$

For circuits with high Q the deviation δ is small to the half-power points and $(2 + \delta)/(1 + \delta)$ is very nearly equal to 2; then $Z = R + 2jRQ\delta$. For half power, the j term is equal to the resistance R (see Fig. 7.6). Therefore $2RQ\delta = R$.

For f_H, δ is positive. Hence $R = +2RQ\delta$. (vi)
For f_L, δ is negative. Hence $R = -2RQ\delta$. (vii)

Dividing equation (vi) throughout by R $\quad 1 = +2Q\delta$
Dividing equation (vii) throughout by R $\quad 1 = -2Q\delta$, so that

$$\delta = \pm\frac{1}{2Q}.$$

But $\delta = \dfrac{\omega - \omega_0}{\omega_0}$. Therefore $\dfrac{\omega_H - \omega_0}{\omega_0} = +\dfrac{1}{2Q}$ (viii)

and $\dfrac{\omega_L - \omega_0}{\omega_0} = -\dfrac{1}{2Q}$. (ix)

Subtracting equation (ix) from equation (viii)

$$\frac{\omega_H - \omega_0}{\omega_0} - \frac{\omega_L - \omega_0}{\omega_0} = \frac{1}{2Q} - \left(\frac{-1}{2Q}\right)$$

$$\frac{\omega_H - \omega_L}{\omega_0} = \frac{2}{2Q}$$

$$\omega_H - \omega_L = \frac{\omega_0}{Q}.$$

But $(\omega_H - \omega_L)$ is the bandwidth expressed in rad/s, therefore

$$\text{bandwidth} = \frac{\omega_0}{Q}, \text{ or in terms of frequency, } \frac{f_0}{Q} \text{ Hz.} \qquad (7.6)$$

Equation (7.6) is applicable also to the parallel resonant circuit. It may be proved in exactly the same manner commencing with the impedance of the circuit for any frequency.

WORKED EXAMPLE 7.1

A series resonant circuit has the following parameters: frequency at resonance, $5000/2\pi$ Hz; impedance at resonance, 56 Ω; Q-factor, 25.

(a) Assuming that the capacitor is pure, calculate (i) the capacitance of the capacitor, (ii) the inductance of the inductor.
(b) Assuming that these values are independent of frequency, determine the two frequencies at which the circuit impedance has a phase angle of $\pi/4$ radian.

Since $f_0 = \dfrac{5000}{2\pi}$ Hz, $\omega_0 = 2\pi f_0 = 5000$ rad/s.

(a) At resonance, the circuit impedance is that of the resistance alone, $R = 56\ \Omega$.

Now $Q = \dfrac{\omega_0 L}{R}$ (equation 7.3), therefore

$$25 = \frac{5000L}{56}$$

$$L = 0.28 \text{ H.}$$

At resonance, $\omega_0 L = \dfrac{1}{\omega_0 C}$, therefore

$$5000 \times 0.28 = \frac{1}{5000C}$$

$$C = \frac{1}{(5000)^2 \times 0.28} = 0.143\ \mu\text{F.}$$

(b) For a phase angle of $\pi/4$ rad (45°), $(\omega L - 1/\omega C) = R$ (Fig. 7.6)

$$\omega \times 0.28 - \frac{1}{\omega \times 0.143 \times 10^{-6}} = 56.$$

Multiplying throughout by $0.143 \times 10^{-6}\ \omega$ gives

$$4.004 \times 10^{-8}\ \omega^2 - 1 = 8.008 \times 10^{-6}\ \omega.$$

Dividing throughout by 4.004×10^{-8} and transposing

$$\omega^2 - 200\omega = 24\,975\,025.$$

This quadratic equation may be solved by any known method. Completing the square

$$\omega^2 - 200\omega + (-100)^2 = 24\,975\,025 + 100^2$$
$$(\omega - 100)^2 = 24\,985\,025$$
$$(\omega - 100) = \pm 4998.5$$
$$\omega = 4998.5 + 100 = 5098.5 \text{ rad/s}$$
$$\text{or } \omega = -4998.5 + 100 = -4898.5 \text{ rad/s.}$$

The second result is impossible practically since it is negative. However starting from $(1/\omega C - \omega L) = R$ to determine the lower frequency, it will be found that the results are the same with the signs changed. The value 4898.5 rad/s is the required lower angular velocity. The reader might care to verify this result for himself.

The two frequencies for 45° phase shift and half power (3 dB) points are therefore

$$\frac{5098.5}{2\pi} = 811.4 \text{ Hz} \quad \text{and} \quad \frac{4898.5}{2\pi} = 779.6 \text{ Hz.}$$

An alternative approach is to use equation (7.6).

$$\text{Bandwidth} = \frac{f_0}{Q} = \frac{5000/2\pi}{25} = 31.83 \text{ Hz.}$$

Assuming symmetry about f_0, the bandwidth is from $(f_0 - 31.83/2)$ up to $(f_0 + 31.83/2)$

$$f_L = \frac{5000}{2\pi} - 15.9 = 779.8 \text{ Hz;} \qquad f_H = \frac{5000}{2\pi} + 15.9 = 811.7 \text{ Hz.}$$

The slight differences between the answers using the two methods are due to the assumptions made during the proof of equation (7.6).

WORKED EXAMPLE 7.2

An inductor has a resistance of 20 Ω and when connected across a 120 V, 50 Hz supply draws a current of 1 A. For a supply voltage of 85 V, calculate (a) the frequency of a supply to cause parallel resonance with a capacitor of capacitance 30 μF, (b) the dynamic impedance of the parallel circuit, (c) the value of the make-up current, (d) the capacitor current, and (e) the circuit Q-factor.

The impedance of the inductor at 50 Hz $= \dfrac{V}{I} = \dfrac{120}{1} = 120\ \Omega$.

$$\text{Now, } Z = \sqrt{(R^2 + X_L^2)}$$
$$120 = \sqrt{(20^2 + X_L^2)}, \text{ hence, } X_L = 118.32\ \Omega.$$

Therefore, $L = \dfrac{118.32}{2\pi} = 0.3766$ H.

(a) For parallel resonance,

$$f_0 = \frac{1}{2\pi}\sqrt{\left(\frac{1}{LC} - \frac{R^2}{L^2}\right)} \qquad \text{(equation 7.4)}$$

$$f_0 = \frac{1}{2\pi}\sqrt{\left(\frac{1}{0.3766 \times 30 \times 10^{-6}} - \frac{20^2}{0.3766^2}\right)}$$

$$= 46.58 \text{ Hz.}$$

(b) $Z_D = \dfrac{L}{CR}$ (equation 7.5)

$$= \frac{0.3766}{30 \times 10^{-6} \times 20} = 627.7\ \Omega.$$

(c) Supply (make-up) current $= \dfrac{V}{Z_D} = \dfrac{85}{627.7} = 0.135$ A.

(d) $I_C = \dfrac{V}{X_C} = V\omega C$ A

$$= 85 \times 2\pi \times 46.58 \times 30 \times 10^{-6} = 0.746 \text{ A.}$$

(e) $Q\text{-factor} = \dfrac{I_C}{I_{\text{make-up}}} = \dfrac{0.746}{0.135} = 5.5.$

SELF-ASSESSMENT EXAMPLE 7.3

A coil with inductance 2.5 mH and resistance 15 Ω is connected in parallel with a capacitor of capacitance 0.001 πF and fed from a 1.5 V source. Determine (a) the frequency at which the current drawn from the source is a minimum, (b) the rms value of the make-up current, (c) the rms value of the capacitor current, (d) the Q-factor of the circuit, (e) the values of f_L, f_H and the bandwidth.

7.4 Coupled coils

7.4.1 Mutual inductance and coupling coefficient

Two circuits are said to possess mutual inductance if, when a current changes in one of them, the resultant flux change links at least in part with a second circuit so inducing a voltage in it. The two circuits are said to have a mutual inductance of one henry if, when a current changes in one circuit at a rate of one ampere per second, one volt is induced in the second circuit. In symbols

$$e_2 = M\frac{di_1}{dt}$$

the subscripts 1 and 2 denote the first and second circuit respectively, M is the mutual inductance in henry.

For alternating currents of sine form, $i_1 = I_{m(1)} \sin \omega t$, so that

$$e_2 = M\frac{dI_{m(1)} \sin \omega t}{dt} = M\omega I_{m(1)} \cos \omega t \text{ V.} \quad \text{(i)}$$

The maximum value of e_2 is $E_{m(2)}$ and this occurs when $\cos \omega t = 1$.

The voltage wave (equation i) is a cosine function whereas the current was a sine function. The voltage wave is therefore 90° ahead of the current wave and this phase shift is indicated by using +j.

Hence

$$E_{m(2)} = jM\omega I_{m(1)}.$$

Dividing both sides of the equation by $\sqrt{2}$ reduces both $E_{m(2)}$ and $I_{m(1)}$ to rms values when

$$E_2 = j\omega M I_1. \quad (7.7)$$

This is similar to the familiar result associated with self-inductance, namely

$$E_1 = j\omega L_1 I_1 \quad \text{or} \quad E_1/I_1 = j\omega L_1 \ \Omega. \quad (7.8)$$

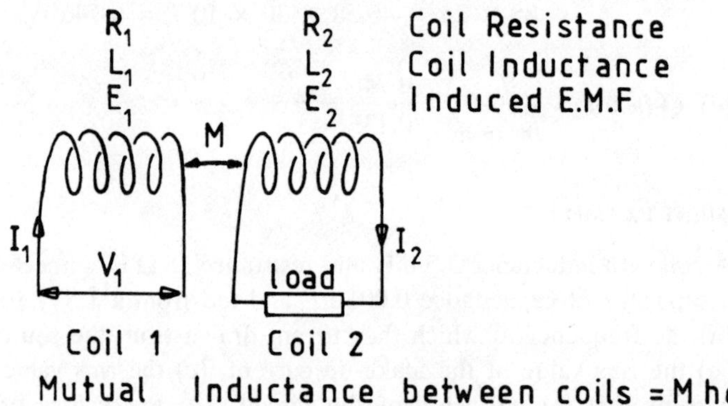

Fig. 7.7

Equations (7.7) and (7.8) are illustrated in Fig. 7.7. Assuming that all the flux created by the current I_1 in coil 1 links with coil 2, i.e. assuming perfect or tight coupling, the mutual inductance has the highest value possible, M_{max}

$$M_{max} = \sqrt{(L_1L_2)}. \quad (7.9)$$

If there is leakage flux so that only part of the flux from coil 1 links with coil 2, then

$$M = k\sqrt{M_{max}} = k\sqrt{(L_1L_2)} \quad (7.10)$$

where k is the coupling coefficient.

For a derivation of equations (7.9) and (7.10), see appendix 3.

7.4.2 *The effect of secondary loading on the input impedance of the primary*

Figure 7.8 shows the circuit diagram of two mutually coupled coils, including the resistances of both the primary and secondary coils, R_1 and R_2 respectively. The secondary is connected to a load with impedance Z_L.

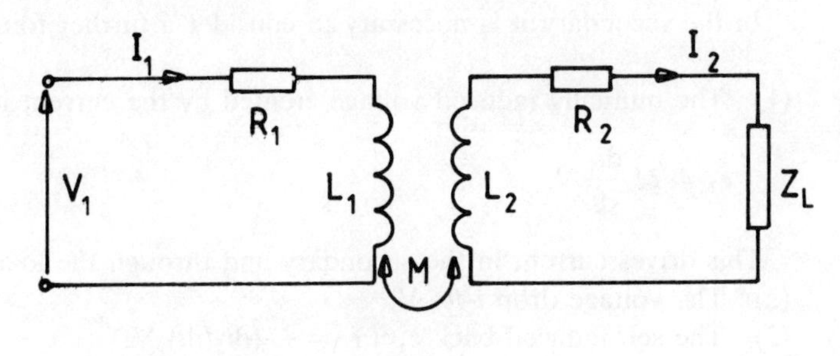

Fig. 7.8

Suppose the applied voltage V_1 is of sinusoidal form. Then, assuming linearity between current and flux and using instantaneous values, the equation for the primary may be drawn up.

There are four voltages acting in the primary

(1) $v_1 = V_{m(1)} \sin \omega t$ V. This will drive a current $i_1 = I_{m(1)} \sin (\omega t - \phi)$ A.
(2) The resistive volt drop, $i_1 R_1$ V.
(3) The back e.m.f. e induced by the self-inductance L_1

$$e_1 = L_1 \frac{di_1}{dt} \text{ V.}$$

By Lenz's law this will oppose V_1.

(4) The mutually induced e.m.f. $e_{1(M)}$ in the primary set up by current i_2 in the secondary

$$e_{1(M)} = M \frac{di_2}{dt}.$$

It is necessary to decide whether the mutually induced voltage helps or opposes V_1. With no current in the secondary, i.e. the load disconnected, there will be no primary mutually induced voltage and I_1 will have a certain value which depends on the primary parameters. Connecting the load and considering the effect of increasing the magnitude of the secondary current, there will be power developed in the secondary so that the power input to the primary must increase. This will occur only if the voltage $e_{1(M)}$ is helping v_1. If $e_{1(M)}$ opposed v_1, for increasing secondary current, the input current would fall, which is clearly a nonsense.

The equation for the primary in instantaneous values becomes

$$v_1 + M\frac{di_2}{dt} - i_1R_1 - L_1\frac{di_1}{dt} = 0.$$

Transposing and substituting the steady-state values for sinusoidal input from equations (7.7) and (7.8)

$$V_1 = I_1R_1 + j\omega L_1I_1 - j\omega MI_2$$

$$V_1 = I_1\,(R_1 + j\omega L_1) - j\omega MI_2. \qquad (7.11)$$

In the secondary it is necessary to consider a further four voltages.

(1) The mutually induced voltage created by the current in the primary

$$e_2 = M\frac{di_1}{dt}\text{ V.}$$

This drives current in the secondary and through the load.
(2) The voltage drop i_2R_2 V.
(3) The self-induced back e.m.f. $= L_2(di_2/dt)$ V.
(4) The voltage drop across the load, i_2Z_L V.

In instantaneous values the secondary equation becomes

$$M\frac{di_1}{dt} - i_2R_2 - L_2\frac{di_2}{dt} - i_2Z_L = 0.$$

Transposing and substituting the steady-state values for sinusoidal input from equations (7.7) and (7.8)

$$j\omega MI_1 = I_2R_2 + j\omega L_2I_2 + I_2Z_L$$

$$j\omega MI_1 = I_2(R_2 + j\omega L_2 + Z_L). \qquad (7.12)$$

Note that in equation (7.12), Z_L will have to be expressed in rectangular form and added correctly to R_2 and $j\omega L_2$.

Consider the case where the load is inductive $Z_L = R_L + j\omega L_L$.

Total secondary impedance $= (R_2 + R_L) + j\omega(L_2 + L_L)$.

Let $(R_2 + R_L) = R_s$ and $(L_2 + L_L) = L_s$. Equation (7.12) becomes

$$j\omega MI_1 = I_2(R_s + j\omega L_s). \qquad (7.13)$$

This yields a pair of simultaneous equations (7.11 and 7.13). A solution by eliminating the I_2 term involves multiplying equation (7.11) by $(R_s + j\omega L_s)$ to form equation (7.14), and equation (7.13) by $j\omega M$ to form equation (7.15).

$$V_1(R_s + j\omega L_s) = I_1(R_1 + j\omega L_1)\,(R_s + j\omega L_s) - j\omega MI_2(R_s + j\omega L_s) \qquad (7.14)$$

$$j\omega MI_1 \times j\omega M = j\omega MI_2(R_s + j\omega L_s)$$

$$-\omega^2M^2I_1 = j\omega MI_2(R_s + j\omega L_s). \qquad (7.15)$$

Adding equations (7.14) and (7.15) gives

$$V_1(R_s + j\omega L_s) - \omega^2 M^2 I_1 = I_1(R_1 + j\omega L_1)(R_s + j\omega L_s)$$

dividing throughout by $(R_s + j\omega L_s)$

$$V_1 - \frac{\omega^2 M^2 I_1}{R_s + j\omega L_s} = I_1(R_1 + j\omega L_1)$$

$$V_1 = I_1\left[(R_1 + j\omega L_1) + \frac{\omega^2 M^2}{R_s + j\omega L_s}\right]$$

$$\frac{V_1}{I_1} = (R_1 + j\omega L_1) + \frac{\omega^2 M^2}{R_s + j\omega L_s} \quad (7.16)$$

$V_1/I_1 = Z_{in}$, the input impedance of the primary coil.

Without the effect of mutual inductive coupling to the secondary the input impedance would be simply $(R_1 + j\omega L_1)$.

Hence, bringing a second coil close to the first coil, such as to establish mutual inductance M, has the effect of modifying the input impedance by a term

$$\frac{\omega^2 M^2}{R_s + j\omega L_s}.$$

Rationalising this term gives

$$\frac{\omega^2 M^2(R_s - j\omega L_s)}{R_s^2 + (\omega L_s)^2}.$$

Since the secondary inductive reactance is now multiplied by −j it appears to be capacitive at the input terminals. Thus the reactance at the input becomes the difference between

$$j\omega L_1 \quad \text{and} \quad \frac{j\omega L_s\, \omega^2 M^2}{R_s^2 + (\omega L_s)^2}$$

which results in a lowering of the input impedance from that in the uncoupled state (see worked example 7.4).

WORKED EXAMPLE 7.4

Determine the value of input current to the primary coil in the circuit shown in Fig. 7.9.

From equation (7.10) $M = k\sqrt{(L_1 L_2)}$
$M = 0.2\sqrt{(50 \times 10^{-3} \times 50 \times 10^{-3})} = 10$ mH.

For each coil, at 1 kHz $X_L = 2\pi \times 1000 \times 50 \times 10^{-3} = 314.16\ \Omega$.

For the load, $X_{L(load)} = 2\pi \times 1000 \times 10 \times 10^{-3} = 62.83\ \Omega$.

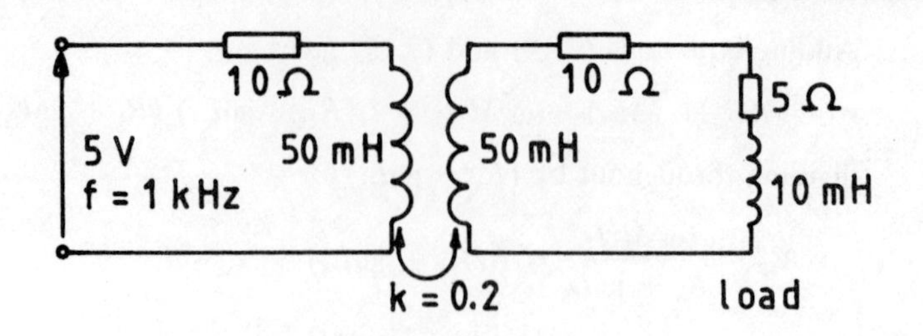

Fig. 7.9

$$\text{Total secondary impedance} = (10 + 5) + j(314.16 + 62.83) = (15 + j377)\ \Omega.$$

$$\text{Primary input impedance} = R_1 + j\omega L_1 + \frac{\omega^2 M^2}{R_s + j\omega L_s} \quad \text{(equation 7.16)}$$

$$= 10 + j314.16 + \frac{(2\pi \times 1000)^2\ (10 \times 10^{-3})^2}{(15 + j377)}$$

$$= 10 + j314.16 + \frac{3947.8(15 - j377)}{15^2 + 377^2}$$

$$= 10 + j314.16 + \frac{59\,217.6 - j1\,488\,336}{142\,354}$$

$$= 10.416 + j303.7\ \Omega.$$

Observe that this value is less than $(R_1 + j\omega L_1)$.

$$|I_1| = \frac{5}{10.416 + j303.7} = \frac{5}{303.88} = 16.45 \text{ mA}.$$

Without the presence of the secondary, I_1 would have been

$$\frac{5}{10 + j314.16} = 15.9 \text{ mA}.$$

The effect of the secondary is small since the degree of coupling is small ($k = 0.2$).

Generally speaking when a capacitor is connected in the secondary it will be employed to cause the secondary to be purely resistive, i.e. to tune it to resonance. Rarely, if ever, will sufficient capacitance be used so as to cause the secondary impedance to become capacitive overall. The solution of problems will therefore be as outlined in worked example 7.4 or with Z_s purely resistive.

WORKED EXAMPLE 7.5

Two similar coils have resistance 20 Ω and inductance 25 mH. They are mutually coupled and $k = 0.6$.

The input voltage to the first coil is 20 V at a frequency of 1 kHz. A load of $50\angle -36.87°$ Ω is connected to the second coil. Determine (a) the power input to the first coil, (b) the power developed by the load.

From equation (7.10) $M = 0.6\sqrt{[(25 \times 10^{-3})(25 \times 10^{-3})]} = 15$ mH.

For both coils $X_L = 2\pi \times 1000 \times 25 \times 10^{-3} = 157\ \Omega$.

The load $= 50\angle -36.87° = (40 - j30)\ \Omega$.

Using equation (7.16)

$$Z_{in} = 20 + j157 + \frac{(2\pi \times 1000)^2 \times (15 \times 10^{-3})^2}{(20 + j157) + (40 - j30)}$$

$$= 20 + j157 + \frac{8882.6\ (60 - j127)}{60^2 + 127^2} \qquad \text{(see part (b) (ii) below)}$$

$$= 47 + j100.2\ \Omega.$$

$$I_1 = \frac{V}{Z_{in}} = \frac{20}{47 + j100.2} = \frac{20}{110.67\angle 64.87°} = 0.181\angle -64.87°\ \text{A}.$$

(a) Power input $= VI \cos\phi = 20 \times 0.181 \times \cos 64.87°$
$= 1.54$ W.

or $|I|^2 \times$ total effective resistance in the primary $= 0.181 \times 47$
$= 1.54$ W.

(b) To calculate the power developed by the load there are two ways to progress.

(i) Determine the secondary induced voltage and hence the secondary current and power.

Using equation (7.7) $E_s = j\omega M I_1$
$= j2\pi \times 1000 \times 15 \times 10^{-3} \times 0.181\angle -64.87°$
$= 17.06\angle 25.13°$ V (j $= +90°$).

The angle is of no significance in this question and will be disregarded.

$$I_2 = \frac{E_s}{Z_s} = \frac{17.06}{(20 + j157) + (40 - j30)}$$

$$|I_2| = \frac{17.06}{140.46} = 0.12\ \text{A}.$$

Power in the load $= |I_2|^2 \times R_{load} = 0.12^2 \times 40 = 0.59$ W.

(ii) The resistance referred into the primary because of the total secondary resistance of 60 Ω is expressed as

$$\frac{8882.6 \times 60}{60^2 + 127^2} = 27\ \Omega. \qquad \text{(see above)}$$

The power in the primary due to this referred resistance is

$$|I_1|^2 \times 27 = 0.18^2 \times 27 = 0.8845\ \text{W}.$$

The secondary resistance consists of 20 Ω in the coil and 40 Ω in the load.

The load resistance = 40/60 = 2/3 of the total.

The power developed in the primary due to the load resistance is therefore two-thirds of that due to the total secondary resistance, i.e. two-thirds of 0.8845 = 0.59 W.

SELF-ASSESSMENT EXAMPLE 7.6

Two coils each with resistance 20 Ω and inductance 25 mH are mutually coupled so that the coupling coefficient is 0.8. The secondary is connected to a load of (40 − j30) Ω. A capacitor with capacitance 2.87 μF is connected in series with the primary coil. For a supply voltage of 20 V at 1 kHz, determine the power developed in the load resistance.

7.4.3 *Critical coupling*

From the work on maximum power transfer (chapter 2) it is known that maximum power is transferred from a resistive source of e.m.f. into a resistive load when the load resistance is equal to the source resistance.

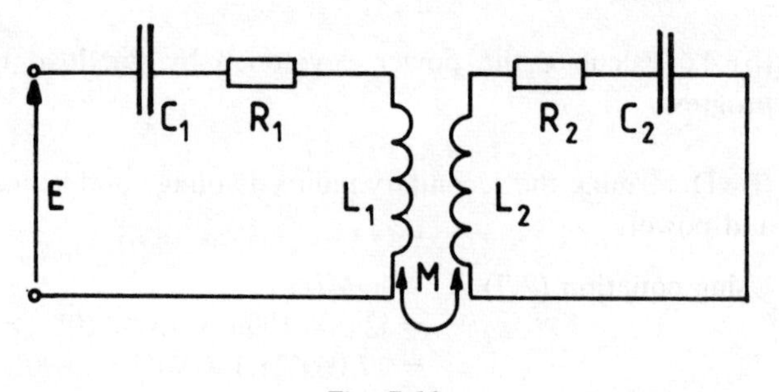

Fig. 7.10

By individually tuning both primary and secondary using capacitors C_1 and C_2 as shown in Fig. 7.10, equation (7.16) for the input impedance reduces to

$$Z_{in} = R_1 + \frac{\omega_0^2 M^2}{R_2}.$$

The primary coil and capacitor may now be considered as a generator with an e.m.f. of E volts and an internal resistance of R_1 ohms. Maximum power will be transferred into the secondary when the secondary resistance referred into the primary has the same value as R_1. Hence for maximum power transfer

$$\frac{\omega_0^2 M^2}{R_2} = R_1. \qquad \text{(i)}$$

The particular value of mutual inductance which will cause this to occur is known as the critical value M_{CR}. Transposing (i) gives

$$\omega_0^2 M_{CR}^2 = R_1 R_2 \qquad (7.17)$$

but $M = k\sqrt{(L_1L_2)}$, (equation 7.10)

or $M_{CR} = k_{CR}\sqrt{(L_1L_2)}$

where k_{CR} is the corresponding value of coupling coefficient to give maximum power transfer.

$$M_{CR}^2 = k_{CR}^2 L_1L_2.$$

Substituting for M_{CR}^2 in equation (7.17)

$$\omega_0^2 k_{CR}^2 L_1L_2 = R_1R_2.$$

Transposing gives

$$k_{CR}^2 = \frac{R_1R_2}{\omega_0L_1 \times \omega_0L_2}.$$

Now from equation (7.3)

$$Q = \frac{\omega_0L}{R}, \qquad \text{so that } \frac{R}{\omega_0L} = \frac{1}{Q}.$$

Hence $\dfrac{R_1}{\omega_0L_1} = \dfrac{1}{Q_1}$ and $\dfrac{R_2}{\omega_0L_2} = \dfrac{1}{Q_2}$. Therefore

$$k_{CR}^2 = \frac{1}{Q_1} \times \frac{1}{Q_2}$$

$$k_{CR} = \sqrt{\left(\frac{1}{Q_1\,Q_2}\right)}. \tag{7.18}$$

WORKED EXAMPLE 7.7

Two identical coils each with resistance 10 Ω and self-inductance 75 mH are individually tuned to resonance using capacitors connected as shown in Fig. 7.11. The first coil is supplied with 5 V at a frequency of 2.5 kHz.

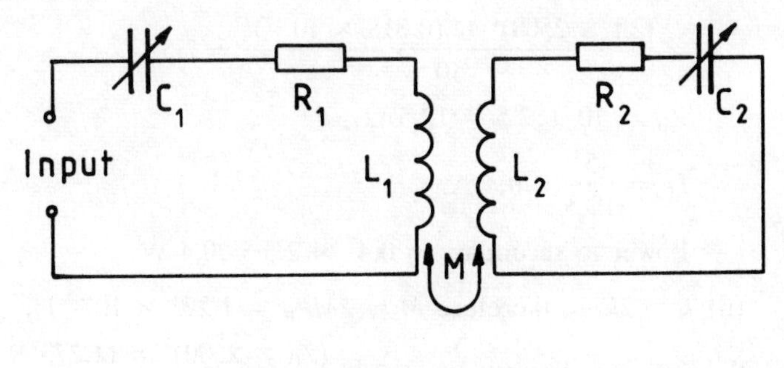

Fig. 7.11

(a) Determine (i) the required mutual inductance between the coils to achieve maximum power transfer into the secondary, (ii) the value of k_{CR} for (i) above, and (iii) the value of the power transferred into the secondary.

(b) Determine the power transferred into the secondary for (i) $k = \frac{1}{2}k_{CR}$, (ii) $k = 2k_{CR}$.

(a) (i) From equation (7.17), for critical coupling $R_1 = \dfrac{\omega_0^2 M_{CR}^2}{R_2}$.

Transposing gives

$$M_{CR} = \sqrt{\left(\frac{R_1 R_2}{\omega_0^2}\right)},$$

or since the coils are identical

$$M_{CR} = \frac{R_1}{\omega_0} = \frac{10}{2\pi \times 2500} = 0.637 \text{ mH}.$$

(ii) From equation (7.10) $M_{CR} = k_{CR} \surd(L_1 L_2)$, hence

$$k_{CR} = \frac{M_{CR}}{\surd(L_1 L_2)}$$

$$k_{CR} = \frac{0.637 \times 10^{-3}}{75 \times 10^{-3}} = 8.49 \times 10^{-3}.$$

(iii) Since there is critical coupling, then by definition R_2 is equal to R_1. Therefore

total resistance referred to the primary $= 10 + 10 = 20\ \Omega$.

$$I_1 = \frac{V}{Z_{in}} = \frac{5}{20} = 0.25 \text{ A}.$$

The secondary power is equal to the power input to the primary due to the referred secondary resistance (see worked example 7.5(b) (ii)).

Secondary power $= 0.25^2 \times 10 = 0.625$ W.

(b) (i) $k = \frac{1}{2}k_{CR}$, therefore $M = \frac{1}{2}M_{CR} = 0.318 \times 10^{-3}$ H.
Secondary resistance referred into the primary $= \omega^2 M^2 / R_2$

$$= \frac{(2\pi \times 2500)^2 \times (0.318 \times 10^{-3})^2}{10} = 2.5\ \Omega.$$

$$Z_{in} = 10 + 2.5 = 12.5\ \Omega.$$

$$I_1 = \frac{5}{12.5} = 0.4 \text{ A}.$$

Power to secondary $= 0.4^2 \times 2.5 = 0.4$ W.

(ii) $k = 2k_{CR}$, therefore $M = 2M_{CR} = 1.273 \times 10^{-3}$ H.

Secondary resistance referred $= \dfrac{(2\pi \times 2500)^2 \times (1.273 \times 10^{-3})^2}{10}$

$= 40\ \Omega$.

$$I_1 = \frac{5}{10 + 40} = 0.1 \text{ A}.$$

Secondary power $= 0.1^2 \times 40 = 0.4$ W.

This example demonstrates the fact that the power transferred to the secondary falls from a relatively high value at k_{CR} to lower values as k is both increased and decreased. It does not conclusively prove the presence of a maximum value of k_{CR}. This is verified by differentiating the expression for secondary current I_2, with respect to ωM.

$$I_2 = \frac{j\omega M I_1}{R_2} \quad \text{and} \quad I_1 = \frac{E}{R_1 + (\omega^2 M^2/R_2)}.$$

7.4.4 *Coupled tuned circuits and bandwidth*

In the series or parallel tuned circuit the bandwidth is quite narrow. The circuits are highly selective, that is to say they have their maximum voltage or current magnification at a particular frequency, the response falling by 3 dB quite rapidly as the frequency deviates either way from the resonant value.

Where, for example, speech, music or television signals are to be transmitted, much wider bandwidths are required.

The ear responds to frequencies between about 40 Hz and 18 kHz. All these frequencies are contained in music either as fundamental or harmonic frequencies. To be able to listen to music in anything like the original clarity from a radio receiver or gramophone record player the equipment must be able to deliver this range of frequencies. From 40 Hz to 18 000 Hz is a bandwidth of 18 000 − 40 = 17 960 Hz. The medium wave radio transmissions have an 8.5 kHz bandwidth whereas VHF transmissions have a bandwidth in excess of 15 kHz. When listening to both of these carrying the same programme, switching from one to the other will show the difference quite clearly.

For television receivers the bandwidth is in the region of 8 MHz since the signal not only provides the capability for high-quality sound but must be capable of adjusting the electron beam intensity with great rapidity as it scans the vision tube forming light and dark squares upon its surface.

At the other end of the bandwidth spectrum there is the telephone system operating with a bandwidth of 3 kHz (400 to 3400 Hz) which is adequate for speech but sometimes leads to recognition problems.

The wide bandwidth required in radio, etc. is achieved by using mutual inductance coupling between two tuned circuits such as has already been considered in this chapter, but with the degree of coupling considered from a bandwidth point of view and not from that for maximum power transfer.

In Fig. 7.11 both primary and secondary are tuned to a common resonant angular velocity ω_0. This case has already been considered under 'critical coupling' when a single frequency was supplied. Now consider the connection of the input to a source which provides a range of frequencies, as would be obtained from a radio aerial for example.

The aerial carries signals of all frequencies from transmitters within its range.

For angular velocities above ω_0, ωL_1 is greater than $1/\omega C_1$, and ωL_2 is greater than $1/\omega C_2$. Both primary and secondary are therefore predominantly inductive. Using equation (7.16)

$$Z_{in} = R_1 + j(\omega L_1 - 1/\omega C_1) + \frac{\omega^2 M^2}{R_2 + j(\omega L_2 - 1/\omega C_2)}$$

$$= R_1 + j(\omega L_1 - 1/\omega C_1) + \frac{\omega^2 M^2 (R_2 - j(\omega L_2 - 1/\omega C_2))}{R_2^2 + (\omega L_2 - 1/\omega C_2)^2} \quad \text{(i)}$$

Thus the excess inductive reactance in the secondary appears to be capacitive ($-j$) to the primary. There will be one particular value of ω for which this apparent capacitance cancels the inductance of the primary, i.e. when

$$+j(\omega L_1 - 1/\omega C_1) + \left[\frac{-j\omega^2 M^2(\omega L_2 - 1/\omega C_2)}{R_2^2 + (\omega L_2 - 1/\omega C_2)}\right] = 0 \quad \text{(ii)}$$

At this angular velocity the circuit as a whole appears to be resistive and there is resonance at an angular velocity other than ω_0.

At angular velocities less then ω_0, $1/\omega C_1$ is greater than ωL_1 and $1/\omega C_2$ is greater than ωL_2. Both primary and secondary circuits are predominantly capacitive. However, capacitance in the secondary appears to be inductive to the primary. There will be one particular value of ω for which this apparent inductance cancels the excess capacitance of the primary, resulting in a further resonant frequency. The coupled circuit as a whole therefore has three resonant angular velocities, ω_0, ω_H and ω_L.

As the coil resistances are small they can be neglected and equation (ii) reduces to

$$j(\omega L_1 - 1/\omega C_1) + \frac{-j\omega^2 M^2(\omega L_2 - 1/\omega C_2)}{(\omega L_2 - 1/\omega C_2)^2} = 0$$

Dropping the j in each term and transposing gives

$$\omega L_1 - \frac{1}{\omega C_1} = \frac{\omega^2 M^2}{(\omega L_2 - 1/\omega C_2)}.$$

Cross multiplying $(\omega L_1 - 1/\omega C_1)(\omega L_2 - 1/\omega C_2) = \omega^2 M^2$

$$\omega^2 L_1 L_2 - \frac{L_1}{C_2} - \frac{L_2}{C_1} + \frac{1}{\omega^2 C_1 C_2} = \omega^2 M^2 .$$

Multiplying by C_1C_2/ω^2 throughout

$$\frac{\omega^2}{\omega^2} L_1 L_2 C_1 C_2 - \frac{C_1 C_2 L_1}{\omega^2 C_2} - \frac{C_1 C_2 L_2}{\omega^2 C_1} + \frac{C_1 C_2}{\omega^2 \omega^2 C_1 C_2} = \frac{\omega^2 M^2 C_1 C_2}{\omega^2}$$

$$L_1 L_2 C_1 C_2 - \frac{C_1 L_1}{\omega^2} - \frac{C_2 L_2}{\omega^2} + \frac{1}{\omega^4} = M^2 C_1 C_2. \quad \text{(iii)}$$

Now $M = k\sqrt{(L_1 L_2)}$ (equation 7.11), therefore $M^2 = k^2 L_1 L_2$ and $\omega_0^2 = 1/LC$ for each circuit (equation 7.1). Therefore

$$\frac{1}{\omega_0^2} = L_1C_1 = L_2C_2. \tag{iv}$$

Substituting equation (iv) in equation (iii) and also for M^2 gives

$$\frac{1}{\omega_0^2}\frac{1}{\omega_0^2} - \frac{1}{\omega_0^2\omega^2} - \frac{1}{\omega_0^2\omega^2} + \frac{1}{\omega^4} = k^2L_1L_2C_1C_2 = \frac{k^2}{\omega_0^4}$$

$$\frac{1}{\omega_0^4} - \frac{2}{\omega_0^2\omega^2} + \frac{1}{\omega^4} = \frac{k^2}{\omega_0^4}$$

$$\frac{1}{\omega^4} - \frac{2}{\omega_0^2\omega^2} = \frac{k^2-1}{\omega_0^4}. \tag{v}$$

Let $\frac{1}{\omega^2} = x$, then equation (v) becomes

$$x^2 - \frac{2x}{\omega_0^2} = \frac{k^2-1}{\omega_0^4}.$$

This is a quadratic equation which may be solved using any known method. Completing the square

$$x^2 - \frac{2x}{\omega_0^2} + \left(-\frac{1}{\omega_0^2}\right)^2 = \frac{k^2-1}{\omega_0^4} + \left(\frac{1}{\omega_0^2}\right)^2$$

$$\left(x - \frac{1}{\omega_0^2}\right)^2 = \frac{k^2}{\omega_0^4}$$

$$\left(x - \frac{1}{\omega_0^2}\right) = \pm\frac{k}{\omega_0^2}$$

$$x = \frac{k}{\omega_0^2} + \frac{1}{\omega_0^2} = \frac{(k+1)}{\omega_0^2}$$

$$\text{or } x = -\frac{k}{\omega_0^2} + \frac{1}{\omega_0^2} = \frac{(1-k)}{\omega_0^2}.$$

But $x = \frac{1}{\omega^2}$, therefore

$$\frac{1}{\omega^2} = \frac{(k+1)}{\omega_0^2} \quad \text{or } \omega = \frac{\omega_0}{\sqrt{(1+k)}} \tag{7.19(a)}$$

$$\text{or} \quad \frac{1}{\omega^2} = \frac{(1-k)}{\omega_0^2} \quad \omega = \frac{\omega_0}{\sqrt{(1-k)}} \tag{7.19(b)}$$

Although these two solutions have been obtained from equation (ii) which supposed a higher angular velocity than ω_0, they are both valid. Equation (7.19(b)) gives ω_H, the denominator being less than 1, while equation (7.19(a)) gives ω_L, the denominator being greater than 1. Reversing the positions of

$(1/\omega C_1$ and $\omega L_1)$ and of $(1/\omega C_2$ and $\omega L_2)$ to establish the lower angular velocity for resonance yields the same roots.

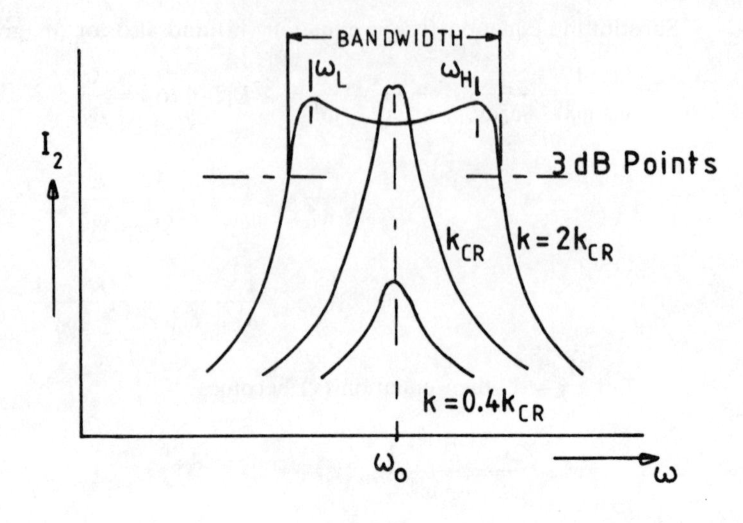

Fig. 7.12

For values of k less than k_{CR}, the secondary current is small and the peak response is small. For values of k from k_{CR} and upwards, the secondary current develops the 'double hump' form shown in Fig. 7.12 which is drawn for two identical coils each with Q in the region of 50. For $k = 2k_{CR}$, the extra bandwidth over the simple response curve (Fig. 7.5) is clear. ω_L and ω_H and the bandwidth in Fig. 7.12 are for this value of coupling coefficient only. Ideally the response curve should be rectangular giving a level response between the two peaks. This is never realised in practice, but by careful adjustment of the two resistances and by fine adjustment of the coupling coefficient it may be approached. From worked example 7.7 it can be seen that using $k = 2k_{CR}$ the power transferred to the secondary is less than that for k_{CR}, but for this application power is sacrificed for bandwidth.

WORKED EXAMPLE 7.8

Two air-cored coils are coupled magnetically, one being connected to an alternating supply and the other short circuited. Calculate (a) the primary input impedance in the coupled condition given that $R_1 = 5\ \Omega$, $R_2 = 2.5\ \Omega$, $L_1 = L_2 = 10$ mH and M = 2 mH. Both primary and secondary are tuned using series connected capacitors to a frequency of $10\,000/2\pi$ Hz. (b) The additional resonant frequencies created by the coupling. (c) The ratio k/k_{CR} for this arrangement of coils.

(a) Since both primary and secondary are tuned to resonance, only the resistances need to be considered. From equation (7.16)

$$Z_{in} = R_1 + \frac{\omega_0^2 M^2}{R_2}$$

$$= 5 + \frac{(10\,000)^2 \times (2 \times 10^{-3})^2}{2.5}$$

$$= 165\ \Omega.$$

(b) Equations (7.19) $\omega_H = \dfrac{\omega_0}{\sqrt{(1-k)}}$; $\quad \omega_L = \dfrac{\omega_0}{\sqrt{(1+k)}}$.

Equation (7.10) $M = k\sqrt{(L_1L_2)}$, transposing gives $k = \dfrac{M}{\sqrt{(L_1L_2)}}$

or for identical inductances $k = \dfrac{M}{L} = \dfrac{2\text{ mH}}{10\text{ mH}} = 0.2.$

Therefore $\omega_H = \dfrac{10\,000}{\sqrt{(1-0.2)}} = 11\,180$ rad/s; $\quad f_H = \dfrac{11\,180}{2\pi}$ Hz.

$\omega_L = \dfrac{10\,000}{\sqrt{(1+0.2)}} = 9129$ rad/s; $\quad f_L = \dfrac{9129}{2\pi}$ Hz.

(c) From equation (7.17), for k_{CR}, $R_1 = \dfrac{\omega_0^2 M^2}{R_2}$. Transposing

$$M_{CR} = \sqrt{\frac{R_1R_2}{\omega_0^2}} = \sqrt{\frac{5 \times 2.5}{10\,000^2}} = 0.35 \times 10^{-3}\ \text{H}.$$

$$k_{CR} = \frac{M_{CR}}{L} = \frac{0.35 \times 10^{-3}}{10 \times 10^{-3}} = 0.035.$$

Therefore $k/k_{CR} = 0.2/0.035 = 5.66.$

SELF-ASSESSMENT EXAMPLE 7.9

Two coils are coupled by mutual inductance. Both primary and secondary are tuned to resonance. Given that $L_1 = L_2 = 238\ \mu$H, $C_1 = C_2 = 106$ pF and $k = 0.02$, (a) determine the three frequencies, f_0, f_H and f_L at which resonance occurs. (b) What value of k would double the range of frequencies between f_H and f_L?

7.5 The power transformer

7.5.1 The ideal transformer

The power transformer consists of two coils wound on a laminated iron core in the simplest single-phase type. The core is designed to have the smallest possible leakage flux and therefore to approach the perfect case of mutual coupling when $k = 1$ and $M = \sqrt{(L_1 L_2)}$ (equation 7.9).

Consider a perfectly coupled pair of coils as shown in Fig. 7.13. The primary coil has N_p turns and has a self-inductance of L_p henry. The secondary has N_s turns and a self-inductance L_s henry. The resistances of both windings are ignored for the moment as they are small compared with ωL_p and ωL_s. The secondary is supplying current to a load with impedance Z_2 ohms.

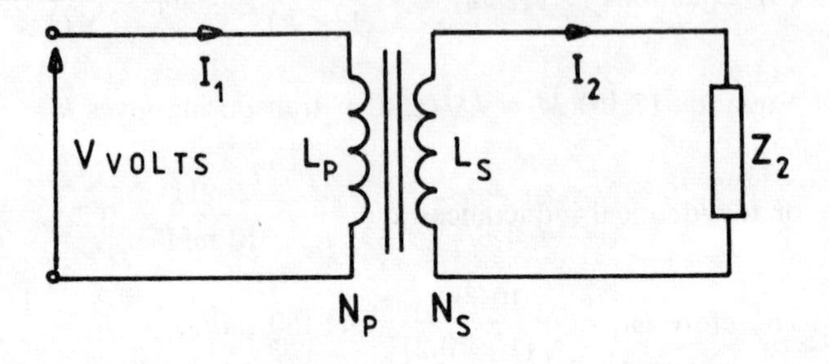

Fig. 7.13

$$I_1 = \frac{V}{Z_{in}} = \frac{V}{j\omega L_p + \dfrac{\omega^2 M^2}{j\omega L_s + Z_2}} \qquad \text{(using equation 7.16 for } Z_{in}\text{).}$$

Multiply throughout by $(j\omega L_s + Z_2)$

$$I_1 = \frac{V(j\omega L_s + Z_2)}{j\omega L_p(j\omega L_s + Z_2) + \omega^2 M^2}$$

$$= \frac{V(j\omega L_s + Z_2)}{-\omega^2 L_p L_s + j\omega L_p Z_2 + \omega^2 M^2}. \qquad \text{(i)}$$

This is a general equation, no assumptions at all having been made with respect to coupling, etc. As already stated, with a power transformer tight coupling is obtained, so that $M = \sqrt{(L_p L_s)}$ (very closely), hence

$$M^2 = L_p L_s \quad \text{and} \quad \omega^2 M^2 = \omega^2 L_p L_s.$$

Hence in the denominator of equation (i) $-\omega^2 L_p L_s + \omega^2 M^2 = 0$ when

$$I_1 = \frac{V(j\omega L_s + Z_2)}{j\omega L_p Z_2}$$

$$= V\left(\frac{j\omega L_s}{j\omega L_p Z_2}\right) + V\left(\frac{Z_2}{j\omega L_p Z_2}\right)$$

$$= V\left(\frac{L_s}{L_p} \times \frac{1}{Z_2}\right) + V\left(\frac{1}{j\omega L_p}\right)$$

$$= \begin{bmatrix}\text{a current present}\\ \text{due to the load}\\ \text{impedance } Z_2\end{bmatrix} + \begin{bmatrix}\text{a current lagging } V \text{ by } 90^\circ\\ \text{limited only by } L_p\text{, i.e. the}\\ \text{magnetising current } I_m\end{bmatrix}.$$

$1/Z_2 = Y_2$, the admittance of the load, hence the admittance of the load referred into the primary is

$$\frac{L_s}{L_p} \times \frac{1}{Z_2}.$$

Notice that the effect of Z_2 in the primary is determined by the ratio L_p/L_s. It is no longer necessary to consider $\omega^2 M^2$ with tight-coupled coils. It is general to consider impedances when working with power transformers so that inverting gives

$$\text{effective primary impedance} = Z_2 \frac{L_p}{L_s} \ \Omega.$$

From equation (5.13) $L = N \dfrac{d\phi}{di}$ H. Now $d\phi \propto N\, di$, so that

$$L \propto \frac{N \times N\, di}{di} \qquad \text{or } L \propto N^2.$$

Therefore $L_p \propto N_p^2$ and $L_s \propto N_s^2$, and $\dfrac{L_p}{L_s} = \left(\dfrac{N_p}{N_s}\right)^2$.

$$\text{Hence, the effective impedance in the primary} = Z_2 \left(\frac{N_p}{N_s}\right)^2 \Omega. \qquad (7.20)$$

The equivalent circuit for the ideal transformer is, therefore, as shown in Fig. 7.14.

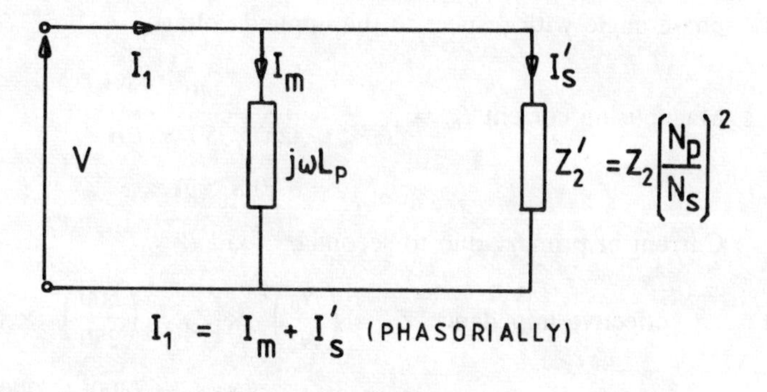

Fig. 7.14

The magnetising current I_m is independent of transformer loading and flows at all the times the transformer is connected to the supply. I'_s is the secondary current referred to the primary, or balancing current, i.e. the current which flows in the primary because I_s flows in the secondary and in the load.

In Fig. 7.14

$$\text{VA developed in } Z'_2 = (I'_s)^2\, Z_2 \left(\frac{N_p}{N_s}\right)^2.$$

This must be the same as that actually developed in the load Z_2 by current I_s if this circuit is to represent the transformer accurately. Therefore

$$I_s^2 Z_2 = (I_s')^2\ Z_2\left(\frac{N_p}{N_s}\right)^2$$

Transposing

$$\left(\frac{I_s}{I_s'}\right)^2 = \frac{Z_2}{Z_2}\left(\frac{N_p}{N_s}\right)^2.$$

Cancelling Z_2 and taking the square root of both sides gives

$$\frac{I_s}{I_s'} = \frac{N_p}{N_s} \qquad \text{and } I_s' = I_s\frac{N_s}{N_p}\ \text{A.} \qquad (7.21)$$

The current input to the primary is the phasor sum of I_m and I_s'.

WORKED EXAMPLE 7.10

An iron-cored transformer has primary and secondary coils with 500 and 250 turns respectively. The inductance of the primary coil is 1.6 H. It is connected to a 440 V, 50 Hz supply and the secondary supplies a load with impedance (100 + j50) Ω.

Assuming ideal operation, determine the value of the primary current and its phase angle with respect to the applied voltage.

$$\text{Magnetising current } I_m = \frac{V}{j\omega L_p} = \frac{440}{j2\pi \times 50 \times 1.6}$$

$$= -j0.875 \text{ A.}$$

Current in primary due to secondary load I_s'

$$\text{effective impedance } Z_2' = \left(\frac{N_p}{N_s}\right)^2 \times Z_2 = \left(\frac{500}{250}\right)^2 \times (100 + j50)$$

$$= (400 + j200)\ \Omega.$$

$$I_s' = \frac{440}{(400 + j200)} = \frac{440\ (400 - j200)}{400^2 + 200^2}$$

$$= (0.88 - j0.44) \text{ A.}$$

$$I_1 = I_m + I_s'$$

$$= -j0.875 + (0.88 - j0.44) = (0.88 - j1.315) \text{ A}$$

$$= 1.582\angle -56.2° \text{ A.}$$

I_1 = 1.582 A, the current lags the supply voltage by 56.2°.

7.5.2 *The practical transformer*

In the previous section ideal operation was considered. There are a number of imperfections which must be considered.

(1) The core being made from iron suffers hysteresis and eddy current losses and so there must be a power input from the supply to provide these. In the equivalent circuit a resistance R_0 is drawn in parallel with $j\omega L_p$. This resistor has a value calculated from open-circuit test results on the transformer which determine the core loss. For example, the core loss is equal to 60 W at a full rated voltage of 440 V. It would appear that this loss is taking place in an effective resistance calculated from

$$\text{power} = \frac{V^2}{R_0}, \qquad \text{hence } R_0 = \frac{440^2}{60} = 3227\ \Omega\,.$$

(2) However carefully designed the transformer is, there will in fact be some leakage flux, so that the primary coil is shown in two sections. The main section of N_p turns is considered to be perfectly coupled to the secondary while a very small section is considered to be uncoupled and it is this small section which sets up the leakage flux. The reactance of the uncoupled section is called leakage reactance and is marked X_p in Fig. 7.15. A similar argument may be applied to the secondary and its leakage reactance is marked X_s in Fig. 7.15. (See also appendix 3 and section 7.4.1.)

(3) Each winding has a small resistance and where efficiency and volt-drop calculations are to be performed these resistances have to be considered.

The overall equivalent circuit diagram becomes as shown in Fig. 7.15.

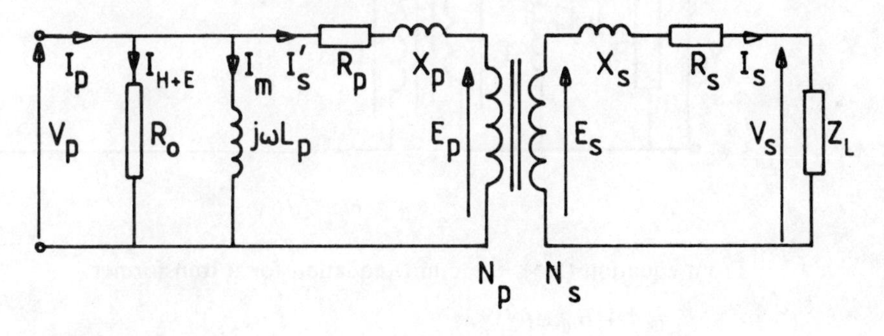

Fig. 7.15

In Fig. 7.15
R_p and X_p = primary winding resistance and leakage reactance respectively.
R_s and X_s = secondary winding resistance and leakage reactance respectively.
E_p = primary induced voltage which opposes V_p.
E_s = secondary induced voltage which drives the secondary current I_s through the load impedance such that the load p.d. equals V_s volts.
I_{H+E} = current to provide hysteresis and eddy current losses of the core.
I_p = total input current to the primary = phasor sum of I'_s, I_m and I_{H+E}.

It might be argued that R_p should be on the supply side of $j\omega L_p$ and R_0 since I_m and I_{H+E} have to flow in the primary winding and hence its resistance. This is indeed true, but in this position, as I_s' flows, the voltage drop across R_p would seem to cause the magnetising current and core losses to fall. These are in fact essentially constant so that the best equivalent circuit is as shown. It gives good results when predicting efficiency and voltage changes with load.

Using equation (7.20), all the secondary values of impedance may be referred into the primary as in Fig. 7.14. However, it is often more convenient to refer the series impedance from the primary into the secondary. Equation (7.20) gives

$$\text{primary impedance} = \left(\frac{N_p}{N_s}\right)^2 \times \text{secondary impedance}$$

$$\text{primary impedance} \times \left(\frac{N_s}{N_p}\right)^2 = \text{secondary impedance.}$$

$$\text{Hence } R_p \text{ referred to the secondary} = R_p' = R_p\left(\frac{N_s}{N_p}\right)^2$$

$$\text{and } X_p \text{ referred to the secondary} = X_p' = X_p\left(\frac{N_s}{N_p}\right)^2$$

when the equivalent circuit diagram becomes as shown in Fig. 7.16.

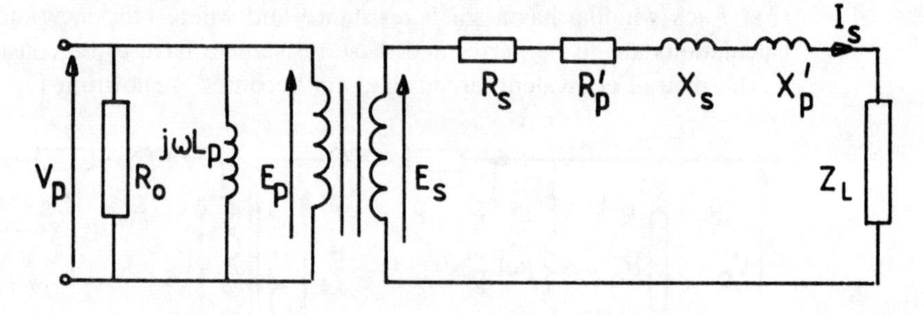

Fig. 7.16

From equation (5.5), the e.m.f. equation for a transformer

$$E = 4.44 B_m A f N \text{ V.}$$

Therefore $E_p = 4.44 B_m A f N_p$ and $E_s = 4.44 B_m A f N_s$.

Dividing E_p by E_s gives
$$\frac{E_p}{E_s} = \frac{N_p}{N_s}. \qquad (7.22)$$

WORKED EXAMPLE 7.11

The ratio of turns of a single-phase transformer is 8:1. The resistances of primary and secondary windings are 0.85 Ω and 0.012 Ω respectively. The corresponding leakage reactances are 4.8 Ω and 0.07 Ω.

Determine the value of voltage to be applied to the primary to cause 165 A to flow in a short circuit across the secondary terminals. (Ignore magnetising current and core losses.)

The transformer as described in the question may be drawn as shown in Fig. 7.17(a).

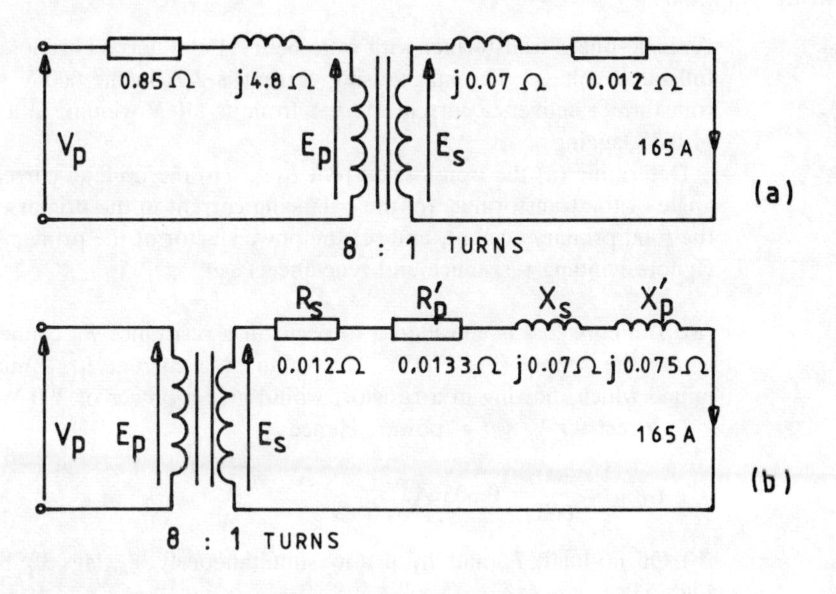

Fig. 7.17

Referring primary resistance and reactance to the secondary yields

$$R'_p = R_p \times \left(\frac{N_s}{N_p}\right)^2 = 0.85 \times \left(\frac{1}{8}\right)^2 = 0.0133\ \Omega$$

$$X'_p = X_p \times \left(\frac{N_s}{N_p}\right)^2 = 4.8 \times \left(\frac{1}{8}\right)^2 = 0.075\ \Omega\,.$$

Observe that when referring values from a winding with a large number of turns into one with less turns, the impedance values decrease, and vice versa.

In Fig. 7.17(b), the total impedance of the transformer referred to the secondary $Z_T^{(S)}$ is given by

$$Z_T^{(S)} = \sqrt{[(0.012 + 0.0133)^2 + (0.07 + 0.075)^2]} = 0.147\ \Omega.$$

$$E_s = IZ_T^{(S)} = 165 \times 0.147 = 24.26\ \text{V}.$$

Note that if the output terminals were not short circuited, the load impedance would have to be included in $Z_T^{(S)}$.

$$\text{Now } \frac{E_p}{E_s} = \frac{N_p}{N_s}\,. \qquad \text{(equation 7.22)}$$

$$\text{Therefore } \frac{E_p}{24.26} = \frac{8}{1}$$

$$E_p = 8 \times 24.26 = 194.1 \text{ V.}$$

Since the primary impedance is zero, $V_p = E_p = 194.1$ V.

WORKED EXAMPLE 7.12

A single-phase transformer with ratio 660 V:110 V has a core loss of 200 W at full rated voltage. The magnetising current is 2 A in the 660 V winding. The transformer delivers a current of 65 A from its 110 V winding at a power factor of 0.75 lagging.

Determine (a) the iron loss current I_{H+E}, (b) the no-load current and phase angle of the transformer, (c) the balancing current in the primary winding, (d) the total primary current, and (e) the power factor of the primary at this load. (Ignore winding resistance and reactance.)

(a) The core loss is considered to occur in a resistance R_0 connected directly across the supply (Fig. 7.15). The value of a current I_{H+E} must be determined which, flowing in a resistor, would give a power of 200 W at 660 V.

In a resistor $V \times I$ = power. Hence

$$I_{H+E} = \frac{200}{660} = 0.303 \text{ A.}$$

(b) On no-load, I_m and I_{H+E} flow simultaneously. I_m lags V_p by 90°. From Fig. 7.18 (a)

$$\text{no-load current } I_0 = \sqrt{(I_{H+E}^2 + I_m^2)} = \sqrt{(0.303^2 + 2^2)}$$
$$= 2.023 \text{ A.}$$

$$\text{Also } \tan \phi_0 = \frac{I_m}{I_{H+E}}, \qquad \text{hence } \phi_0 = 81.38°.$$

(c) From equation (7.21)

$$I_s' = I_s \frac{N_s}{N_p} = I_s \frac{E_s}{E_p} \qquad \text{(using equation 7.22)}$$

$$I_s' = 65 \times \frac{110}{660} = 10.83 \text{ A.}$$

Since at all instants I_s' 'balances' I_s, I_s' has the same phase angle with respect to E_p that I_s has with respect to E_s. The load power factor is 0.75. Therefore I_s lags E_s by $\cos^{-1} 0.75 = 41.4°$ and $E_s = V_s$, since winding impedance is ignored.

It follows that I_s lags E_p (and V_p) by 41.4°.

(d) The total primary current is equal to the phasor sum of I_m, I_{H+E} and I_s'. In Fig. 7.18(b), resolving I_s' into vertical horizontal components

horizontal component = 10.83 sin 41.4 = 7.16 A
vertical component = 10.83 cos 41.4 = 8.12 A.

Adding horizontal components: 7.16 + 2 = 9.16 A.
Adding vertical components: 8.12 + 0.303 = 8.423 A.

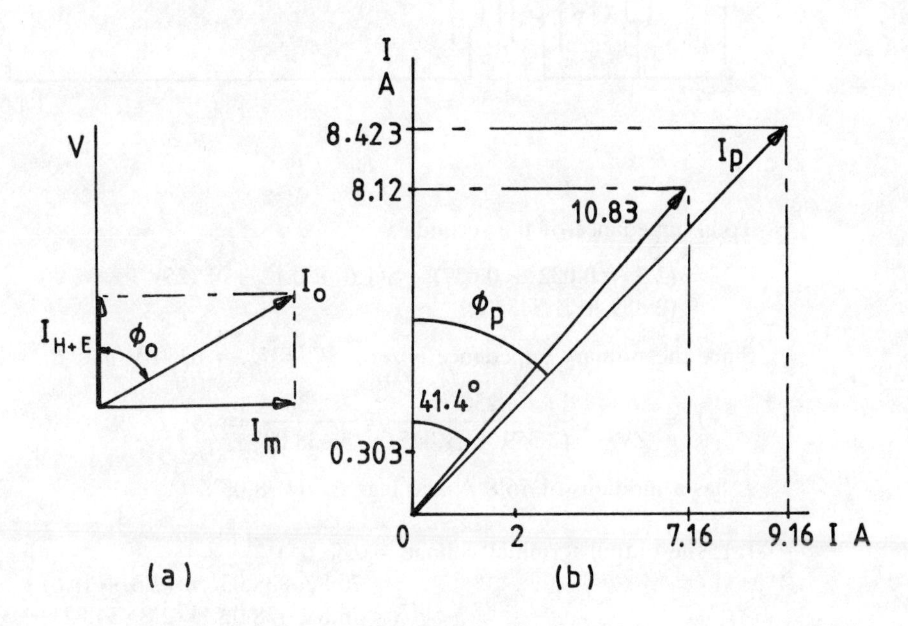

Fig. 7.18

By Pythagoras' theorem $I_p = \sqrt{(9.16^2 + 8.423^2)} = 12.44$ A.

(e) Power factor $= \cos \phi_p = \dfrac{8.423}{12.44} = 0.677 \qquad (\phi_p = 47.4°)$.

WORKED EXAMPLE 7.13

A single-phase, 50 Hz transformer with voltage ratio 6350 V:230 V has the following details: $R_p = 20.7\ \Omega$, $R_s = 0.032\ \Omega$, $X_p = 95\ \Omega$, $X_s = 0.12\ \Omega$. The core loss is 2 kW. The secondary is connected to a load of (2.3 + j1.6) Ω.

Calculate, for this condition, assuming the primary to be connected to a 6350 V supply, (a) the output terminal voltage, and (b) the operating efficiency.

Refer the primary resistance and reactance into the secondary.

$$R'_p = 20.7\left(\frac{230}{6350}\right)^2 = 0.027\ \Omega; \qquad X'_p = 95\left(\frac{230}{6350}\right)^2 = 0.125\ \Omega.$$

The equivalent circuit is shown in Fig. 7.19.

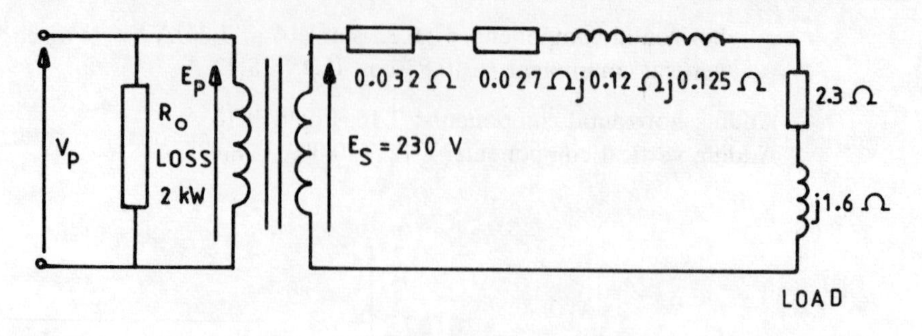

Fig. 7.19

Total impedance of the secondary

$$= (2.3 + 0.032 + 0.027) + j(1.6 + 0.12 + 0.125)$$
$$= (2.359 + j1.845)\ \Omega.$$

Since the primary impedance is zero, $V_p = E_p = 6350$ V and $E_s = 230$ V.

$$I_s = \frac{E_s}{Z_T^{(S)}} = \frac{230}{(2.359 + j1.845)} = \frac{230}{3\angle 38.03^\circ} = 76.8\angle -38.03^\circ \text{ A}.$$

I_s has a modulus of 76.8 A and lags E_s by 38.03°.

(a) The output terminal voltage $= IZ_{\text{load}}$

$$= 76.8\angle -38.03^\circ \times (2.3 + j1.6)$$
$$= 76.8\angle -38.03^\circ \times 2.8\angle 34.82^\circ$$
$$= 215.18\angle -3.2^\circ \text{ V}.$$

V_s has a modulus of 215.18 V and lags E_s by 3.2°.

(b) Load power $= I_s^2 R_L = 76.8^2 \times 2.3 = 13\,566$ W.

Power loss in windings $= I_s^2 \times$ winding resistance

$$= 76.8^2\,(0.032 + 0.027)$$
$$= 348 \text{ W}.$$

Core loss = 2 kW.

$$\text{Efficiency} = \frac{\text{output}}{\text{input}} = \frac{\text{output}}{\text{output} + \text{losses}}$$
$$= \frac{13\,566}{13\,566 + 348 + 2000}$$
$$= 0.852 \text{ p.u.}$$

SELF-ASSESSMENT EXAMPLE 7.14

A single-phase transformer has its primary connected to a 220 V supply. It has total equivalent resistance referred to the primary of 0.1 Ω and total reactance referred into the primary of 0.5 Ω. The secondary is connected to a load of resistance 200 Ω and reactance +j100 Ω. The turns ratio is 1:5 from primary to secondary. The core loss is 60 W.

Determine (a) the total resistance referred into the secondary, (b) the total reactance referred into the secondary, (c) the secondary current I_s, (d) the secondary terminal voltage, and (e) the efficiency.

There is a considerable amount of further work which can be undertaken on power transformers including some 'short-cuts' in working allowed in the relevant British Standards. These give close approximate answers with substantially reduced working. These are to be considered as applications and not principles.

This chapter has attempted to show that mutual inductance is employed to obtain several different results, maximum power transfer, wide bandwidth or, in the case of power transformers, the greatest possible efficiency while changing voltages from one level to another. Each function overlaps the others to some extent. There is a great tendency to treat mutual inductance for light-current applications as something quite different to that required for heavy-current applications which is, of course, quite unjustified.

FURTHER PROBLEMS

7.15 A capacitor with capacitance 53 μF is connected in series with a coil of inductance 0.0955 H. The maximum current drawn from a 200 V, variable-frequency supply is 5 A. Determine (a) the circuit current, (b) the power factor, when supplied at 50 Hz.

7.16 What is meant by the magnification factor of a series resonant circuit?

Sketch a graph of current against frequency for a series circuit containing inductance and capacitance to which a constant voltage, variable-frequency supply is connected. Explain how the shape of the curve would be affected by a change in the circuit resistance.

A circuit consists of a capacitor with capacitance 0.02 μF in series with a coil possessing resistance and inductance. The resonant frequency is 5 kHz. The magnification factor Q = 110. Determine (a) the resistance of the coil, (b) the inductance of the coil, (c) the lower 3 dB frequency.

7.17 (a) Explain what is meant by (i) the dynamic impedance of a parallel resonant circuit, (ii) the voltage magnification of a series resonant circuit.

(b) A circuit consists of a capacitor with capacitance 20 μF and a coil with resistance 120 Ω and inductance 0.5 H. Determine the resonant frequency when (i) the capacitor is connected in series with the coil, (ii) the capacitor is connected in parallel with the coil.

7.18 A coil and a capacitor are connected in series to a source of e.m.f. of 10 mV and a frequency of 1 MHz. If the coil has an effective resistance of 20 Ω and an inductance of 159 μH, determine (a) the value of capacitance to tune the circuit to resonance, (b) the value of circuit current at resonance, (c) the p.d. across the capacitor at resonance, and (d) the bandwidth and frequencies for half power.

7.19 An a.c. circuit consists of a coil with inductance 0.1 H and resistance 50 Ω shunted by a capacitor having capacitance 10 μF. The supply frequency is adjusted until the supply current and voltage are in phase. Calculate (a) the supply frequency for this condition, (b) the value of the dynamic impedance of the circuit.

7.20 The secondary of the circuit shown in Fig. 7.20 is at resonance when f = 50 kHz. (M = 0.1 mH and R_s = 10 Ω.) Calculate the effective impedance coupled into the primary from the secondary at this frequency.

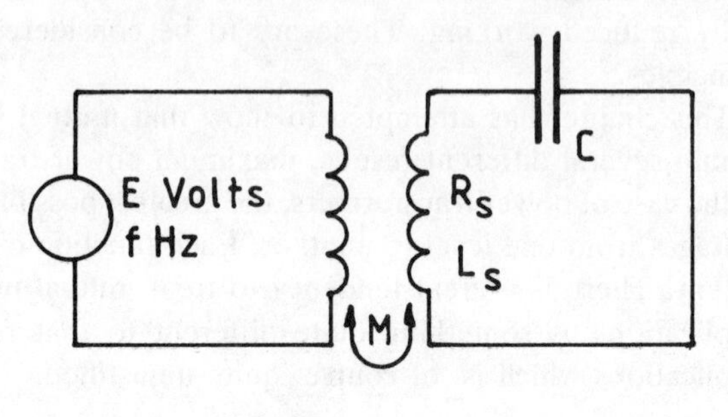

Fig. 7.20

7.21 A coil having inductance 5 mH and resistance 20 Ω is connected to a source of 2 V at a frequency of 10 kHz. It is mutually coupled to a second coil having inductance 20 mH and resistance 100 Ω. The coupling coefficient is 0.4. The second coil is connected to a variable capacitor which is adjusted to make the secondary circuit resonant. Calculate (a) the value of the primary current, (b) the capacitor voltage under these conditions.

7.22 Two air-cored coils possessing resistance and self-inductance are coupled by mutual inductance. The primary is connected to an alternating power supply while the secondary is short circuited. (a) Derive an expression for the effective impedance of the primary under these conditions. Given that R_p = 10 Ω, L_p = 20 mH, R_s = 5 Ω. L_s = 20 mH, k = 0.15 and f = 2 kHz, calculate this value of impedance. (b) If capacitors are now connected in series with both primary and secondary windings such that both circuits are resonant at $10\,000/2\pi$ Hz when in isolation, calculate the two additional frequencies created by the mutual coupling.

7.23 (a) Two coils having self-inductance L_1 and L_2 henry respectively have mutual inductance M henry between them. Define the coupling coefficient k between these two coils in terms of L_1, L_2 and M. (b) Given that the values of L_1 and L_2 are 0.1 mH and 0.3 mH respectively and that the corresponding resistances are 5 Ω and 21 Ω, and that each coil is connected in series with a capacitor which causes resonance in each circuit when isolated at 50 kHz, calculate (i) the value of the critical coupling coefficient and the corresponding value of mutual inductance, (ii) the power transferred into the secondary circuit for the condition in (i), the primary being fed at 5 V.

7.24 Two coils coupled by mutual inductance are used to couple a load of impedance $100\angle 0°$ Ω to a source of e.m.f. having internal impedance of $200\angle 0°$ Ω operating at 0.8 MHz. The primary winding has inductance 100 μH and is connected to the source via a capacitor C. The secondary winding with inductance 25 μH is connected directly to the load.

Calculate the degree of coupling and the value of C which gives maximum power in the load. The resistances of both coils may be ignored. (*Hint*: begin by referring the whole of the secondary impedance into the primary and then equate resistive parts of primary and secondary referred impedances.)

7.25 Upon what factors does the no-load current of a transformer depend?

A transformer draws a magnetising current of 9.9 A and an in-phase current of 1.1 A to supply hysteresis and eddy current losses when fed at its rated voltage. The ratio is 4:1 step-down. Determine the value of the total primary current and the power factor when the secondary supplies a load with 200 A at 0.8 power factor lagging.

7.26 A single-phase transformer with ratio 3300 V: 240 V has the following parameters: $R_p = 4.5$ Ω, $X_p = 22.6$ Ω, $R_s = 0.026$ Ω, $X_s = 0.12$ Ω. The core loss is 325 W. A load of (10 + j5) Ω is connected to the secondary (240 V) terminals.

For the primary connected to a 3300 V supply, determine (a) the secondary current, (b) the balancing current in the primary, (c) the output terminal voltage, and (d) the operating efficiency.

Answers

7.3 (a) 100 654 Hz (b) 9 μA (c) 0.949 mA (d) $Q = 105.4$
(e) 100 176 Hz; 101 131 Hz; bandwidth, 955 Hz

7.6 2.8 W

7.9 (a) $f_0 = 1.002$ MHz; $f_H = 1.012$ MHz; $f_L = 0.992$ MHz (b) $k = 0.039$

7.14 (a) $R_T^{(S)} = 2.5$ Ω (b) $X_T^{(S)} = 12.5$ Ω
(c) $I_s = 1100/231.65 \angle 29° = 4.748\angle -29°$ A (d) $V_s = 1061$ V
(e) Power out = 4508.7 W; losses = 56.4 + 60 W; efficiency: 0.975

7.15 (a) 4 A (b) 0.8

7.16 (a) 14.5 Ω (b) 0.0507 H (c) 4.977 kHz

7.17 (b) (i) 50.3 Hz (ii) 32.7 Hz

7.18 (a) 159 pF (b) 500 μA (c) 0.5 V (d) 20 kHz; 0.99 MHz; 1.01 MHz

7.19 (a) 137.8 Hz (b) 200 Ω

7.20 98.7 Ω

7.21 (a) 2.76 mA (b) 8.72 V ($I_s = 6.94$ mA)

7.22 (a) (10.1 + j245.7) Ω (b) 1484 Hz; 1726 Hz

7.23 (b) (i) $k = 0.188$; $M = 32.6$ μH (ii) 1.25 W

7.24 From secondary: (244.9 − j307.8) k^2 Ω. To match resistances, $k = 0.904$.
Capacitor: $X_C = -j251.1$ Ω; $C = 792$ pF

7.25 57.6 A; 0.71 lag

7.26 (a) 21.18 A (b) 1.54 A (c) 236.7 V (d) 0.928 p.u.

Chapter 8
Laplace Transforms and the Solution of Circuit Transients

This chapter is not intended to be a comprehensive treatment of Laplace transforms but rather an introduction to the topic for those whose mathematical studies have not reached this stage and as an encouragement to study further this very useful tool for solving differential equations, involved amongst other things in circuit transients and transmission lines with which this book deals. The solution of differential equations by other means is difficult or even impossible.

When a voltage is applied to a circuit consisting of elements which store energy, i.e. inductance, capacitance, or a combination of both, there is a period of fluctuation of current and energy preceding the settling down to the final steady-state conditions. Likewise when the voltage is removed or changed there is a period of transient disturbance.

To analyse the transient performance, the Laplace transform is used. This is a development of work carried out by a British electrical engineer, Oliver Heavyside.

8.1 Comparison with logarithms

When an operation such as multiplication, division or raising a number to a power is to be undertaken the direct approach may be difficult. For example, consider raising 1.96 to the power 0.54 by direct mathematical methods. The case of division employing logarithms with the aid of a flow diagram will be examined.

The quotient N_1/N_2 is to be determined using logarithms to a base 'b'. The transforms are

$$\log_b N_1 = x, \text{ hence } b^x = N_1,$$

$$\log_b N_2 = y, \text{ hence } b^y = N_2.$$

Therefore $\dfrac{N_1}{N_2} = \dfrac{b^x}{b^y} = b^{(x-y)}$.

Hence the logarithm of $\dfrac{N_1}{N_2} = (x - y)$.

To find the value of N_1/N_2 the inverse transform is performed by looking up $(x - y)$ in the anti-logarithm tables.

The procedure is illustrated in the flow diagram (Fig. 8.1).

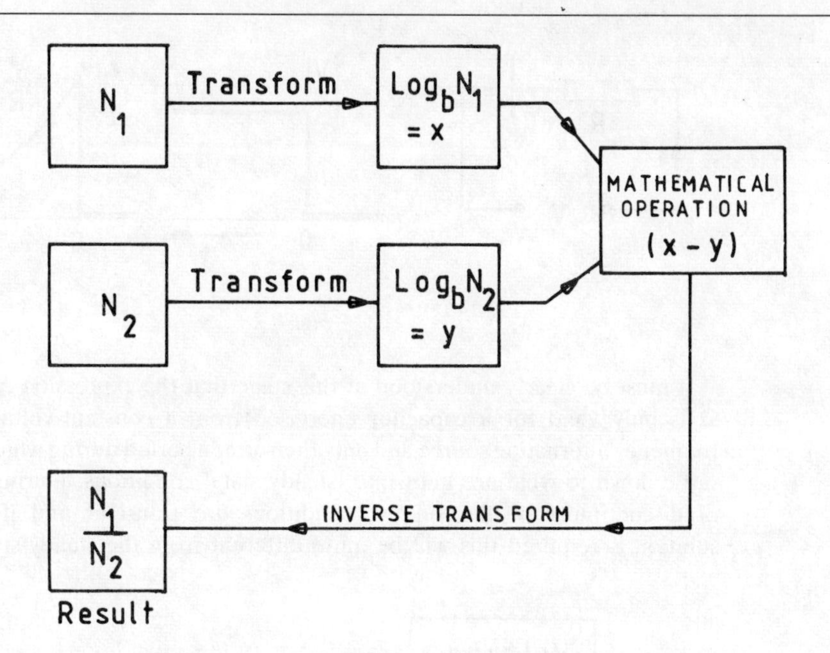

Fig. 8.1

8.2 The Laplace transform

8.2.1 Definition

The Laplace transform changes the expression for a quantity from that involving time, $f(t)$, to a complex function, $F(s)$, where s is a complex quantity, $(\alpha + j\omega)$. α is a real number and ω is the angular velocity in rad/s.

As an example, consider a direct voltage being suddenly applied to a capacitive circuit which is initially uncharged. The current falls exponentially from a maximum value given by $I = V/R$ amperes to zero. The stimulus is a suddenly applied steady voltage, or step voltage as it is termed. The voltage is a function of time in that

at $t < 0$, $V = 0$
$t = 0$, $V = 0$
$t > 0$, $V = V$ volts.

The opposition to current flow or apparent impedance of the circuit increases with time, see Fig. 8.2. If $f(t)$ is the impedance, at any instant in time

$$i = \frac{V}{\text{apparent impedance}}.$$

To eliminate time from the mathematical operations, both voltage and the circuit constants R and C are transformed into their Laplace transforms. The flow diagram is then as shown in Fig. 8.3.

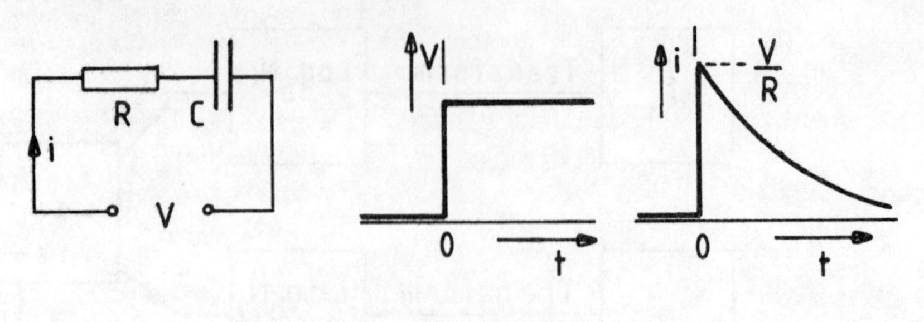

Fig. 8.2

It must be clearly understood at this stage that the expression $X_C = (1/\omega C)$ Ω is only valid for a capacitor energised from a constant-voltage constant-frequency alternating source and only then after a period during which conditions settle down to what are known as 'steady-state' conditions. During any period of discontinuity of the supply, conditions are transient and if a transient solution is required this will be quite different from the steady-state solution

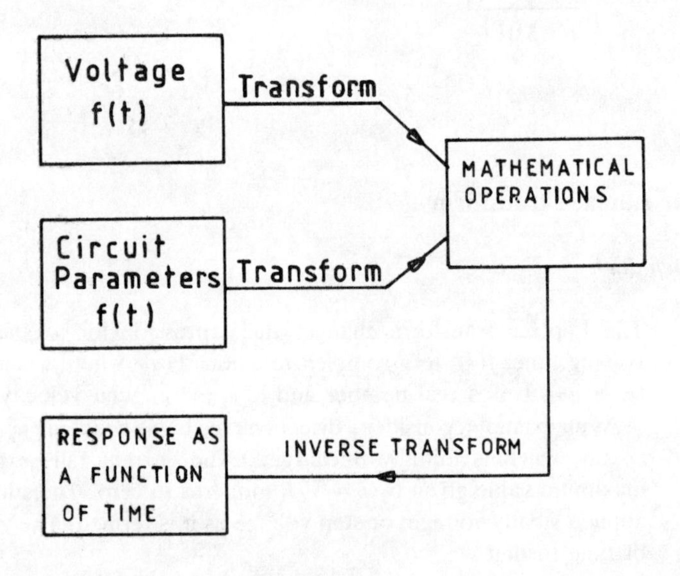

Fig. 8.3

and will involve differential equations which may be solved using Laplace transforms.

If $f(t)$ is a function of time then its Laplace transform $\mathscr{L}[f(t)]$ is defined as

$$\int_0^\infty f(t)\mathrm{e}^{-st}\,\mathrm{d}t, \text{ and is written as } F(s). \tag{8.1}$$

In words, the Laplace transform of a function is the integral between zero and infinity with respect to time of that function multiplied by natural e raised to the power $(-st)$. The integration eliminates the time variable so that the resulting transform is a function of s. The only requirement for the value of s is that the integral (equation 8.1) must be convergent as $t \to \infty$. If the integral is divergent then the particular time function has no Laplace transform.

Logarithms do not need to be derived from first principles each time they are used as they are readily available in books of tables. It is the same with Laplace transforms; they may be referred to in published works. Some of them are listed at the end of this chapter. However, the derivation of some of them will be examined before progressing to their use.

8.2.2 *Derivation of a selection of Laplace transforms*

(1) $f(t) = \text{constant (V)}$.

$$\mathscr{L}V = \int_0^\infty V\mathrm{e}^{-st}\,\mathrm{d}t \qquad \text{(using equation 8.1)}$$

$$= -\left(\frac{V\mathrm{e}^{-st}}{s}\right)_0^\infty = (0) - \left(-\frac{V}{s}\right) = \frac{V}{s}.$$

V/s is the Laplace transform of a constant voltage V suddenly applied to a circuit, i.e. a step function.

$$F(s) = \frac{V}{s}.$$

(2) $f(t) = \mathrm{e}^{at}$.

$$\mathscr{L}\mathrm{e}^{at} = \int_0^\infty \mathrm{e}^{at}\,\mathrm{e}^{-st}\,\mathrm{d}t = \int_0^\infty \mathrm{e}^{(a-s)t}\,\mathrm{d}t.$$

Now since at $t = \infty$, $\mathrm{e}^{(a-s)t}$ must converge, s must be larger than a. The integral is therefore expressed as

$$\mathscr{L}\mathrm{e}^{at} = \int_0^\infty \mathrm{e}^{-(s-a)t}\,\mathrm{d}t$$

$$= \left(\frac{\mathrm{e}^{-(s-a)t}}{-(s-a)}\right)_0^\infty = (0) - \left(-\frac{1}{s-a}\right).$$

$$F(s) = \frac{1}{s-a}.$$

(3) $f(t) = \sin \omega t$.

$$\mathrm{e}^{\mathrm{j}\theta} = \cos\theta + \mathrm{j}\sin\theta \text{ (Euler's formula)} \qquad \text{(i)}$$

$$\mathrm{e}^{-\mathrm{j}\theta} = \cos\theta - \mathrm{j}\sin\theta. \qquad \text{(ii)}$$

Subtracting equation (ii) from equation (i)

$$\mathrm{e}^{\mathrm{j}\theta} - \mathrm{e}^{-\mathrm{j}\theta} = 2\mathrm{j}\sin\theta$$

$$\sin\theta = \frac{\mathrm{e}^{\mathrm{j}\theta} - \mathrm{e}^{-\mathrm{j}\theta}}{2\mathrm{j}}.$$

$$\mathcal{L}(\sin \omega t) = \int_0^\infty \left(\frac{e^{j\omega t} - e^{-j\omega t}}{2j}\right) e^{-st}\, dt \qquad \text{(writing } \omega t \text{ for } \theta\text{)}$$

$$= \int_0^\infty \frac{1}{2j}\left[\left(e^{-(s - j\omega)t} - e^{-(s + j\omega)t}\right)\right] dt$$

$$= \frac{1}{2j}\left[\frac{-e^{-(s - j\omega)t}}{(s - j)} + \frac{e^{-(s + j\omega)t}}{(s + j)}\right]_0^\infty$$

$$= \frac{1}{2j}\left[0 + 0 - \left(-\frac{1}{s - j\omega} + \frac{1}{s + j\omega}\right)\right]$$

$$= \frac{1}{2j}\left[\frac{2j\omega}{s^2 + \omega^2}\right] = \frac{\omega}{s^2 + \omega^2}.$$

$$\mathcal{L}(\sin \omega t) = F(s) = \frac{\omega}{s^2 + \omega^2}.$$

(4) $f(t) = \sinh \omega t$.
Since $\sinh \omega t = \frac{1}{2}(e^{\omega t} - e^{-\omega t})$

$$\mathcal{L}(\sin \omega t) = \int_0^\infty \tfrac{1}{2}(e^{\omega t} - e^{-\omega t})e^{-st}\, dt$$

$$= \frac{1}{2}\int_0^\infty e^{-(s - \omega)t} - e^{-(s + \omega)t}\, dt$$

$$= \frac{1}{2}\left[-\frac{e^{-(s - \omega)t}}{(s - \omega)} + \frac{e^{-(s + \omega)t}}{(s + \omega)}\right]_0^\infty$$

$$= \frac{1}{2}\left(\frac{1}{s - \omega} - \frac{1}{s + \omega}\right) = \frac{1}{2}\left[\frac{s + \omega - (s - \omega)}{s^2 - \omega^2}\right]$$

$$= \frac{\omega}{s^2 - \omega^2}.$$

$$\mathcal{L}(\sinh \omega t) = F(s) = \frac{\omega}{s^2 - \omega^2}.$$

(5) The Laplace transform of derivatives.
A knowledge of the method of integration by parts is required to understand this section.

Given two different functions of t, u and v

$$\frac{d}{dt}(uv) = v\frac{du}{dt} + u\frac{dv}{dt} \qquad \text{(differentiation of a product)}$$

transposing gives

$$v\frac{du}{dt} = \frac{d}{dt}(uv) - u\frac{dv}{dt}.$$

Integrating both sides with respect to t and putting in the limits 0 and ∞

$$\int_0^\infty v\frac{du}{dt}\, dt = \left[uv - \int_0^\infty u\frac{dv}{dt}\, dt\right]_0^\infty. \qquad \text{(i)}$$

Now if $y = f(t)$, the Laplace transform of its derivative, using the defining equation (8.1) is

$$\mathcal{L}\left[\frac{df(t)}{dt}\right] = \int_0^\infty e^{-st}\frac{df(t)}{dt}\,dt. \tag{8.1(a)}$$

Let $e^{-st} = v$. Then $\dfrac{dv}{dt} = -se^{-st}$. (ii), (iii)

Let $\dfrac{df(t)}{dt} = \dfrac{du}{dt}$. Then $u = f(t)$. (iv)

Substituting equations (ii) and (iv) in equation (8.1(a))

$$\mathcal{L}\left[\frac{df(t)}{dt}\right] = \int_0^\infty v\frac{du}{dt}\,dt. \tag{v}$$

Multiplying equation (ii) by equation (iv) $uv - e^{-st} f(t)$. (vi)

Multiplying equation (iii) by equation (iv) $\displaystyle\int_0^\infty u\frac{dv}{dt}\,dt = \int_0^\infty f(t)\,(-se^{-st})\,dt$

$$= -s\int_0^\infty e^{-st} f(t)\,dt. \tag{vii}$$

Substituting equation (vi) for uv and equation (vii) for $\displaystyle\int_0^\infty u\frac{dv}{dt}\,dt$ in equation (i)

$$\int_0^\infty v\frac{du}{dt}\,dt = \left[e^{-st} f(t) - \left(-s\int_0^\infty e^{-st} f(t)\,dt\right)\right]_0^\infty.$$

But $\displaystyle\int_0^\infty v\frac{du}{dt}\,dt = \mathcal{L}\left[\frac{df(t)}{dt}\right]$. (equation v)

Therefore, the Laplace transform of the first derivative of $f(t)$ is

$$\left[e^{-st} f(t) + s\int_0^\infty e^{-st} f(t)\,dt\right]_0^\infty. \tag{viii}$$

Now, in equation (viii) $\int_0^\infty e^{-st} f(t)\,dt$ should be recognised as the Laplace transform of $f(t)$. Therefore, $s\int_0^\infty e^{-st} f(t)\,dt = s\mathcal{L}f(t)$.

Also, $\left[e^{-st} f(t)\right]_0^\infty = 0 - f(t)_0$ (i.e. minus the value of the function at $t = 0$).

Therefore, $\mathcal{L}\left[\dfrac{df(t)}{dt}\right] = -f(t)_0 + s\mathcal{L}f(t)$.

In words, the Laplace transform of the first derivative of a function is equal to s times the Laplace transform of the function itself minus the value of the function itself at time zero.

In symbols $\mathcal{L}\left[\dfrac{df(t)}{dt}\right] = \mathcal{L}f'(t) = sF(s) - f(0)$.

Using a similar method the Laplace transform for the second differential can be arrived at

$$\mathcal{L}\left[\frac{d^2f(t)}{dt^2}\right] = \mathcal{L}f''(t) = s\mathcal{L}f'(t) - f'(0)$$
$$= s^2\mathcal{L}f(t) - sf(0) - f'(0)$$
$$= s^2F(s) - sf(0) - f'(0).$$

In words, the Laplace transform of the second derivative of a function is equal to s^2 times the Laplace transform of the function itself, minus s times the value of the function at time zero, minus the value of the first differential of the function at time zero.

8.3 The exponential factor (shifting theorem)

By definition (equation 8.1), $\mathcal{L}f(t) = F(s) = \int_0^\infty f(t)e^{-st}\,dt$.

Multiplying $f(t)$ by e^{-kt} gives

$$\mathcal{L}\left[f(t) \times e^{-kt}\right] = \int_0^\infty f(t) \times e^{-kt} \times e^{-st}\,dt$$
$$= \int_0^\infty f(t)e^{-(s+k)t}\,dt.$$

To examine what this means, suppose that $f(t)$ is a step function V. Then

$$\mathcal{L}\left[Ve^{-kt}\right] = \int_0^\infty Ve^{-(s+k)t}\,dt = \left[\frac{Ve^{-(s+k)t}}{-(s+k)}\right]_0^\infty$$
$$= \frac{V}{s+k}.$$

It has already been deduced that $\mathcal{L}V = V/s$ (section 8.2.2). The effect of multiplying V by e^{-kt} is to add a k to the s in the Laplace transform. This is a general theorem, a further example being the case of $\mathcal{L}$ (sin ωt).

$$\mathcal{L}(\sin \omega t) = \frac{\omega}{s^2 + \omega^2} \quad \text{and} \quad \mathcal{L}(\sin \omega t \times e^{-kt}) = \frac{\omega}{(s+k)^2 + \omega^2}.$$

8.4 Examples of finding the original function from a given Laplace transform

(using the table of transforms at the end of this chapter)

(1) $F(s) = \dfrac{3}{s+2} = 3\left(\dfrac{1}{s+2}\right)$. Find $f(t)$.

Laplace pair no. 3 $\mathcal{L}e^{-at} = \dfrac{1}{s+a}$.

Therefore, the inverse (working backwards) of $1/(s + 2) = e^{-2t}$ $(a = 2)$

$$\text{and of } 3\left(\frac{1}{s + 2}\right) = 3e^{-2t}.$$

Therefore $$f(t) = 3e^{-2t}.$$

(2) $$F(s) = \frac{3s + 8}{s^2 + 4} = \frac{3s}{s^2 + 4} + \frac{8}{s^2 + 4}. \qquad \text{(i)}$$

Laplace pair no. 6 $\mathcal{L}(\cos \omega t) = \dfrac{s}{s^2 + \omega^2}$.

Laplace pair no. 5 $\mathcal{L}(\sin \omega t) = \dfrac{\omega}{s^2 + \omega^2}$.

It is necessary to rearrange equation (i) so that the denominators are the sum of s^2 and (a number)^2, thus

$$F(s) = 3\left(\frac{s}{s^2 + 2^2}\right) + 4\left(\frac{2}{s^2 + 2^2}\right).$$

Observe that in the second term it has been necessary to leave a 2 in the numerator to match that in the denominator in order that it is similar to Laplace pair no. 5.
With $\omega = 2$, $f(t) = 3 \cos 2t + 4 \sin 2t$.

(3) $$F(s) = \frac{2}{s^2 - 5^2} = 2\left(\frac{1}{s^2 - 5^2}\right).$$

Laplace pair no. 12 $\mathcal{L}(\sinh at) = \dfrac{a}{s^2 - a^2}$.

To give an inverse sinh at, $F(s)$ needs to have a 5 as its numerator to match that in the denominator.

Multiply inside the parentheses by 5 and divide outside by 5 so that the overall value remains unchanged

$$F(s) = \frac{2}{5}\left(\frac{1 \times 5}{s^2 - 5^2}\right).$$

$$f(t) = \frac{2}{5} \sinh 5t \qquad (a = 5).$$

(4) $$F(s) = \frac{s + 1}{s(s^2 + 4s + 8)}.$$

The expression is divided into partial fractions

$$\frac{s + 1}{s(s^2 + 4s + 8)} = \frac{A}{s} + \frac{Bs + C}{s^2 + 4s + 8} \qquad \text{(i)}$$

$$= \frac{A(s^2 + 4s + 8) + (Bs + C)s}{s(s^2 + 4s + 8)}$$

$$= \frac{As^2 + 4As + 8A + Bs^2 + Cs}{s(s^2 + 4s + 8)}.$$

Equating numerators

$$s + 1 = As^2 + 4As + 8A + Bs^2 + Cs.$$

For equality, each type of term must balance numbers, s terms and s^2 terms, hence

$$s^2 \text{ terms } 0 = A + B \quad \text{(ii)}$$
$$s \text{ terms } 1 = 4A + C \quad \text{(iii)}$$
$$\text{numbers } 1 = 8A. \quad \text{(iv)}$$

From equation (iv) $A = 1/8$.

Substituting the value for A in equation (iii) $1 = 4(1/8) + C$
$C = 1/2$.

Substituting the value of A in equation (ii) $B = -1/8$.
Substituting the values of A, B and C in equation (i)

$$F(s) = \frac{(1/8)}{s} + \frac{-(1/8s + 1/2)}{s^2 + 4s + 8}.$$

From Laplace pair no. 1 the inverse for the first term is obtained

$$\mathcal{L}\frac{1}{8} = \frac{1/8}{s}, \text{ therefore } f(t) = 1/8. \quad \text{(v)}$$

Now consider the second term. There are no transforms with fractional values of s in them. Multiply numerator and denominator by -8 to produce a single s

$$F(s) = -\frac{1}{8}\left[\frac{-8(-1/8 + 1/2)}{s^2 + 4s + 8}\right] = -\frac{1}{8}\left[\frac{(s - 4)}{s^2 + 4s + 8}\right]. \quad \text{(vi)}$$

To give a perfect inverse it is necessary to have a term similar to $(s^2 + \omega^2)$ or $((s + k)^2 + \omega^2)$ as the denominator.

Now $(s + 2)^2 = s^2 + 4s + 4$
so that $(s^2 + 4s + 8) = s^2 + 4s + 4 + 4 = (s + 2)^2 + 4$
$= (s + 2)^2 + 2^2$.

Equation (vi) becomes $F(s) = -\frac{1}{8}\left[\frac{(s - 4)}{(s + 2)^2 + 2^2}\right]$. (vii)

Laplace pair no. 8 $\mathcal{L}(e^{-kt} \cos \omega t) = \frac{s + k}{(s + k)^2 + \omega^2}$.

Laplace pair no. 7 $\mathcal{L}(e^{-kt} \sin \omega t) = \frac{\omega}{(s + k)^2 + \omega^2}$.

From Laplace pair no. 8, it is observed that to give a perfect inverse

$$F(s) = \frac{s + 2}{(s + 2)^2 + 2^2}.$$

In equation (vii) the numerator is $(s - 4)$. This may be rewritten as $(s + 2) - 6$ without changing its value. Equation (vii) becomes

$$F(s) = -\frac{1}{8}\left[\frac{(s + 2)}{(s + 2)^2 + 2^2} - \frac{6}{(s + 2)^2 + 2^2}\right]$$
$$= -\frac{1}{8}\left[\frac{(s + 2)}{(s + 2)^2 + 2^2} - 3\left(\frac{2}{(s + 2)^2 + 2^2}\right)\right].$$

Therefore, $f(t) = -(1/8)\,(\mathrm{e}^{-2t}\cos 2t - 3\mathrm{e}^{-2t}\sin 2t)$. (viii)

Bringing this together with the first term (equation v) gives

$$f(t) = (1/8)\,(1 - \mathrm{e}^{-2t}\cos 2t + 3\mathrm{e}^{-2t}\sin 2t).$$

Similar methods apply to types 5, 6, 7 and 8 following (only the partial fractions are quoted).

(5) $F(s) = \dfrac{E}{s(s + a)}$. The partial fractions are $\dfrac{A}{s} + \dfrac{B}{(s + a)}$.

(6) $F(s) = \dfrac{1}{(s + p)^2\,(s^2 + q)}$. The partial fractions are

$$\frac{A}{s + p} + \frac{B}{(s + p)^2} + \frac{Cs + D}{(s^2 + q)}.$$

(7) $F(s) = \dfrac{1}{s^2(xs + y)}$. The partial fractions are $\dfrac{A}{s} + \dfrac{B}{s^2} + \dfrac{C}{xs + y}$.

(8) $F(s) = \dfrac{E}{(s^2 + p^2)\,(s + a)}$. The partial fractions are $\dfrac{A}{s + a} + \dfrac{Bs + C}{s^2 + p^2}$.

8.5 Circuit transforms

8.5.1 Resistance

$$v_R = iR.$$

Taking Laplace transforms

$$\mathcal{L}v_R = \mathcal{L}iR$$

$$R = \frac{\overline{v_R}}{\overline{i}}$$

where $\overline{v_R} = \mathcal{L}v_R$ and $\overline{i} = \mathcal{L}i$. (Fig. 8.4(a))

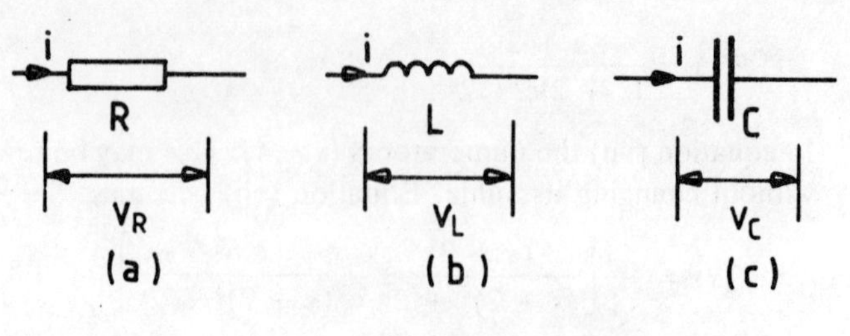

Fig. 8.4

8.5.2 Pure inductance

$$v_L = L\frac{di}{dt}.$$

Taking Laplace transforms

$$\mathscr{L}v_L = L(s\bar{i} - i_{(0)}). \qquad \text{(Laplace pair no. 9)}$$

Again writing $\bar{i}$ for $\mathscr{L}i = F(s)$. $i_{(0)}$ is the value of current at $t = 0$.
If the current is zero at $t = 0$ then

$$\mathscr{L}v_L = Ls\bar{i}$$

$$\frac{\overline{v_L}}{\bar{i}} = sL. \qquad \text{(Fig. 8.4(b))}$$

8.5.3 Pure capacitance

$$i = C\frac{dv}{dt}.$$

$$\mathscr{L}i = C(s\overline{v_C} - v_{(0)}) \qquad \text{(Laplace pair no. 9)}$$

writing $\overline{v_C}$ for $\mathscr{L}v_C = F(s)$.
If the capacitor is initially uncharged, i.e. $v_{(0)} = 0$, then

$$\bar{i} = Cs\overline{v_C}$$

$$\frac{\overline{v_C}}{\bar{i}} = \frac{1}{Cs}. \qquad \text{(Fig. 8.4(c))}$$

With circuit transforms it will be observed that s has replaced (jω) in the steady-state reactances jωL and 1/jωC respectively. This is true in all cases where the initial conditions are zero. When simplifying complicated networks it is convenient to use conventional methods involving j and then substitute s in the relevant places in the final expression for impedance when a transient solution is required.

When solving equations, $\bar{i}$ or $\bar{v}$ is written to indicate the value of $F(s)$ being sought as it is more convenient than using $\mathscr{L}i$ or $\mathscr{L}v$.

WORKED EXAMPLE 8.1

Derive expressions for the current and capacitor voltage in an RC circuit suddenly connected to a direct source of V volts.

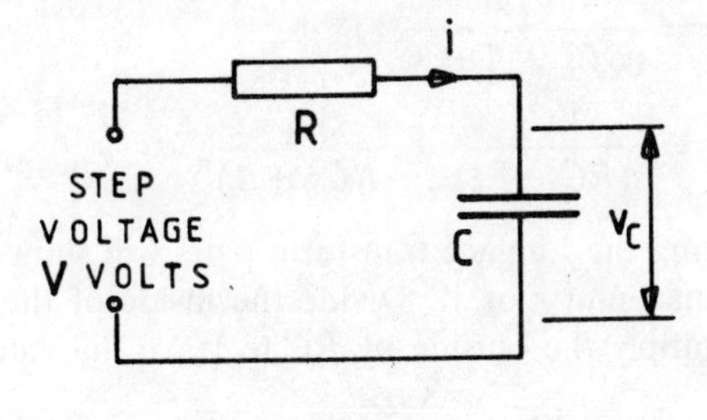

Fig. 8.5

$$V = iR + v_C.$$

Now $dq = C\,dv_C$, and transposing $dv_C = \frac{dq}{C} = \frac{i\,dt}{C}$. The voltage V is applied to the circuit at $t = 0$. The capacitor is initially uncharged.

After time t, capacitor voltage $v_C = \int dv_C = \int_0^t \frac{i\,dt}{C}$.

Therefore $V = iR + \int_0^t \frac{i\,dt}{C}$.

Taking Laplace transforms

Laplace pair 1 $\mathcal{L}V = \frac{V}{s}$.   Laplace pair 11 $\mathcal{L}\left[\int_0^t i\,dt\right] = \frac{\bar{i}}{s}$

(writing $\bar{i}$ for $\mathcal{L}i$).

$$\frac{V}{s} = \bar{i}R + \frac{\bar{i}}{Cs}$$

$$= \bar{i}\left(R + \frac{1}{Cs}\right).$$

This result could have been obtained directly by at first writing the solution in steady-state form

$$V = IR + \frac{I}{j\omega C} = I\left(R + \frac{1}{j\omega C}\right)$$

then substituting the voltage and circuit transforms

$$\frac{V}{s} = \bar{i}\left(R + \frac{1}{Cs}\right).$$

Transposing to obtain $\bar{i}$ gives

$$\bar{i} = \frac{(V/s)}{R + (1/Cs)}$$

$$= \frac{(V/s)}{(RCs + 1)/Cs}$$

$$= \frac{VCs}{s(RCs + 1)} = \frac{VC}{(RCs + 1)}.$$

Inspecting the Laplace transform pairs will show that no transform exists with other than unit s or s^2. Divide the inside of the denominator bracket by RC and multiply the outside by RC to leave the value unchanged

$$\bar{i} = \frac{VC}{RC(s + 1/RC)} = \frac{V}{R}\frac{1}{(s + 1/RC)}.$$

Compare this with Laplace pair 3 $\mathscr{L}e^{-at} = \dfrac{1}{s + a}$

by letting $a = 1/RC$, i = inverse of $\dfrac{V}{R}\dfrac{1}{(s + 1/RC)} = \dfrac{V}{R}\,e^{-t/RC}$.

But $v_C = \displaystyle\int_0^t \frac{i\,dt}{C}$. Substituting the value for i gives

$$v_C = \int_0^t \frac{(V/R)\,e^{-t/RC}}{C}\,dt$$

$$= \left[\frac{(V/R)\,e^{-t/RC}}{-C\,(1/RC)}\right]_0^t$$

$$= (-V\,e^{-t/RC} + V\,e^{-0})$$

$$v_C = V(1 - e^{-t/RC}) \text{ volts.}$$

At $t = \infty$, $e^{-t/RC} = 0$ and $v_C = V$. The value of V is the final steady-state value of capacitor voltage. The steady-state solution is $v_C = V$, the supply voltage.

At any time t, v_C = steady-state voltage $-(Ve^{-t/RC})$.

Generally it may be stated that

condition at time t = final steady-state condition + transient response.

WORKED EXAMPLE 8.2

A capacitor C carrying a charge q_0 and with a standing voltage V volts is suddenly connected to a resistance R Ω at $t = 0$. Using Laplace transforms develop an expression for the current in the resistor at any time t. (The instantaneous capacitor voltage is v_C volts.)

At $t = 0$, $q_0 = CV$, or transposing $q_0/C = V$. (i)

Charge leaves the capacitor, flowing through the resistor.

Charge leaving $= \int_0^t i\,dt$ coulombs.

After time t, charge remaining on the capacitor, $q = q_0 - \int_0^t i\,dt\;C$ (ii)

Since $q = Cv_C$, $v_C = q/C$. (iii)

Substituting equation (ii) in equation (iii)

$$v_C = \frac{q_o - \int_0^t i\,dt}{C} = \frac{q_o}{C} - \int_0^t \frac{i\,dt}{C}. \quad \text{(iv)}$$

v_C is the potential difference at the capacitor terminals and, since it is connected to a resistor, it must be the p.d. across the resistor. Also,

p.d. across the resistor $= iR$ volts.

Therefore $v_C = iR$. (v)

Substituting equations (v) and (i) in equation (iv) gives

$$iR = V - \int_0^t \frac{i\,dt}{C}.$$

Taking Laplace transforms of both sides (the individual transforms are as used in worked example 8.1)

$$\bar{i}R = \frac{V}{s} - \frac{\bar{i}}{Cs}$$

$$\frac{V}{s} = \bar{i}(R + 1/Cs)$$

$$\bar{i} = \frac{V}{s(R + 1/Cs)}.$$

This is identical to the expression found for the charging current in worked example 8.1 and the solution is

$$i = \frac{V}{R}\,\mathrm{e}^{-t/RC}\ \mathrm{A}.$$

The final steady-state current is zero when the capacitor is fully discharged.

Current at time t = steady-state current + transient responses.

WORKED EXAMPLE 8.3

A coil with resistance $R\ \Omega$ and inductance L henry is suddenly connected to a direct voltage of V volts at time $t = 0$. Deduce an expression for the circuit current i amperes at any time t seconds after switching on.

In the loop $V - iR - L\dfrac{\mathrm{d}i}{\mathrm{d}t} = 0$

Laplace pair 1 $\mathscr{L}V = \dfrac{V}{s}$ Laplace pair 9 $\mathscr{L}L\dfrac{\mathrm{d}i}{\mathrm{d}t} = s\bar{i} - i_{(0)}$

$i_{(0)}$ is the value of current at $t = 0$ $\therefore i_{(0)} = 0$.
Hence, the Laplace transform of the loop equation is

$$\frac{V}{s} - \bar{i}R - Ls\bar{i} = 0 \quad \text{or} \quad \frac{V}{s} = \bar{i}(R + sL).$$

As in worked example 8.1, this equation could have been written by using the voltage and component transforms directly.

Transposing $\bar{i} = \dfrac{V}{s(R + sL)}$ dividing top and bottom by L, $\bar{i} = \dfrac{V/L}{s(s + R/L)}$

Using partial fractions (Section 8.4, Type 5)

$$\frac{A}{s} + \frac{B}{R/L + s} = \frac{V/L}{s(s + R/L)} \tag{i}$$

$$A\,\{R/L + \mathrm{s}\} + \mathrm{sB} = V/L$$

Equate numbers $AR/L = V/L$ $\therefore A = \dfrac{V}{R}$

Equate values of s $As + Bs = 0$ $\therefore B = -V/R$

Substitute values in equation (i) above

$$\bar{i} = \frac{V}{Rs} - \frac{V}{R(R/L + s)}$$

Taking inverses:

Laplace pair 1 Inverse $\dfrac{V}{R}\dfrac{1}{\mathrm{s}} = \dfrac{V}{R}$

Laplace pair 3 Inverse $\dfrac{V}{R}\dfrac{1}{(s + R/L)} = \dfrac{V}{R}\mathrm{e}^{-Rt/L}$

$$\therefore i = \frac{V}{R}(1 - \mathrm{e}^{(-Rt/L)})\ A$$

SELF-ASSESSMENT EXAMPLE 8.4

A steady current I amperes is established in a coil with resistance R Ω and inductance L henry. At time $t = 0$, the supply is disconnected and the input terminals are simultaneously short circuited forming the circuit shown in Fig. 8.6. Deduce that the expression for the current i amperes t seconds after shorting out the terminals is given by

$$i = I\,\mathrm{e}^{-Rt/L}\mathrm{A}$$

(Hints: In the loop equation in worked example 8.3, $V = 0$. The driving voltage is now $L(\mathrm{d}i/\mathrm{d}t)$ so that the polarity of the voltage across the inductor has reversed. Also to be considered is the fact that the circuit current is falling so that di/dt is negative. $i_{(0)}$ is *not* zero in this case.)

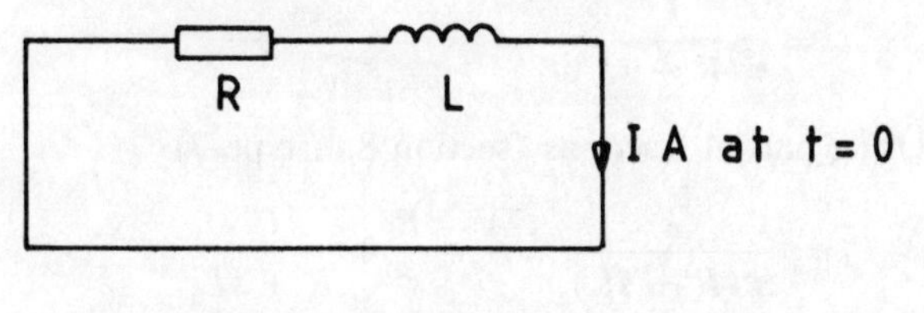

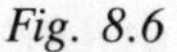
Fig. 8.6

The general forms for currents and voltages in inductive and capacitive circuits dealt with in working examples 8.1 to 8.4 are shown in Fig. 8.7.

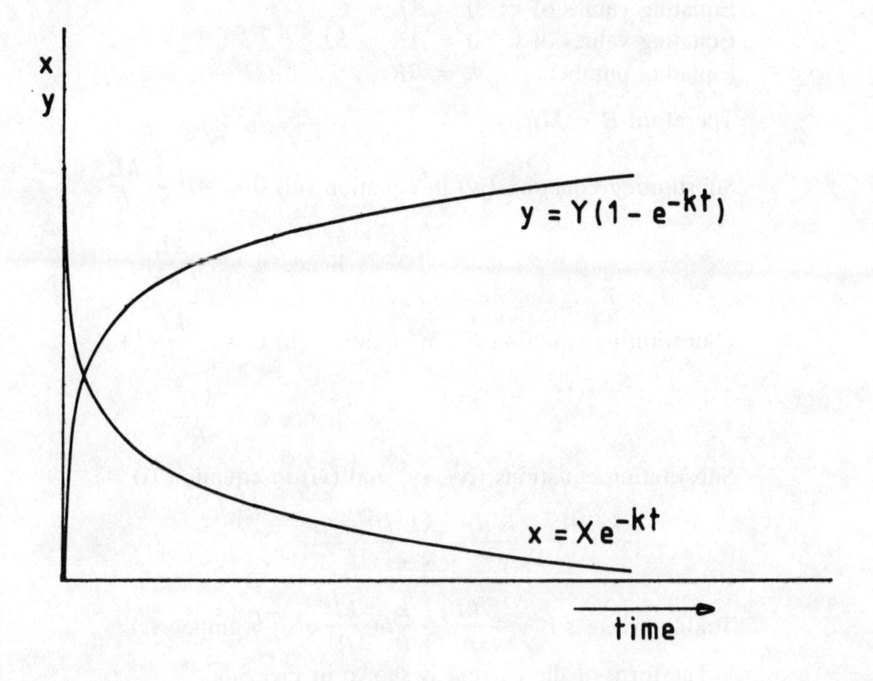

Fig. 8.7

WORKED EXAMPLE 8.5

Determine an expression for the current at time t seconds when an RL circuit is fed with a ramp voltage given by $v = kt$ volts. The initial conditions are zero.

Initially the circuit carries no current so the circuit transforms can be used directly.

Steady-state $V = I(R + j\omega L)$ and $V = kt$.

$$\mathcal{L}kt = \frac{k}{s^2} \text{ (Laplace pair 4).}$$

Therefore $\frac{k}{s^2} = \bar{i}(R + sL)$

$$\bar{i} = \frac{k}{s^2(R + sL)}.$$

Using partial fractions (section 8.4, type 7)

$$\bar{i} = \frac{k}{s^2(R + sL)} = \frac{A}{s} + \frac{B}{s^2} + \frac{C}{R + sL} \quad \text{(i)}$$

$$= \frac{As(R + sL) + B(R + sL) + Cs^2}{s^2(R + sL)}.$$

Hence $k = ARs + ALs^2 + BR + BLs + Cs^2$.

Equating values of s^2 $\quad 0 = AL + C$ (ii)
Equating values of s $\quad 0 = AR + BL$ (iii)
Equating numbers $\quad k = BR$.

Therefore $B = k/R$. (iv)

Substituting equation (iv) in equation (iii) $0 = AR + \frac{kL}{R}$

hence $A = -\frac{kL}{R^2}$. (v)

Substituting equation (v) in equation (ii) $0 = -\frac{kL^2}{R^2} + C$

hence $C = \frac{kL^2}{R^2}$. (vi)

Substituting equations (iv), (v) and (vi) in equation (i)

$$i = \frac{-kL/R^2}{s} + \frac{k/R}{s^2} + \frac{kL^2/R^2}{R + sL}.$$

Taking inverses $i = -\frac{kL}{R^2} + \frac{k}{R}t + \frac{kL}{R^2}\,\mathrm{e}^{-Rt/L}$ amperes.

The form of the current is shown in Fig. 8.8.

SELF-ASSESSMENT EXAMPLE 8.6

Determine an expression for the input current to the parallel circuit shown in Fig. 8.9. The input voltage is a ramp given by $v = kt$. The initial conditions are zero.

There are two methods, both of which could be attempted to advantage.
(i) From worked example 8.5 the current in R and L is known for a ramp input. Determine an expression for the current in C alone and then add this to the former value.
(ii) The impedance of a parallel circuit is given by

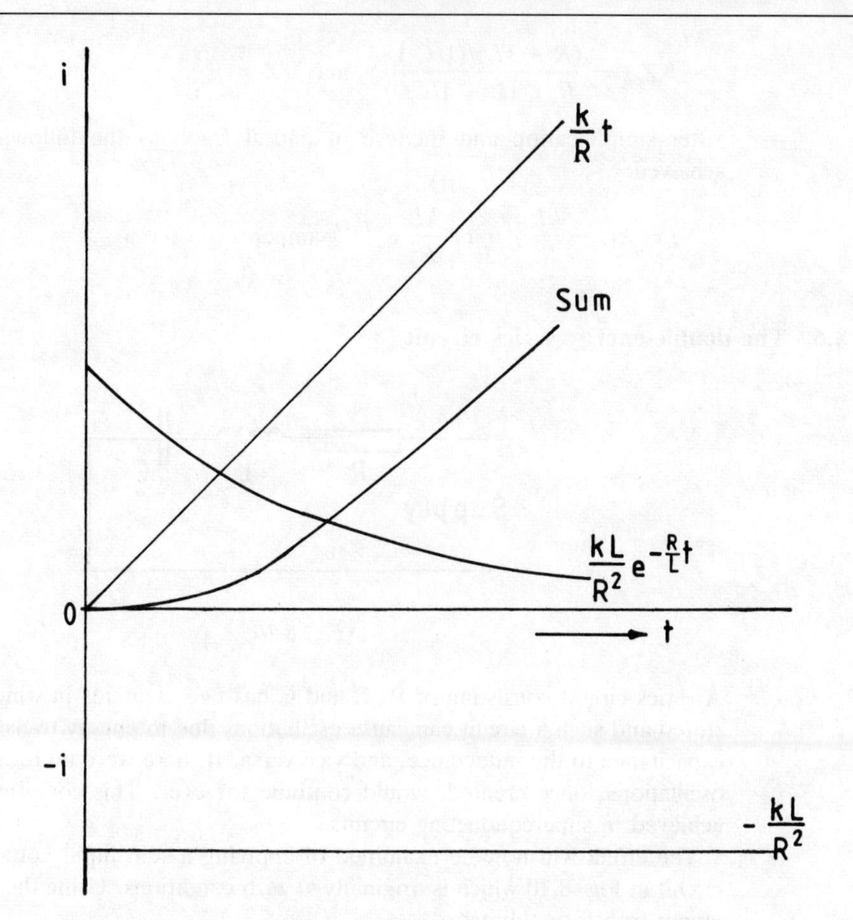

Fig. 8.8

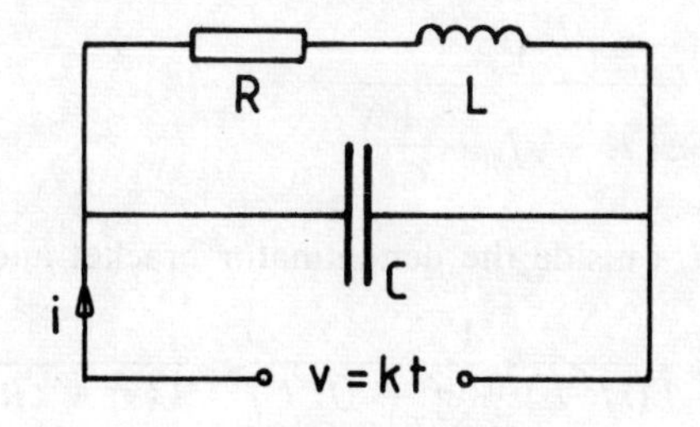

Fig. 8.9

$$Z_{eq} = \frac{Z_1 Z_2}{Z_1 + Z_2}.$$

Writing in the impedances in their steady-state form, and then substituting s for $(j\omega)$ gives

$$\mathcal{L}Z_{eq} = \frac{(R + sL)/(1/Cs)}{R + sL + 1/Cs} \qquad \text{and } \mathcal{L}Z = \mathcal{L}\frac{v}{i}.$$

After simplification and the use of partial fractions the following result is achieved:

$$i = kC - \frac{kL}{R^2} + \frac{k}{R}t + \frac{kL}{R^2}\,e^{-Rt/L} \text{ amperes.}$$

8.6 The double-energy series circuit

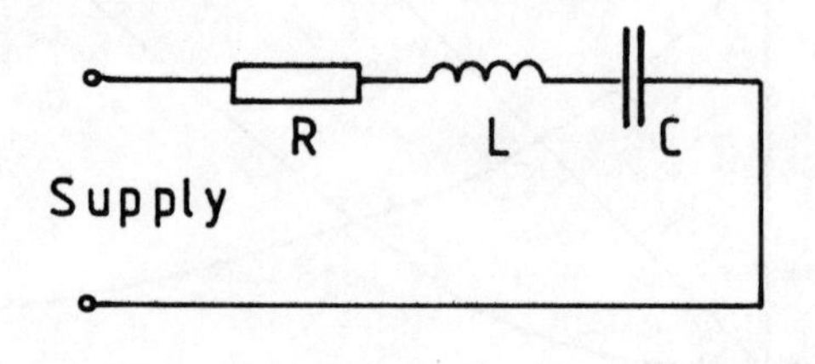

Fig. 8.10

A series circuit consisting of R, L and C has two elements in which energy is stored and such a circuit can cause oscillations due to energy transfer from the capacitance to the inductance, and vice versa. If there were no resistance, such oscillations, once created, would continue for ever. This condition has been achieved in superconducting circuits.

The effect will now be examined of applying a step input voltage V to the circuit in Fig. 8.10 which is originally at zero conditions. Using the voltage and circuit transforms direct

$$\frac{V}{s} = \bar{i}\left(R + sL + \frac{1}{Cs}\right)$$

$$\bar{i} = \frac{V}{s\left(R + sL + \dfrac{1}{Cs}\right)}.$$

Taking L outside the denominator bracket and s inside the bracket gives

$$\bar{i} = \frac{V}{L(sR/L) + s^2 + 1/LC)} = \frac{V}{L(s^2 + (R/sL) + 1/LC)}.$$

By adding $\left(\dfrac{R}{2L}\right)^2$ to $\left(s^2 + \dfrac{R}{L}s\right)$, a perfect square is created since

$$\left[s^2 + \frac{R}{L}s + \left(\frac{R}{2L}\right)^2\right] = \left(s + \frac{R}{2L}\right)^2.$$

To maintain the value of the denominator the value $(R/2L)^2$ must be subtracted from $1/LC$, hence

$$\bar{i} = \frac{V}{L[s^2 + (sR/L) + (R/2L)^2 + 1/LC - (R/2L)^2]}$$

$$= \frac{V}{L[(s + R/2L)^2 + 1/LC - (R/2L)^2]}.$$

This transform is of the type $\dfrac{V}{L[(s + k)^2 + p^2]}$

where $p^2 = \dfrac{1}{LC} - \left(\dfrac{R}{2L}\right)^2$ or $p = \sqrt{\left[\dfrac{1}{LC} - \left(\dfrac{R}{2L}\right)^2\right]}$.

$$\text{Therefore } \bar{i} = \frac{V}{L[(s + R/2L)^2 + \{\sqrt{[1/LC - (R/2L)^2]}\}^2]}. \tag{8.2}$$

Equation (8.2) has four solutions

(1) Undamped, when $R = 0$.
(2) Underdamped, when $(R/2L)^2$ is less than $1/LC$.
(3) Critically damped when $(R/2L)^2 = 1/LC$ or, by transposing, $R = 2\sqrt{(L/C)}$.
(4) Overdamped, when $(R/2L)^2$ is greater than $1/LC$.

Each of these solutions will be examined in turn.

(1) Undamped, with $R = 0$ in equation (8.2)

$$\bar{i} = \frac{V}{L(s^2 + (1/LC))}$$

from work on series resonance

$$\omega_0 = \sqrt{\left(\frac{1}{LC}\right)} \text{ (equation 7.1),} \qquad \text{hence } \frac{1}{LC} = \omega_0^2$$

$$\text{and } \bar{i} = \frac{V}{L(s^2 + \omega_0^2)}.$$

To make a perfect inverse, ω_0 is needed in the numerator. Multiply and divide by ω_0

$$\bar{i} = \frac{V}{\omega_0 L}\left(\frac{\omega_0}{s^2 + \omega_0^2}\right).$$

From Laplace pair 5

$$i = \frac{V}{\omega_0 L} \sin \omega_0 t \text{ amperes.} \tag{8.3}$$

This is a sine wave of amplitude $V/\omega_0 L$ amperes and angular velocity ω_0 rad/s (see Fig. 8.11).

(2) Underdamped, $(R/2L)^2$ less than $1/LC$.

Equation (8.2) stands unchanged, and it may be seen to be in the form

$$\bar{i} = \frac{V}{L[(s + k)^2 + \omega^2]} \qquad \text{(Laplace pair 7)}$$

$$k = \frac{R}{2L} \quad \text{and } \omega = \sqrt{\left[\frac{1}{LC} - \left(\frac{R}{2L}\right)^2\right]}.$$

It requires ω in the numerator for a perfect inverse. Multiply and divide by ω

$$\bar{i} = \frac{V}{\omega L}\left[\frac{\omega}{(s + k)^2 + \omega^2}\right].$$

$$i = \frac{V}{\omega L} e^{-kt} \sin \omega t.$$

Substituting for k and ω

$$i = \frac{V}{L\sqrt{[1/LC - (R/2L)^2]}} e^{-Rt/2L} \sin \sqrt{[1/LC - (R/2L)^2]}t \text{ A.} \qquad (8.4)$$

This is an oscillatory current with angular velocity ω, which is decaying exponentially. The rate of decay is determined by the ratio of R to L (see Fig. 8.11).

(3) Critically damped, $(R/2L)^2 = 1/LC$.

Equation (8.2) reduces to

$$\bar{i} = \frac{V}{L(s + R/2L)^2}.$$

The inverse of $\frac{1}{s^2} = t$ (Laplace pair 4).

Adding $R/2L$ to the s involves the exponential shift theorem (section 8.3) and this multiplies t by $e^{-Rt/2L}$. Therefore

$$\text{inverse of } \left\{\frac{1}{(s + R/2L)^2}\right\} = te^{-Rt/2L}$$

$$\text{and inverse of } \left[\frac{V}{L(s + R/2L)^2}\right] = \frac{V}{L} te^{-Rt/2L} \text{ A.} \qquad (8.5)$$

This is non-oscillatory (see Fig. 8.11).

(4) Overdamped, $(R/2L)^2$ greater than $1/LC$.

In equation (8.2), $1/LC - (R/2L)^2$ will be negative and it is therefore necessary to write the second term in the denominator as

$$-\{\sqrt{[(R/2L)^2 - 1/LC]}\}^2.$$

$$\bar{i} = \frac{V}{L[(s + R/2L)^2 - \{\sqrt{[(R/2L)^2 - 1/LC]}\}^2]}.$$

This is of the form $\dfrac{V}{L[(s+k)^2 - a^2]}$. (Laplace pair 14)

However, an a is needed in the numerator to make a perfect inverse. Multiplying and dividing by a gives

$$\bar{i} = \frac{Va}{aL[(s+k)^2 + a^2]}$$

where $a = \sqrt{\left[\left(\frac{R}{2L}\right)^2 - \frac{1}{LC}\right]}$ and $k = \frac{R}{2L}$.

Taking the inverse $i = \dfrac{V}{aL}\,e^{-kt}\sinh at$

and substituting for a and k gives

$$\bar{i} = \frac{V}{L\sqrt{[(R/2L)^2 - 1/LC]}}\,e^{-Rt/2L}\sinh\sqrt{[(R/2L)^2 - 1/LC]}\;t\ \text{A}. \qquad (8.6)$$

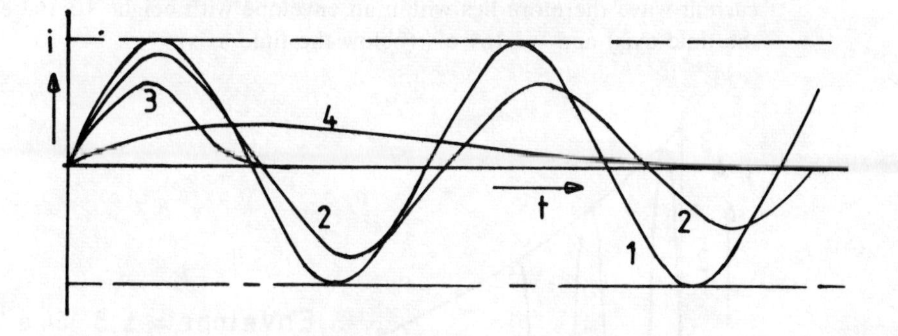

Fig. 8.11

Figure 8.11 shows typical response curves for the four conditions covered in this section (labelled 1 to 4).

WORKED EXAMPLE 8.7

A circuit consisting of a coil with resistance 1 Ω and inductance 0.5 H in series with a capacitor with capacitance 2000 μF is connected at $t = 0$ to a direct voltage of 50 V. Determine an expression for the current at time t and sketch the waveform over the first 2 seconds.

Firstly it is necessary to determine the state of damping of the circuit.

Evaluate $\left(\dfrac{R}{2L}\right)^2$: $\quad \left(\dfrac{1}{2 \times 0.5}\right)^2 = 1$

Evaluate $\dfrac{1}{LC}$: $\quad \dfrac{1}{0.5 \times 2000 \times 10^{-6}} = 1000$

Hence $(R/2L)^2$ is less than $1/LC$ and the circuit is underdamped. For the underdamped circuit, the solution is given in equation (8.4)

$$i = \frac{V}{L\,\sqrt{[1/LC - (R/2L)^2]}}\,e^{-Rt/2L}\sin\sqrt{[1/LC - (R/2L)^2]}\,t$$

$$= \frac{50}{0.5\sqrt{(1000 - 1)}}\,e^{-t}\sin\sqrt{(1000 - 1)}\,t$$

$$i = 3.164\,e^{-t}\sin 31.61t \text{ amperes.}$$

Consider values of $3.164\,e^{-t}$:

at $t = 0$ value $= 3.164$
at $t = 0.5$ s value $= 3.164 \times 0.6065 = 1.92$
at $t = 1$ s value $= 3.164 \times 0.368 = 1.164$
at $t = 2$ s value $= 3.164 \times 0.135 = 0.428$.

During positive half-cycles, sin ωt is positive and the product ($3.164\,e^{-t}$ sin $31.61t$) is positive. During negative half-cycles the product is negative. The current wave therefore lies within an envelope with height $+3.164\,e^{-t}$ (above the time axis) and $-3.164\,e^{-t}$ (below the time axis).

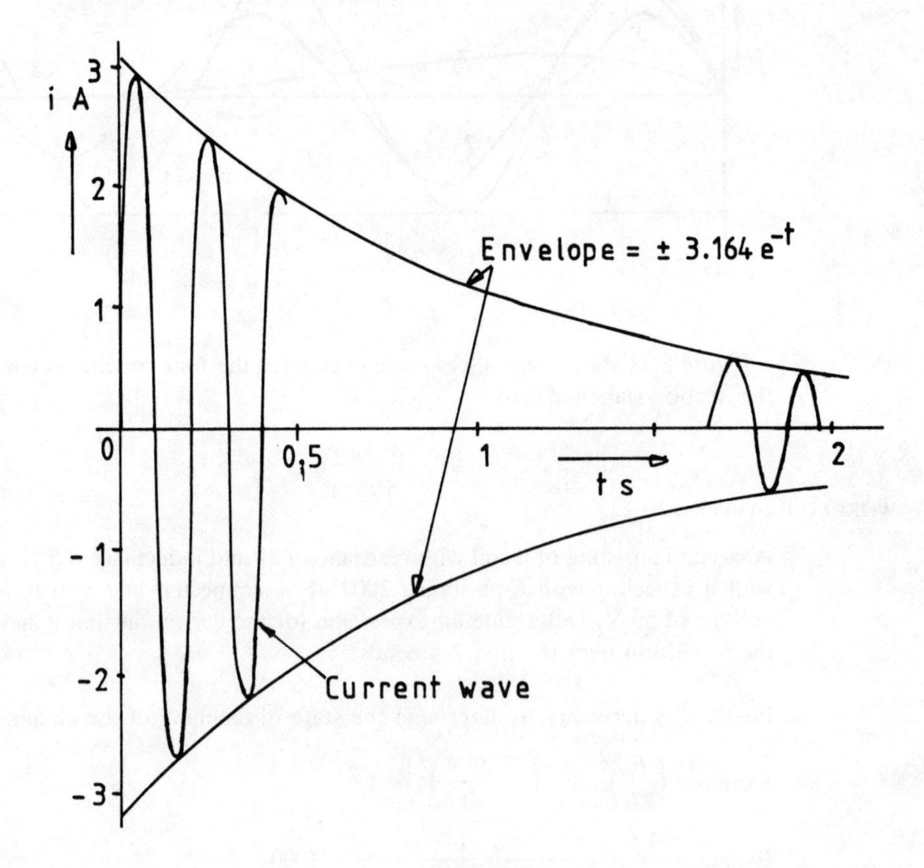

Fig. 8.12

At 31.61 rad/s, $f = 5.031$ Hz. The periodic time = 0.198 s. One quarter cycle takes 0.0497 s.

The resultant current wave is shown in Fig. 8.12.

SELF-ASSESSMENT EXAMPLE 8.8

To the circuit used in worked example 8.7 is added a variable resistor connected in series with the coil.

Determine the value to which the resistor must be set to give critical damping.

Sketch the waveform of the current with this value of resistance in circuit when the supply is at a steady 25 V and is suddenly connected to the circuit at $t = 0$.

8.7 Inductive circuits supplied with a sinusoidal voltage

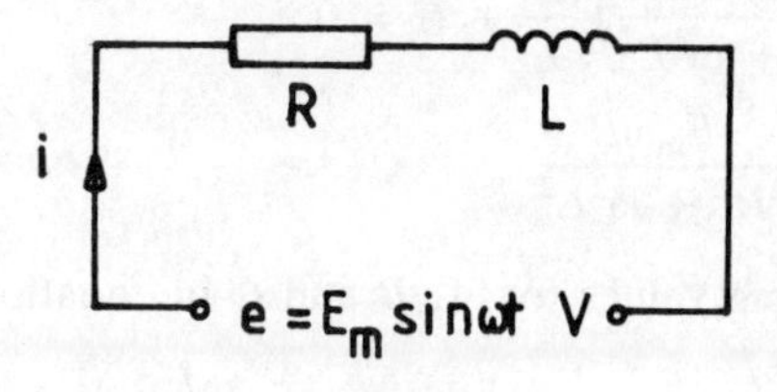

Fig. 8.13

Consider an inductive circuit as shown in Fig. 8.13. It is connected to a source of e.m.f. given by $e = E_m \sin \omega t$ volts, at $t = 0$. An expression for the instantaneous circuit current at time t seconds is required.

At $t = 0$, the e.m.f. is zero and is about to go positive.

Laplace pair 5 $\mathcal{L}(E_m \sin \omega t) = \dfrac{E_m\omega}{s^2 + \omega^2}$.

$$\mathcal{L}(R + j\omega L) = (R + sL).$$

$$\text{Therefore } \frac{E_m\omega}{s^2 - \omega^2} = \bar{i}\,(R + sL).$$

Transposing

$$\bar{i} = \frac{E_m\omega}{(s^2 + \omega^2)\,(R + sL)} = \frac{E_m\omega/L}{(s^2 + \omega^2)\,(s + R/L)}.$$

Using partial fractions (section 8.4, type 8)

$$\frac{E_m\omega/L}{(s^2 + \omega^2)\,(s + R/L)} = \frac{A}{s + R/L} + \frac{Bs + C}{s^2 + \omega^2} \qquad \text{(i)}$$

$$E_m\omega/L = As^2 + A\omega^2 + Bs^2 + (BR/L)s + Cs + C(R/L).$$

Equating values of s^2 $\quad A + B = 0$, therefore $A = (-B)$. (ii)

Equating values of s $\dfrac{BR}{L} + C = 0.$ (iii)

Substituting equation (ii) in equation (iii) $-\dfrac{AR}{L} + C = 0, \quad C = \dfrac{AR}{L}.$ (iv)

Equating numbers $A\omega^2 + C\dfrac{R}{L} = \dfrac{E_m\omega}{L}.$ (v)

Substituting equation (iv) in equation (v) $A\omega^2 + \dfrac{AR^2}{L^2} = \dfrac{E_m\omega}{L}$

$$A = \frac{E_m\omega}{L(\omega^2 + R^2/L^2)} = \frac{E_m\omega}{R^2/L + \omega^2 L} = \frac{E_m\omega L}{R^2 + \omega^2 L^2}.$$

From equation (ii) $B = \dfrac{-E_m\omega L}{R^2 + \omega^2 L^2}.$

Substituting the value of B in equation (iii)

$$\left(\frac{-E_m\omega\, L}{R^2 + \omega^2 L^2}\right)\frac{R}{L} + C = 0$$

$$C = \frac{E_m\omega R}{R^2 + \omega^2 L^2}.$$

Substituting values for A, B and C in equation (i)

$$\bar{i} = \frac{\left(\dfrac{E_m\omega L}{R^2 + \omega^2 L^2}\right)}{s + R/L} - \frac{\left(\dfrac{E_m\omega L}{R^2 + \omega^2 L^2}\right) s + \left(\dfrac{E_m\omega R}{R^2 + \omega^2 L^2}\right)}{s^2 + \omega^2}$$

$$= \frac{E_m\omega}{R^2 + \omega^2 L^2}\left(\frac{L}{s + R/L} - \frac{sL}{s^2 + \omega^2} + \frac{R}{s^2 + \omega^2}\right).$$

Taking inverses

$$i = \frac{E_m\omega}{R^2 + \omega^2 L^2}\left(Le^{-Rt/L} - L\cos\omega t + \frac{R}{\omega}\sin\omega t\right)$$

$$= \frac{E_m\omega}{R^2 + \omega^2 L^2}Le^{-Rt/L} + \frac{E_m\omega}{R^2 + \omega^2 L^2}\left(-L\cos\omega t + \frac{R}{\omega}\sin\omega t\right).$$

The term in brackets is the rectangular form of the phasor shown in Fig. 8.14.

The resultant $= \sqrt{\left[L^2 + \left(\dfrac{R}{\omega}\right)^2\right]} = \dfrac{1}{\omega}\sqrt{(\omega^2 L^2 + R^2)}$ and $\tan\phi = \dfrac{\omega L}{R}.$

$$i = \frac{E_m\omega}{R^2 + \omega^2 L^2}Le^{-Rt/L} + \frac{E_m\omega}{R^2 + \omega^2 L^2} \times \frac{1}{\omega}\sqrt{(\omega^2 L^2 + R^2)}\sin(\omega t - \phi)\text{ A}$$

$$= \frac{E_m\omega}{R^2 + \omega^2 L^2}Le^{-Rt/L} + \frac{E_m}{\sqrt{(R^2 + \omega^2 L^2)}}\sin(\omega t - \phi)\text{ A.} \quad (8.7)$$

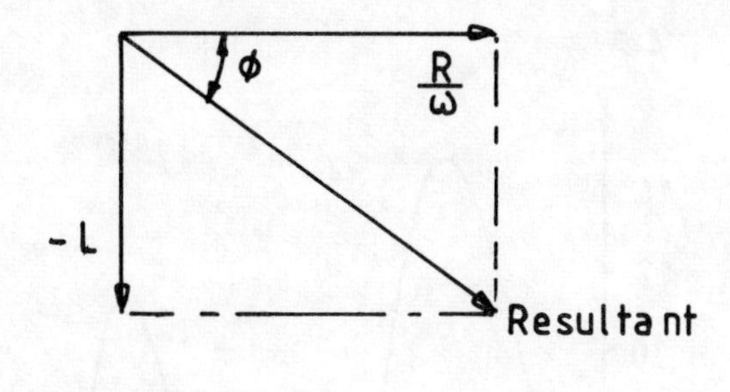

Fig. 8.14

WORKED EXAMPLE 8.9

A coil with resistance 1 Ω and inductance 15.9 mH is connected at time $t = 0$ to a voltage expressed by the equation $v = 50 \sin 314t$ volts. Draw to scale the first two cycles of input current. The secondary is open-circuited (it has no effect).

$$X_L = 2\pi fL = 314 \times 15.9 \times 10^{-3} = 5\ \Omega. \text{ Therefore, } \omega^2L^2 = 25.$$

Substituting values in equation (8.7)

$$i = \frac{50 \times 314}{1 + 25} \times 0.0159\ e^{-1t/0.0159} + \frac{50}{\sqrt{(1^2 + 25)}} \sin(314t - \phi)$$

$$= 9.6\ e^{-62.89t} + 9.805 \sin(314t - \phi)\ \text{A}.$$

$$\phi = \arctan\left(\frac{314 \times 0.0159}{1}\right) = 78.67° = 1.373\ \text{rad}.$$

The following table can be drawn up.

t	$9.6\ e^{-62.89t}$	$9.805 \sin(314t - 1.373)$	i (A)
0	9.6	−9.6	0
0.005	7	1.923	8.923
0.0094	5.32	9.805	15.12 (peak value)
0.01	5.118	9.614	14.73
0.015	3.737	−1.922	1.814
0.02	2.73	−9.61	−6.88
0.025	2	1.923	3.923
0.03	1.455	9.614	11.069
0.035	1.06	−1.922	−0.862
0.04	0.77	−9.61	−8.8

From these results Fig. 8.15 is drawn.

When an inductive circuit is connected to a sinusoidal supply (the voltage wave passing through a zero at the instant the switch is closed), the initial half cycle of current will have a peak value that may be greatly in excess of that of the steady-state sine wave since it contains a transient direct current.

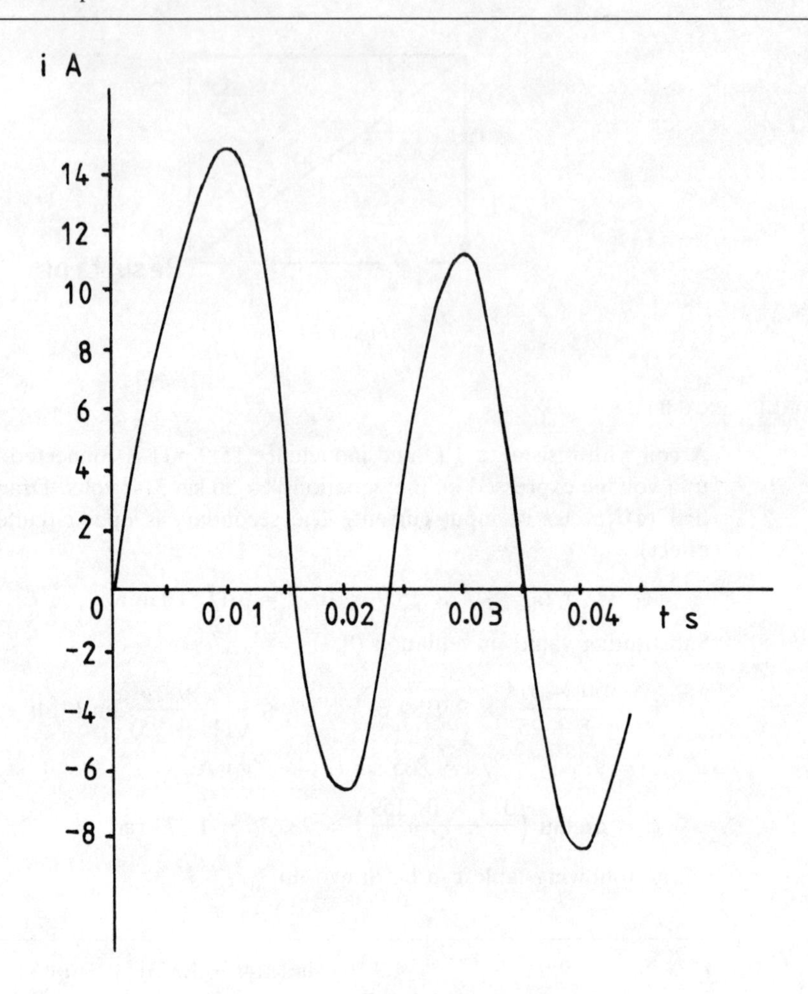

Fig. 8.15

Successive odd numbered half cycles have reduced magnitudes as the transient dies away until the final steady-state sine wave current is left. It should be noted that the right-hand term in equation (8.7) is the normal steady-state solution for this circuit.

In the case of large power transformers the magnitude of the current peak must be allowed for in the design, otherwise the mechanical forces generated may disrupt the end windings. There have been instances in which the windings have been forced up against the transformer casing causing rupture, a leakage of oil and an explosion.

Clearly the point on the voltage waveform at which the switch is closed is a matter of chance, so that it is worth examining what would happen if the switch were closed when the voltage wave was at a maximum value.

SELF-ASSESSMENT EXAMPLE 8.10

A resistive inductive circuit as in Fig. 8.13 is energised from a sinusoidal alternating supply at the instant the voltage wave is at a maximum value.

Therefore at $t = 0$, $e = E_m$. Hence $e = E_m \cos \omega t$ volts. Deduce the following expressions:

(a) $$\bar{i} = \frac{E_m s/L}{(s^2 + \omega^2)(s + R/L)}.$$

(b) Simplifying (a) using partial fractions as in worked example 8.9, deduce that

$$\bar{i} = \frac{\left(\dfrac{-E_m R}{\omega^2 L^2 + R^2}\right)}{s^2 + R/L} + \frac{\left(\dfrac{E_m R\, s}{\omega^2 L^2 + R^2}\right) + E_m\left(\dfrac{\omega^2 L}{\omega^2 L^2 + R^2}\right)}{s^2 + \omega^2}.$$

(c) $$i = \frac{-E_m R}{\omega^2 L^2 + R^2} e^{-Rt/L} + \frac{E_m R}{\omega^2 L^2 + R^2} \cos \omega t + \frac{E_m \omega L}{\omega^2 L^2 + R^2} \sin \omega t$$

$$= \frac{-E_m R}{\omega^2 L^2 + R^2} e^{-Rt/L} + \frac{E_m}{\sqrt{(R^2 + \omega^2 L^2)}} \sin(\omega t + \phi)$$

where $\phi = \arctan(R/\omega L)$.

The form of the resulting current calculated for the circuit parameters in worked example 8.9 is shown in Fig. 8.16. It will be noted that almost perfect symmetry of current waveform is achieved immediately and therefore the ideal time at which to energise an inductive circuit is as the applied voltage passes through a maximum value.

The instant of closing a switch is arbitrary so that conditions vary between the best and worst. When energising a transformer a transient hum is often

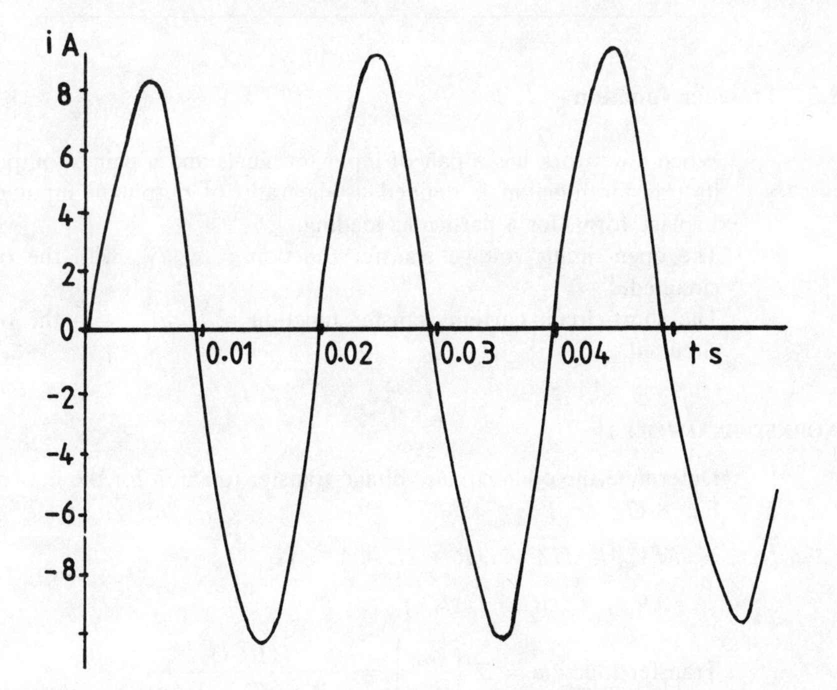

Fig. 8.16

heard for the first 5–10 cycles indicating that the instant of closing the switch was not ideal. When energising a three-phase transformer, since the voltage waveforms are displaced by 120°, at least two phases will be energised at such an instant.

Table for self-assessment example 8.10

$$i = \frac{-50}{1+25}\,\mathrm{e}^{-62.89t} + \frac{50}{\sqrt{(1+25)}}\sin(\omega t + 0.197)$$

$$= 1.92\,\mathrm{e}^{-62.89t} + 9.805\sin(314t + 0.197).$$

t	$-1.92\,\mathrm{e}^{-62.89t}$	$9.805\sin(314t + 0.197)$	i (A)
0	−1.92	1.92	0
0.0044	−1.46	9.805	8.335
0.005	−1.402	9.614	8.216
0.01	−1.024	−1.92	−2.946
0.015	−0.747	−9.614	−10.36
0.02			1.375
0.025			9.216
0.03			−2.21
0.035			−9.826
0.04			1.765
0.045			9.5
0.05			−2.0
0.055			−9.67

8.8 Transfer function

When a network has a pair of input terminals and a pair of output terminals, its transfer function is defined as the ratio of output to input expressed in Laplace form, for a particular loading.

The open-circuit voltage transfer function $= \bar{v}_{out}/\bar{v}_{in}$ with the output open-circuited.

The short-circuit current transfer function $= \bar{i}_{out}/\bar{i}_{in}$ with the output short-circuited.

WORKED EXAMPLE 8.11

Determine the open-circuit voltage transfer function for the network shown in Fig. 8.17.

$$\mathscr{L}V_{in} = \mathscr{L}iZ = \bar{i}(R + sL + 1/Cs)$$

$$\mathscr{L}V_{out} = \mathscr{L}iC = \bar{i}(1/Cs).$$

$$\text{Transfer function} = \mathscr{L}\left[\frac{v_{out}}{v_{in}}\right] = \frac{\bar{i}(1/Cs)}{\bar{i}(R + sL + 1/Cs)}$$

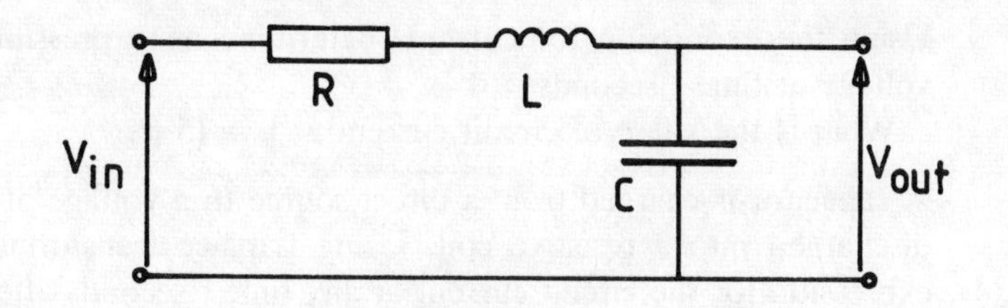

Fig. 8.17

$$= \frac{1}{Cs(R + sL + 1/Cs)}$$

$$= \frac{1}{s^2LC + CRs + 1}$$

$$= \frac{1}{LC(s^2 + (R/L)s + 1/LC)}.$$

SELF-ASSESSMENT EXAMPLE 8.12

Show that the open circuit voltage transfer function of the filter network shown in Fig. 8.18 is given by

$$\frac{1}{LC(s^2 + 1/LC)}.$$

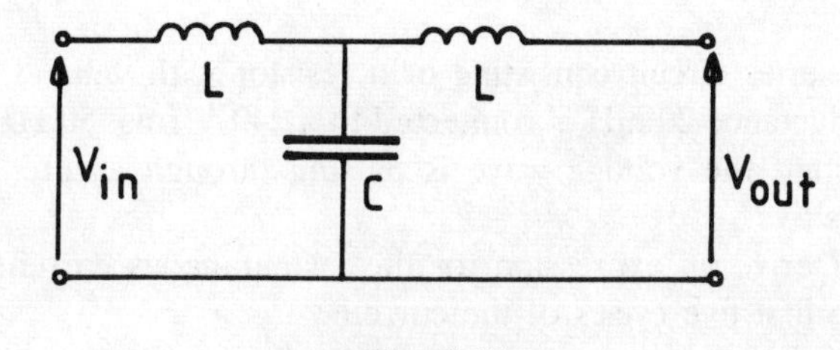

Fig. 8.18

FURTHER PROBLEMS

8.13 A 6 mH inductor having a resistance of 100 Ω is connected in series with a 0.05 μF capacitor. The circuit so formed is suddenly connected to a direct source of 20 V at $t = 0$. Given that the initial conditions are zero, show that the current in the circuit t s after connecting the supply, can be expressed in the form

$$i = I\,e^{-kt} \sin \omega t \text{ A}.$$

Calculate the values of I, k and ω.

8.14 A capacitor with capacitance 0.15 μF in series with a 600 Ω resistor is switched on to a voltage given by $v = 100t$ at $t = 0$. The capacitor is initially uncharged. Deduce an expression for the current at time t seconds after switching on.

Using the expression for current, determine an expression for the capacitor voltage at time t seconds.

What is the value of circuit current at $t = 15\ \mu s$?

8.15 A capacitor is charged from a direct source to a voltage of V_0 volts. It is then discharged into a resistive coil. Using Laplace transforms deduce a general expression for the circuit current at any time t seconds after the capacitor has been connected to the coil.

Given that $C = 0.2\ \mu F$, $L = 0.5$ H and $R = 5\ \Omega$, evaluate the general expression in terms of V_0 and the circuit parameters.

8.16 A ramp voltage $v = kt$ is applied at $t = 0$ to the circuit shown in Fig. 8.19. Given that the capacitor is initially uncharged and the inductor current is zero at $t = 0$, evaluate the value of the input current at $t = 0.04$ second given that $R_1 = 10\ \Omega$, $R_2 = 1000\ \Omega$, $L = 0.35$ H, $C = 0.1\ \mu F$ and $k = 100$.

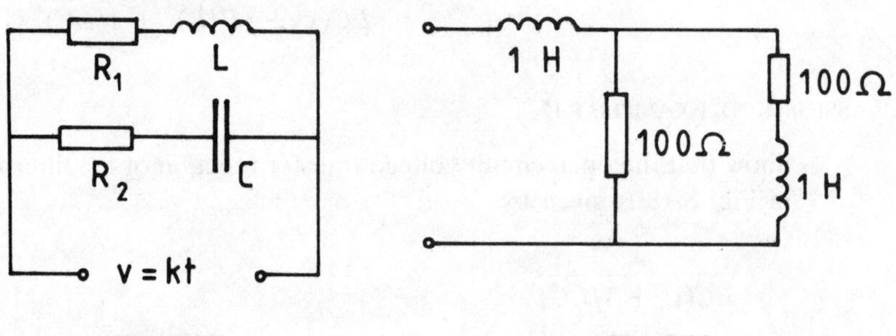

Fig. 8.19

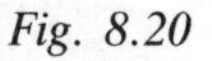

Fig. 8.20

8.17 A series circuit consisting of a resistor with value $1\ \Omega$ and an inductor with inductance 20 mH is connected to a 240 V rms, 50 Hz, sinusoidal supply at the instant the voltage wave is passing through a zero and is about to become positive.

Derive an expression for the instantaneous current and sketch the form of the first five cycles of the current.

8.18 The circuit shown in Fig. 8.20 is connected at $t = 0$ to a 10 V direct supply. Deduce an expression for the input current at any time t seconds after connection.

8.19 A coil has resistance $R\ \Omega$ and inductance L H. A step voltage V volts is applied to it.

Establish and solve the equation for the current at any instant t seconds after the voltage is applied.

If $R = 5\ \Omega$, $L = 5.5$ H and $V = 10$ V, calculate (a) the value of current 0.6 seconds after the voltage is applied, (b) the time taken for the current to reach 1.86 A, (c) the time taken for the current to decay to 0.25 A if, after some considerable time, the coil is shorted-out using a $10\ \Omega$ resistor and the supply is simultaneously removed.

8.20 A $50\ \mu F$ capacitor is charged to 500 V and then connected to a coil of inductance 1 H and resistance $50\ \Omega$. Derive an expression for the current at time t seconds after the connection. Sketch the form of the current wave.

8.21 (a) Derive an expression for the open-circuit voltage transfer function for the circuit shown in Fig. 8.21.
(b) Working from first principles determine (i) the value of resistance which will give critical damping, and (ii) the value of circuit current 10 ms after a constant voltage $v_{in} = 50$ V is suddenly applied to the critically damped circuit.

8.22 (a) What is meant by the term 'critical damping' as applied to the transient response of an electrical circuit?
(b) A coil with inductance L H and resistance R Ω is connected in series with a capacitor of capacitance C F and a variable resistor R_X Ω. Deduce an expression for the value of R_X to give critical damping. Calculate the value of R_X given that $L = 1.5$ H, $C = 50$ μF and $R = 50$ Ω.

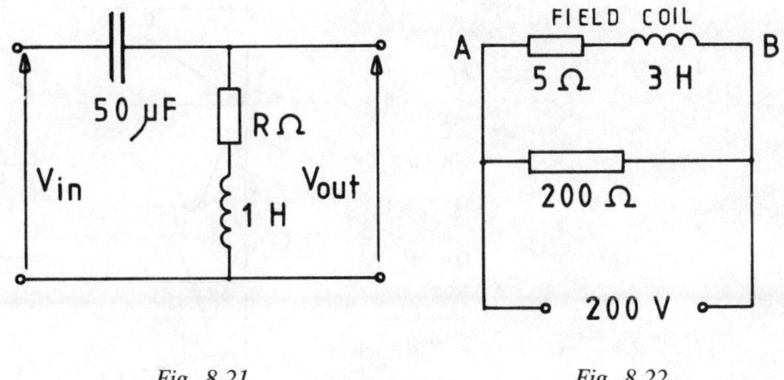

Fig. 8.21

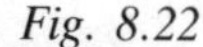

Fig. 8.22

8.23 The field coil of a d.c. motor is shunted by a 200 Ω arc-quenching resistor as shown in Fig. 8.22. Determine (a) an expression for the current in the field coil at time t seconds after connection to the 200 V supply, (b) the total current taken by the parallel components 0.5 seconds after switching on, (c) the peak voltage V_{AB} at the instant the supply is switched off, (d) the time taken for the coil current to fall to 10% of its normal running value after the supply is switched off; both (c) and (d) being considered when steady-state conditions have been achieved after switching on.

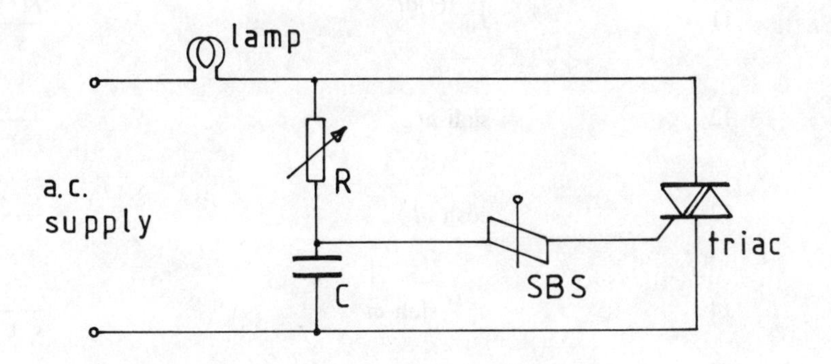

Fig. 8.23

Table of a selection of Laplace transforms

Laplace pair text reference number	$f(t)$	Form of $f(t)$	$\mathcal{L}f(t) = F(s)$
1	V	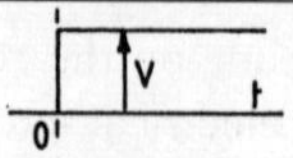	$\frac{V}{s}$
2	e^{at}		$\frac{1}{s - a}$
3	e^{-at}	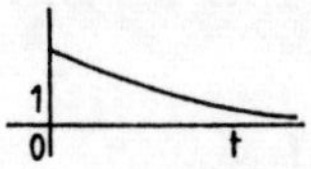	$\frac{1}{s + a}$
4	kt	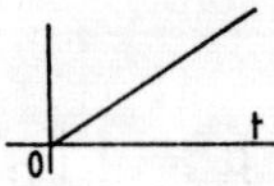	$\frac{k}{s^2}$
5	$\sin \omega t$	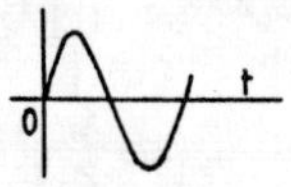	$\frac{\omega}{s^2 + \omega^2}$
6	$\cos \omega t$	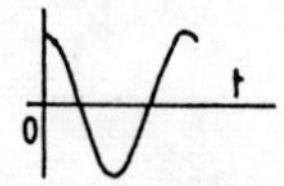	$\frac{s}{s^2 + \omega^2}$
7	$e^{-kt} \sin \omega t$	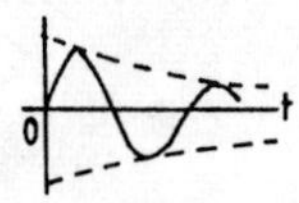	$\frac{\omega}{(s + k)^2 + \omega^2}$
8	$e^{-kt} \cos \omega t$	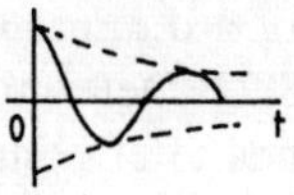	$\frac{s + k}{(s + k)^2 + \omega^2}$
9	$\frac{df(t)}{dt}$		$sF(s) - f_{(0)}$
10	$\frac{d^2f(t)}{dt^2}$		$s^2F(s) - sf_{(0)} - f'_{(0)}$
11	$\int_0^t f(t)dt$		$\frac{F(s)}{s}$
12	$\sinh at$		$\frac{a}{s^2 - a^2}$
13	$\cosh at$		$\frac{s}{s^2 - a^2}$
14	$e^{-kt} \sinh at$		$\frac{a}{s + k^2 - a^2}$
15	$e^{-kt} \cosh at$		$\frac{s + k}{(s + k)^2 - a^2}$

8.24 A transformer primary winding with resistance 2 Ω and inductance 6.6 H is connected to a 50 Hz 440 V sinusoidal supply at the instant the voltage wave is a maximum. Determine the value of the first current peak in the primary winding.

8.25 The transformer in problem 8.24 is energised as the voltage wave passes through a zero. Determine the value of the first current peak in the primary winding.

8.26 Figure 8.23 shows a basic incandescent light dimmer employing a triac and a silicon bilateral switch (SBS). The capacitor C charges through the variable resistor R and at a critical voltage the SBS becomes conducting so switching on the triac. Ideally the triac should be switched on at some point during each half cycle of input to give a flicker-free variable light output from the lamp. A problem with this simple circuit is that if R is sufficiently large the critical voltage is not reached during a particular half cycle and the triac does not become conducting. However during successive half cycles of a.c. supply the capacitor "pumps up" and may reach the critical voltage after several half cycles causing the lamp to flash followed by an off period whilst the capacitor charges once more. This effect creates a very undesirable flicker. The effect is overcome by adding circuitry to discharge the capacitor at the end of every half cycle.
(a) For a supply voltage described by the equation $v = V_{max} \sin \omega t$ volts develop an expression for the current in R in Fig. 8.23. (The lamp resistance when cold will be small compared with that of R so that the effects of R and C alone need to be considered.)
(b) For the sine input, i.e. for $v = 0$ at $t = 0$ determine the magnitude of the residual voltage on the capacitor assuming that the SBS does not become conducting: (i) at the end of the first complete cycle ($\omega t = 2\pi$ rads). (ii) at the end of the second complete cycle ($\omega t = 4\pi$ rads). (iii) Determine the limiting value of capacitor voltage (after a considerable number of cycles). $V_{max} = 240\sqrt{2}$ V, $R = 200$ kΩ, $C = 0.22$ μF, $\omega = 314$ rad/s.

8.27 (a) A resistance R Ω in series with a capacitor with capacitance C F is connected at time instant $t = 0$ to an alternating supply which is described by the equation $v = V_{max} \cos \omega t$ volts. Develop an equation for the current in R at any time t seconds after switching on.
(b) What is the magnitude of this current (i) after 0.01 s (ii) 0.0275 s, given that $v = 100 \cos 314t$ volts, $R = 70.7$ Ω, $C = 45$ μF?
(c) What is the magnitude and phase of the final steady-state current in the RC combination?

Answers

8.8 $R_{total} = 31.6$ Ω, add 30.6 Ω. Current: $t = 0$, $i = 0$; $t = 0.02$ s, $i = 0.53$ A; $t = 0.05$ s, $i = 0.514$ A; $t = 0.1$ s, $i = 0.211$ A; $t = 0.2$ s, $i = 0.018$ A; $t > 0.5$ s, $i = 0$

8.13 $I = 0.058$ A; $k = 8333$; $\omega = 57.1$ rad/s

8.14 $i = 0.15 \times 10^{-4}\,(1 - e^{-11111t})$ A;
$v_C = 100(t + 9 \times 10^{-5}\,(e^{-11111t} - 1))$ V; $i = 2.3$ μ A

8.15 General expression as equation (8.2). Circuit underdamped. Required answer as equation (8.4) with values:

$$i = \frac{V_0}{1.58 \times 10^3} e^{-5t} \sin 3162t \text{ A}$$

8.16 Inductive branch $i = 100\left(\frac{-0.35}{10^2} + \frac{0.04}{10} + \frac{0.35}{100} e^{-10/0.35\times 0.04}\right)$

$$= 1.62 \times 10^{-3} \text{ A}$$

Capacitive branch $i = 10(1 - e^{-10^4 \times 0.04}) = 10\ \mu\text{A}$

Total current 1.63 mA

8.17 $i = 53.5 \sin(\omega t - 80.95°) + 52.96e^{-50t}$ A

8.18 $i = 0.2 - 0.2e^{-150t} \cosh 111.8t - \frac{20}{111.8} \sinh 111.8t.$

This simplifies to: $i = 0.2 - 0.1895e^{-38.2t} - 0.0105e^{-261.8t}$ A

8.19 (a) 0.84 A (b) 2.925 s (c) 0.76 s (don't forget the additional resistance)

8.20 $i = 3.59e^{-25t} \sin 139.2t$ A

8.21 (a) $\frac{R + sL}{R + sL + 1/Cs}$ (b) (i) 282.8 Ω (ii) 0.1215 A

8.22 (b) 296.4 Ω

8.23 (a) $V/R(1 - e^{Rt/L})$ (b) 12.3 A (c) 4000 V (d) 0.034 s

8.24 0.3 A (closely)

8.25 0.6 A (closely)

8.26 (a) $i_{(t)} = \frac{v_{max}\omega C}{1 + \omega^2 C^2 R^2} (\cos \omega t + \omega CR \sin\omega t - e^{-t/CR})$

(b) (i) −8.8 V (ii) −14.59 V (iii) −24.44 V

8.27 (a) $i_{(t)} = \frac{V_{max}}{1 + \omega^2 C^2 R^2} (\omega^2 C^2 R \cos\omega t - \omega C \sin\omega t + (1/R) e^{-t/CR})$

(b) (i) − 0.675 A (ii) −0.99 A (iii) 1∠+45° A (peak)

Chapter 9
Transmission Lines

9.1 Power transmission

9.1.1 Short lines

A three-phase transmission or distribution system is fed at its sending end by a star-connected transformer or generator, the star point of either being connected to earth. Power is transmitted on three wires.

For a star-connected system, the phase and line currents are identical.

$$I_{ph} = I_L$$

The phase voltage is equal to the line voltage divided by the square root of three.

$$V_{ph} = \frac{V_L}{\sqrt{3}}$$

The values of capacitance between lines and between each line and earth of a three-phase overhead line are small per unit length since the spacings between lines are large and so, up to a few kilometres in length, these may be ignored. Only the series resistance and inductance are considered. Calculations are carried out per phase considering the system to be symmetrical so that the line currents in each phase are identical as are the phase (and line) voltages. For asymmetric loading a more complicated analysis is involved using what are called 'symmetrical components'.

One phase of a three-phase system is shown in Fig. 9.1.

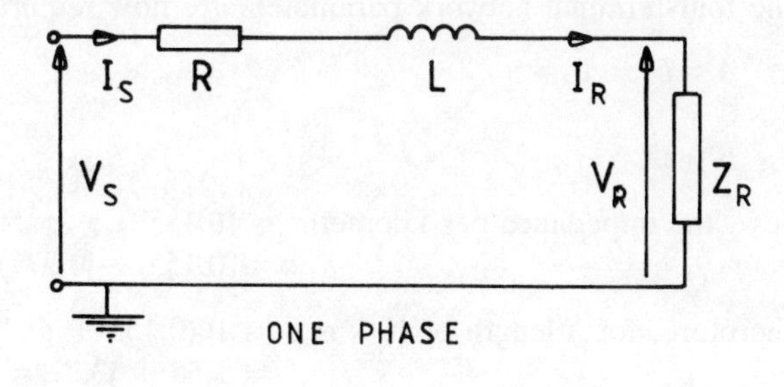

Fig. 9.1

V_S and V_R = phase voltages at the sending end and receiving end, respectively.
I_S and I_R = phase (and line) currents at the sending end and receiving end, respectively.

ϕ_S = phase angle at the sending end.
Z_R and ϕ_R = impedance and phase angle of the load.
R = resistance per phase of the transmission line.
L = series inductance per phase of the transmission line.

Referring to chapter 6:

equation (6.1) $V_S = AV_R + BI_R$, equation (6.2) $I_S = CV_R + DI_R$.

For a network with series impedance only:

equations (6.3) and (6.4) $A = 1$, $B = Z$, $C = 0$ and $D = 1$.

WORKED EXAMPLE 9.1

A three-phase, 50 Hz, 11 kV distribution line 10 km in length has a conductor resistance of 0.155 Ω/km/phase and a series inductance of 1.2 mH/km/phase. A load of 2 MW at 0.75 power factor lagging is supplied at the receiving end of the line at 11 kV. Calculate (a) the sending-end voltage, (b) the sending-end power factor, and (c) the transmission efficiency.

The phase voltage at 11 kV line $= \dfrac{11\,000}{\sqrt{3}} = 6350.8$ volts $(= V_R)$.

Power in a three-phase system $= \sqrt{3} V_L I_L \cos\phi$ watts.

Considering the receiving end

$$\sqrt{3} \times 11\,000 \times I_L \times 0.75 = 2 \times 10^6.$$

Therefore, $I_L\ (= I_R) = \dfrac{2 \times 10^6}{\sqrt{3} \times 11\,000 \times 0.75} = 140$ A.

Taking the receiving-end voltage as reference $V_R = 6350.8\angle 0°$ volts, I_R lags V_R by an angle whose cosine is 0.75. Hence $\phi_R = -41.4°$.

$$I_R = 140\angle -41.4° \text{ A}.$$

The four-terminal network parameters are now required.

$$A = D = 1$$
$$C = 0$$
$$B = Z$$

Now, the impedance per kilometre $= (0.155 + \text{j}\,2\pi\,50 \times 1.2 \times 10^{-3})$
$= (0.155 + \text{j}0.377)\ \Omega$.

Therefore, for a length of 10 km $B = 10(0.155 + \text{j}0.377)$
$= 1.55 + \text{j}3.77 = 4.076\angle 67.65°\ \Omega$.

$$\begin{aligned}(a)\ V_S = AV_R + BI_R &= 1 \times 6350.8\angle 0° + 4.076\angle 67.65° \times 140\angle -41.4° \\ &= 6350.8 + 570.64\angle 26.25° \\ &= 6350.8 + 511.79 + \text{j}252.38 \\ &= 6862.6 + \text{j}252.38 = 6867.2\angle 2.1° \text{ V (phase)}.\end{aligned}$$

Sending-end line voltage = $\sqrt{3} \times 6867.2 = 11\,894$ V.

(b) Since there are no shunt components, $I_S = I_R$. By inspection of Fig. 9.2, it is seen that the angle between I_S and V_S is $(41.4 + 2.1)° = 43.5°$. Therefore, sending-end power factor = cos 43.5° = 0.725.

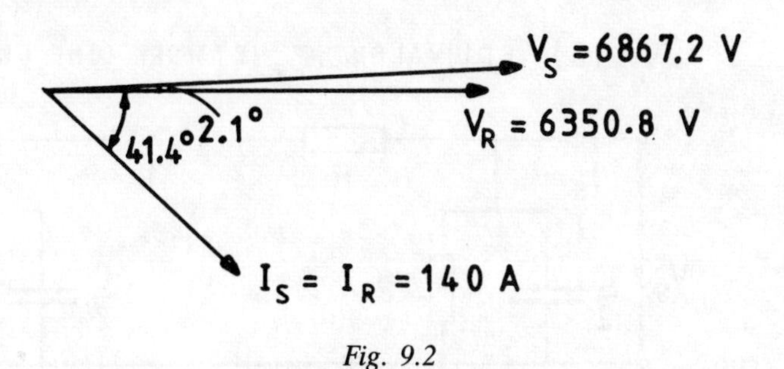

Fig. 9.2

(c) Sending-end power $= \sqrt{3} V_L I_L \cos\phi$ watts (using sending-end quantities)
$= \sqrt{3} \times 11\,894 \times 140 \times 0.725$
$= 2.091$ MW.

$$\text{Efficiency} = \frac{\text{output}}{\text{input}} = \frac{2 \text{ MW}}{2.091 \text{ MW}} = 0.956 \text{ p.u.}$$

SELF-ASSESSMENT EXAMPLE 9.2

A three-phase, 50 Hz transmission line 18 km long has a conductor resistance of 0.2 Ω/km/phase and a series inductance of 1.47 mH/km/phase. Determine the sending-end voltage and its phase with respect to the receiving-end voltage when the line supplies a load of 10 MW at 33 kV and unity power factor.

9.1.2 *Medium length lines*

Increasing the length of an overhead transmission line increases the overall capacitance and also, since for larger distances higher voltages are used, the capacitive or charging current becomes significant and has to be considered. In addition, at higher voltages the leakage current across the insulators may also become significant.

For underground power cables of whatever length, since the cores are generally close together the capacitance is large and must be considered. The leakage current through the insulation may also have an appreciable effect. (See chapter 4, sections 4.14.2 and 4.14.3 on dielectric losses.)

For overhead power lines up to a few hundred kilometres in length and short lengths of interconnecting underground cables, calculations are generally carried out with reasonable accuracy (within 10%) by considering line models made up with 'lumped' constants.

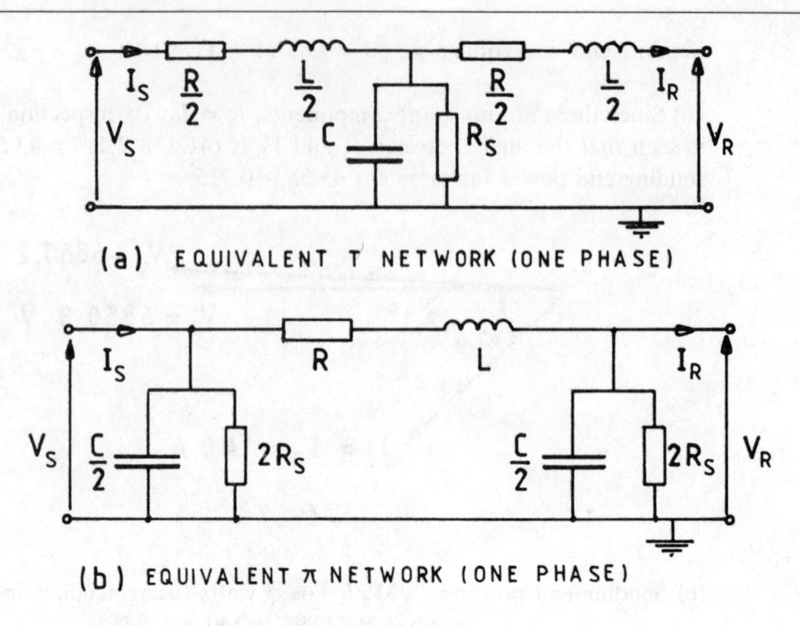

Fig. 9.3

In Fig. 9.3, R and L, V_R, V_S, I_R and I_S are as defined for the short line.

C = total effective capacitance to earth of one line (see also chapter 2, worked example 2.13).

R_s = total shunt resistance to earth of one line.

In the equivalent T-network (Fig. 9.3(a)), the total capacitance and shunt resistance are considered to act midway along the length of the line. On either side of these two components is one half of the line resistance and one half of the line inductance.

$$\text{Shunt admittance } Y_s = \frac{1}{R_s} + \frac{1}{(1/j\omega C)} = (G + j\omega C) \text{ siemens.} \tag{9.1}$$

In the equivalent π-network, one half of the total capacitance is considered to act at each end of the line. The leakage resistance associated with it is doubled so giving one half of the total leakage current at each end of the line. This is shown in Fig. 9.3(b).

$$\text{Shunt admittance at each end of the line} = \frac{Y_s}{2} = \frac{1}{2R_s} + \frac{1}{(1/j\omega(\frac{1}{2}C))}$$

$$= (G/2 + j\omega C/2) \text{ siemens.}$$

Note that G is the total conductance for the line, and C is the total capacitance of the line (both per-phase, as for the short line).

For the equivalent T-network which is symmetrical, the four-terminal network parameters are

$$A = 1 + YZ \qquad B = 2Z + Z^2Y \qquad C = Y \qquad D = 1 + YZ.$$

where Z is the series impedance of *one half* of the line and Y is the total shunt admittance (see Fig. 6.6 and equation (6.9) in chapter 6).

WORKED EXAMPLE 9.3

The per-phase resistance, reactance and capacitive susceptance of a three-phase transmission line are 20 Ω, 50 Ω and $4 \times 10^{-4}\angle 90°$ S respectively. The load at the receiving end of the line is 60 MW at 132 kV line and 0.85 power factor lagging. Considering each phase as a T-network, determine the four-terminal network parameters of the line and hence determine the sending-end power input.

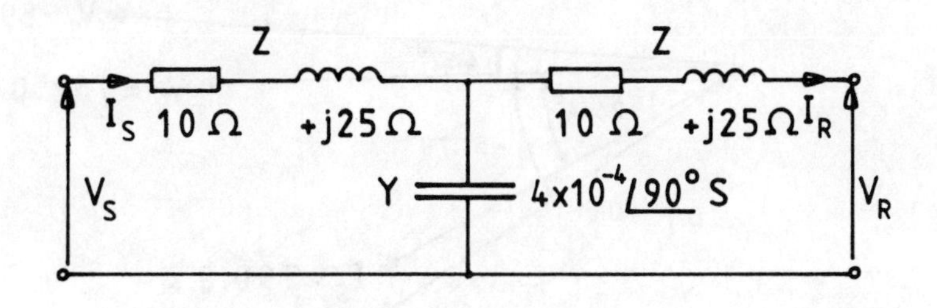

Fig. 9.4

$$V_R = \frac{132\,000}{\sqrt{3}} = 76\,210 \text{ V.}$$

Receiving-end power $= \sqrt{3} V_L I_L \cos \phi$ watts.
Using receiving-end values

$$60 \times 10^6 = \sqrt{3} \times 132\,000 \times I_R \times 0.85$$
$$I_R = 308.7 \text{ A.}$$

Using V_R as reference $V_R = 76\,210\angle 0°$ V.

$$I_R = 308.7\angle \arccos 0.85 = 308.7\angle -31.79° \text{ A}$$

(angle is negative since current lags voltage).

To determine the four-terminal network parameters

$$A = D = 1 + YZ = 1 + 4 \times 10^{-4}\angle 90° \times (10 + j25)$$
$$= 1 + 4 \times 10^{-4}\angle 90° \times 26.93\angle 68.2°$$
$$= 1 + 0.01077\angle 158.2°$$
$$= 1 - 0.01 + j4 \times 10^{-3}$$
$$A = D = 0.99 + j4 \times 10^{-3}$$

$$B = 2Z + Z^2Y = 2(10 + j25) + (10 + j25)^2 \times 4 \times 10^{-4}\angle 90°$$
$$= (20 + j50) + (26.93\angle 68.2°)^2 \times 4 \times 10^{-4}\angle 90°$$
$$= (20 + j50) + 0.29\angle 226.4°$$
$$= 20 + j50 - 0.2 - j0.21$$
$$B = 19.8 + j49.79$$

$$C = Y = 4 \times 10^{-4}\angle 90°$$

Substituting these values in $V_S = AV_R + BI_R$ gives

$$
\begin{aligned}
V_S &= (0.99 + j4 \times 10^{-3})\,(76\,210\angle 0°) + (19.8 + j49.79)\,(308.7\angle -31.79°) \\
&= 0.99\angle 0.23° \times 76\,210\angle 0° + 53.58\angle 68.31° \times 308.7\angle -31.79° \\
&= 75\,447.9 + j302.9 + 13\,292.4 + j9844.5 \\
&= 88\,740 + j10\,147.4 = 89\,318.3\angle 6.52° \text{ V } (\times \sqrt{3} = 154\,703 \text{ V line}).
\end{aligned}
$$

Substituting these values in $I_S = CV_R + DI_R$ gives

$$
\begin{aligned}
I_S &= 4 \times 10^{-4}\angle 90° \times 76\,210\angle 0° + 0.99 + j4 \times 10^{-3} \times 308.7\angle -31.79° \\
&= j30.48 + 260.4 - j159.95 \\
&= 290.8\angle -26.44° \text{ A}.
\end{aligned}
$$

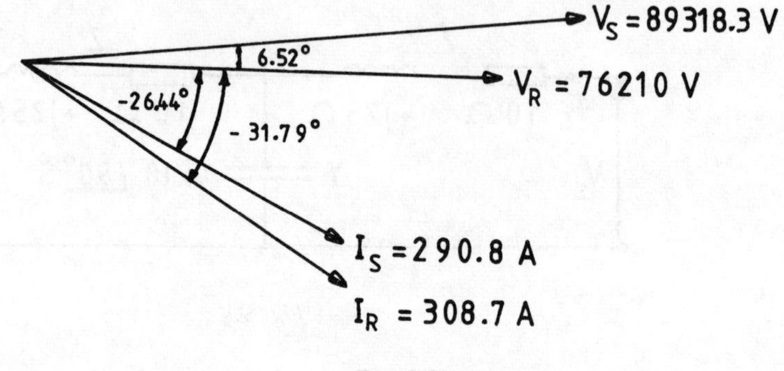

Fig. 9.5

The various quantities are shown in Fig. 9.5.

Angle between V_S and I_S = 26.44° + 6.52° = 32.96°.

$$
\begin{aligned}
\text{Input power} &= \sqrt{3}V_L I_L \cos\phi \text{ watts (using input values)} \\
&= \sqrt{3} \times 154\,703 \times 290.8 \times \cos 32.96° \\
&= 65.38 \text{ MW}.
\end{aligned}
$$

SELF-ASSESSMENT EXAMPLE 9.4

(a) Deduce the values of the four-terminal network parameters A, B, C and D for a symmetrical π-configuration with a total series impedance Z Ω and two shunt admittances each equal to $Y/2$ siemens.
(b) A 160 km length of 132 kV, 3-phase, 50 Hz transmission line has total conductor resistance of 29.5 Ω/phase, inductive reactance of 66.5 Ω/phase and capacitance of 1.44 μF/phase. Leakage current may be ignored. Using the π-representation, determine the A, B, C and D parameters of the line. Hence calculate the required value of the sending-end voltage for a receiving-end power of 50 MVA at 132 kV and 0.8 power factor lagging.

9.2 Transmission lines for high frequencies

From worked example 9.3 and self-assessment example 9.4 it should be observed that there is a phase change between the sending-end voltage V_S, and the receiving-end voltage V_R. In an overhead power line this is generally small, an extremely long length being required to produce a change of 90°. According to spacing this length lies between 2500 km and 6000 km.

The length of line which would be required to cause the receiving-end voltage to be exactly one complete cycle behind the input voltage is known as a wavelength.

Where high frequencies are involved a wavelength may only be a fraction of a metre and in problems where lines have lengths which are significant compared with a wavelength, and especially where small powers are involved so that the criteria for maximum power transfer must be considered, the lumped-parameter approach is not sufficiently accurate. The resistance, reactance and capacitance must be taken as being uniformly distributed along the line.

9.2.1 *The transmission line equations*

Consider a transmission line as shown in Fig. 9.6. It has series impedance $(R + j\omega L)$ Ω per unit length and shunt resistance and capacitance R_s and C respectively per unit length. It is supplied with voltage V_S from an alternating source and the input current is I_S. The receiving-end voltage and current are V_R and I_R respectively.

At a distance x from the sending end an element of line δx in length is considered in which the current is I amperes. At this point the potential difference between line and earth is V volts.

Figure 9.6(a) shows the line complete, length ℓ m. Figure 9.6(b) shows the element of line δx in length. It is represented as a T-equivalent. The impedance of the line $Z = (R + j\omega L)$ Ω per unit length, so that for length δx, the series impedance $= (R + j\omega L)\delta x$ Ω $= Z\delta x$ Ω. Similarly the shunt components are $R_s\delta x$ and $C\delta x$ for the element.

An expression for the characteristic impedance Z_0 of the line (See chapter 6, section 6.8, for a definition of Z_0.)

Consider the small element of line δx in Fig. 9.6(b). Using the iterative method of determination, looking in at terminals AC, the impedance must be equal to Z_0 when terminals BD are connected to an impedance Z_0. Hence, in Fig. 9.7, $Z\delta x/2$ is in series with a parallel combination of the shunt impedance $1/Y_s\delta x$ and $(Z\delta x/2 + Z_0)$ Ω.

$$Z_0 = \frac{Z\,\delta x}{2} + \frac{1}{Y_s\,\delta x}\left(\frac{Z\delta x/2 + Z_0}{1/Y_s\,\delta x + Z\,\delta x/2 + Z_0}\right).$$

Multiplying through by $(1/Y_s\,\delta x + Z\,\delta x/2 + Z_0)$ and expanding the parentheses gives

$$\frac{Z_0}{Y_s\,\delta x} + \frac{Z_0 Z\,\delta x}{2} + Z_0^2 = \frac{Z\,\delta x}{2Y_s\,\delta x} + \frac{Z^2\,(\delta x)^2}{4} + \frac{ZZ_0\,\delta x}{2} + \frac{Z\,\delta x}{2Y_s\,\delta x} + \frac{Z_0}{Y_s\,\delta x}$$

$$Z_0^2 = \frac{Z\,\delta x}{Y_s\,\delta x} + \frac{Z^2\,(\delta x)^2}{4}.$$

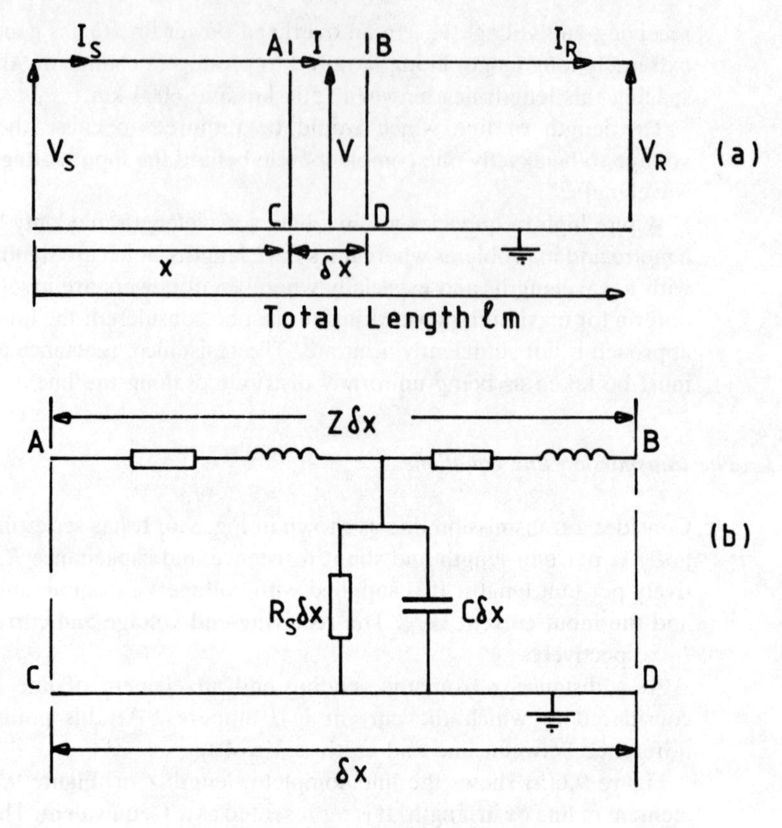

Fig. 9.6

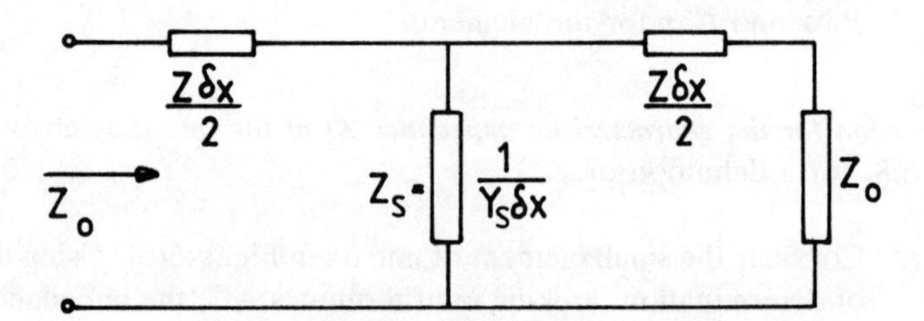

Fig. 9.7

As $\delta x \to 0$, $(\delta x)^2$ may be ignored and $Z_0^2 = \dfrac{Z}{Y_s}$.

But $Z = (R + j\omega L)\ \Omega$ and $Y_s = (G + j\omega C)$ S (section 9.1.2, equation 9.1). Therefore

$$Z_0 = \sqrt{\left[\frac{(R + j\omega L)}{(G + j\omega C)}\right]} \tag{9.2}$$

Now, returning to Fig. 9.6

current drawn by the shunt components in the element δx

$$= VY_s\,\delta x = V(G + j\omega C)\delta x \text{ A}.$$

The current leaving the element of line is less than that entering it by this amount. The change in current through the element equals $-\delta I$, hence

$$-\delta I = V(G + j\omega C)\delta x \text{ A}.$$

In the limit, as $\delta x \to 0$

$$-\frac{dI}{dx} = V(G + j\omega C). \tag{9.3}$$

Owing to the current I A in the series impedance, there will be a fall in line potential of $-\delta V$ volts with respect to earth across the element of line.

$$-\delta V = IZ\,\delta x = I(R + j\omega L)\delta x.$$

In the limit, as $\delta x \to 0$

$$-\frac{dV}{dx} = I(R + j\omega L). \tag{9.4}$$

Partial derivatives are used since both voltage and current are also time-varying.
Differentiating equation (9.3) with respect to x gives

$$-\frac{d^2I}{dx^2} = \frac{dV}{dx}(G + j\omega C). \tag{i}$$

Differentiating equation (9.4) with respect to x gives

$$-\frac{d^2V}{dx^2} = \frac{dI}{dx}(R + j\omega L). \tag{ii}$$

Substituting equation (9.4) in equation (i) gives

$$\frac{d^2I}{dx^2} = I(R + j\omega L)\,(G + j\omega C). \tag{9.5}$$

Substituting equation (9.3) in equation (ii) gives

$$\frac{d^2V}{dx^2} = V(R + j\omega L)\,(G + j\omega C). \tag{9.6}$$

These are the two equations for a transmission line with uniformly distributed parameters.

9.2.2 *A solution of the transmission line equations for a line terminated in its characteristic impedance*

$$\frac{d^2V}{dx^2} = V(R + j\omega L)\,(G + j\omega C). \qquad \text{(equation 9.6)}$$

To simplify the expression, write P^2 for $(R + j\omega L)(G + j\omega C)$ or $P = \sqrt{[(R + j\omega L)(G + j\omega C)]}$. Then equation (9.4) becomes

$$\frac{d^2V}{dx^2} = P^2V.$$

Taking Laplace transforms of both sides (chapter 8, Laplace pair 10)

$$s^2\overline{V} - sV_{(0)} - V'_{(0)} = P^2\overline{V}$$

$$\overline{V}(s^2 - P^2) = sV_{(0)} + V'_{(0)}$$

$$\overline{V} = \frac{sV_{(0)}}{s^2 - P^2} + \frac{V'_{(0)}}{s^2 - P^2}.$$

Taking inverses (Laplace pairs 12 and 13, chapter 8)

$$V = V_{(0)} \cosh Px + \frac{V'_{(0)}}{P} \sinh Px. \qquad \text{(i)}$$

Now $V'_{(0)}$ is the value of $\frac{dV}{dx}$ at $x = 0$.

From equation (9.4), $dV/dx = -I(R + j\omega L)$, and as $x = 0$ is at the sending end of the line, $I = I_S$, the total input current. It is only as progress is made along the line that the value of the current changes due to the shunt components. Therefore

$$V'_{(0)} = -I_S(R + j\omega L).$$

Also in equation (i) above, $V_{(0)}$ is the value of voltage at $x = 0$, which is the sending-end voltage V_S. Therefore $V_{(0)} = V_S$. Writing these values into equation (i)

$$V = V_S \cosh Px - \frac{I_S(R + j\omega L)}{P} \sinh Px. \qquad \text{(ii)}$$

But $P = \sqrt{[(R + j\omega L)(G + j\omega C)]}$, therefore equation (ii) becomes

$$V = V_S \cosh Px - \frac{I_S(R + j\omega L)}{\sqrt{[(R + j\omega L)(G + j\omega C)]}} \sinh Px$$

$$= V_S \cosh Px - I_S\sqrt{\left[\frac{(R + j\omega L)}{(G + j\omega C)}\right]} \sinh Px.$$

Substituting equation (9.2) for Z_0

$$V = V_S \cosh Px - I_S Z_0 \sinh Px. \qquad (9.7)$$

For a correctly terminated line, the input impedance is Z_0 so that $I_S Z_0 = V_S$.

Now, $\cosh Px = \frac{e^{Px} + e^{-Px}}{2}$ and $\sinh Px = \frac{e^{Px} - e^{-Px}}{2}$.

Equation (9.7) can therefore be written as

$$V = V_S\left(\frac{e^{Px} + e^{-Px}}{2}\right) - V_S\left(\frac{e^{Px} - e^{-Px}}{2}\right)$$

$$= \frac{V_S}{2}(e^{Px} + e^{-Px} - e^{Px} + e^{-Px})$$

$$V = V_S e^{-Px} \text{ volts.} \quad (9.8)$$

V is the line voltage at any distance x from the sending end. Transposing equation (9.8)

$$V_S/V = e^{+Px} \quad \text{or } Px = \log_e (V_S/V).$$

In chapter 6, section 6.12.2, the propagation coefficient of a network γ is defined as

$$\gamma = \log_e \frac{I_1}{I_2} = \log_e \frac{V_1}{V_2}$$

where subscripts 1 and 2 denote input and output conditions respectively.

Therefore the quantity P is in fact the propagation coefficient of the transmission line per unit length, and equation (9.8) may be written as

$$V = V_S e^{-\gamma x} \text{ volts.} \quad (9.9)$$

And from equation (6.38), equation (9.9) becomes

$$V = V_S e^{-\alpha x} \angle -\beta x \text{ volts} \quad (9.10)$$

α = the attenuation coefficient in nepers per unit length
β = the phase-change coefficient per unit length
γ = propagation coefficient per unit length = $\sqrt{[(R + j\omega L)(G + j\omega C)]}$.

At the receiving end of the line, $x = \ell$, and equation (9.10) becomes

$$V_R = V_S e^{-\alpha\ell} \angle -\beta\ell.$$

In exactly the same manner, writing I for V starting from equation (9.5), a solution may be arrived at for the current I at any distance x from the sending end. An outline proof is given here.

$$\frac{d^2I}{dx^2} = P^2 I.$$

Using Laplace transforms

$$I = I_{(0)} \cosh Px + \frac{I'_{(0)}}{P} \sinh Px$$

$$I_{(0)} = \text{current when } x \text{ is zero} = I_S.$$

$$I'_{(0)} = \frac{dI}{dx} (\text{at } x = 0) = -V_Z(G + j\omega C). \quad \text{(equation 9.3)}$$

$$\text{Therefore } I = I_S \cosh Px - \frac{V_S(G + j\omega C)}{\sqrt{[(R + j\omega L)(G + j\omega C)]}} \sinh Px$$

$$= I_S \cosh Px - (V_S/Z_0) \sinh Px. \quad (9.11)$$

For a correctly terminated line $I_S = V_S/Z_0$.

Therefore $I = \frac{I_S}{2}(e^{Px} + e^{-Px} - e^{Px} + e^{-Px})$

$= I_S e^{-Px}$

$$I = I_S e^{-\alpha x} \angle -\beta x \text{ A.} \qquad (9.12)$$

At the receiving end of the line, $x = \ell$, and $I_R = I_S e^{-\alpha \ell} \angle -\beta \ell$.

9.2.3 *The four-terminal network parameters of a transmission line with uniformly distributed constants*

From equations (6.13) and (6.14), and Fig. 6.8(b)

$$V_2 = DV_1 - BI_1 \qquad \text{and } -I_2 = CV_1 - AI_1.$$

Treating V_1 as the input voltage V_S, and I_1 as the input current I_S, V_2 and I_2 become the voltage and current respectively, at any point on the line (V and I in section 9.2.2). Hence

$$V = DV_S - BI_S.$$

From equation (9.7)

$$V = V_S \cosh \gamma x - I_S Z_0 \sinh \gamma x$$

it is seen that $D = \cosh \gamma x$, and $B = Z_0 \sinh \gamma x$. Also

$$-I = CV_S - AI_S$$
$$I = -CV_S + AI_S.$$

From equation (9.11)

$$I = -(V_S/Z_0) \sinh \gamma x + I_S \cosh \gamma x$$

it is seen that $C = (\sinh \gamma x)/Z_0$ and $A = \cosh \gamma x$.

These A, B, C and D parameters are for a length x of line. The parameters for the whole line are obtained by writing ℓ for x.

WORKED EXAMPLE 9.5

Determine the values of (a) Z_0, (b) α, and (c) β, for a cable which has the following primary constants per kilometre length.

$R = 46.8\ \Omega$, $G = 38\ \mu S$, $L = 1.1$ mH, $C = 0.06\ \mu F$ and $\omega = 10\,000$ rad/s.

$(R + j\omega L) = 46.8 + j10000 \times 1.1 \times 10^{-3} = 46.8 + j11 = 48.075\angle 13.2°\ \Omega$.

$(G + j\omega C) = 38 \times 10^{-6} + j10000 \times 0.06 \times 10^{-6}$

$= 38 \times 10^{-6} + j6 \times 10^{-4} = 6.01 \times 10^{-4} \angle 86.37°$ S.

(a) $Z_0 = \sqrt{\left(\frac{R + j\omega L}{G + j\omega C}\right)} = \sqrt{\left(\frac{48.075\angle 13.2°}{6.01 \times 10^{-4}\angle 86.37°}\right)} = \sqrt{(79\,991.7\angle -73.17°)}$

$= 282.8\angle -36.58°\ \Omega$.

(b) and (c) $\gamma = \sqrt{[(R + j\omega L)(G + j\omega C)]}$

$$= \sqrt{(48.075\angle 13.2° \times 6.01 \times 10^{-4}\angle 86.37°)}$$

$$= \sqrt{(0.02889\angle 99.57°)} = 0.17\angle 49.78° = 0.109 + j0.13.$$

$\alpha = 0.109$ nepers per km, $\beta = 0.13$ radians per km.

9.2.4 *The logarithmic spiral*

Using the data from worked example 9.5 equation (9.12) may be written in the form

$$I = I_S e^{-0.109x} \angle -0.13x \text{ A (where } x \text{ is measured in km).}$$

Or using degrees instead of radians

$$I = I_S e^{-0.109x} \angle -7.45x° \text{ A.}$$

It follows that for increasing values of x the modulus of the current will be decreasing while its phase with respect to I_S will be increasingly more lagging. This is best demonstrated using specific values.

Let us consider an applied voltage of $282.8\angle -36.58°$ V ($= Z_0$). Then

$$I_S = \frac{V_S}{Z_0} = \frac{282.8\angle -36.58°}{282.8\angle -36.58°} = 1\angle 0° \text{ A.}$$

At $x = 1$ km

$$I = 1e^{-0.109 \times 1} \angle -7.45° \times 1$$
$$= 0.897\angle -7.45° \text{ A.}$$

The current after a length of 1 km has a modulus of 0.897 A and is lagging I_S by 7.45°. For 2 km of line

$$I = 1e^{-0.109 \times 2} \angle -7.45° \times 2$$
$$= 0.804\angle -14.9° \text{ A.}$$

After 2 km of line the modulus of the current is 0.804 A and is lagging I_S by 14.9°.

Similar calculations may be made for any length of line required. The following table is drawn up for lengths of up to 20 km.

Length (km)	Current (A)	Phase angle (degrees)
1	0.897	−7.45
2	0.804	−14.9
3	0.72	−22.35
4	0.646	−29.8
5	0.58	−37.25
10	0.336	−74.5
20	0.113	−149

These results are plotted in Fig. 9.8. The line joining the ends of the phasors is known as the logarithmic spiral.

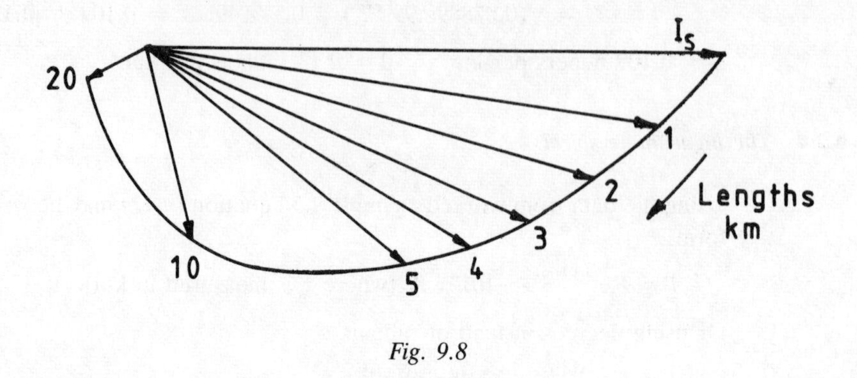

Fig. 9.8

A similar spiral may be drawn for the voltage using the equation

$$V = V_S e^{-0.109x} \angle -7.45x^\circ \text{ V} \qquad \text{(equation 9.10)}$$

at any point on the line $V/I = Z_0$, and the phase angle between V and I will be that of Z_0. In the example used this is 36.58°.

9.2.5 *Wavelength* (λ)

That length of line which will cause the voltage or current phasor to be retarded by one complete revolution is known as the wavelength.

Since β is the phase change in radians per unit length and one revolution equals 2π radians, it will require $2\pi/\beta$ unit lengths to give the required phase change.

$$\lambda = \frac{2\pi}{\beta}. \qquad (9.13)$$

9.2.6 *Phase velocity* (V_p)

Consider a cosine wave input to a transmission line expressed by the equation, $v = V_m \cos \omega t$ volts. It is switched on to the line at $t = 0$, i.e. when $v = V_m$ (Fig. 9.9(a)). The sending end of the line is at V_m volts at $t = 0$.

It takes a finite time for the effect of the applied voltage to be felt further along the line so that after time interval t_1 the effect is felt at x_1 (Fig. 9.9(b)), by which time the input wave has fallen from its peak value. As the voltage wave travels along the line it is attenuated according to equation (9.10). At $t_2 = \pi/2\omega$

$$v = V_m \cos \omega\left(\frac{\pi}{2\omega}\right) = 0 \text{ volts.}$$

The wave has travelled a distance x_2 (Fig. 9.9(c)).

Progressing through the cycle to $t_3 = 2\pi/\omega$

$$v = V_m \cos \omega\left(\frac{2\pi}{\omega}\right) = V_m$$

the input is again at value V_m and the effect has been felt at a distance x_3 along the line (Fig. 9.9(d)). At this point the effect is being felt of a wave which was impressed on the line one cycle beforehand, i.e. the effect has been retarded by exactly one complete cycle. The length x_3 is therefore one wavelength (section 9.2.5).

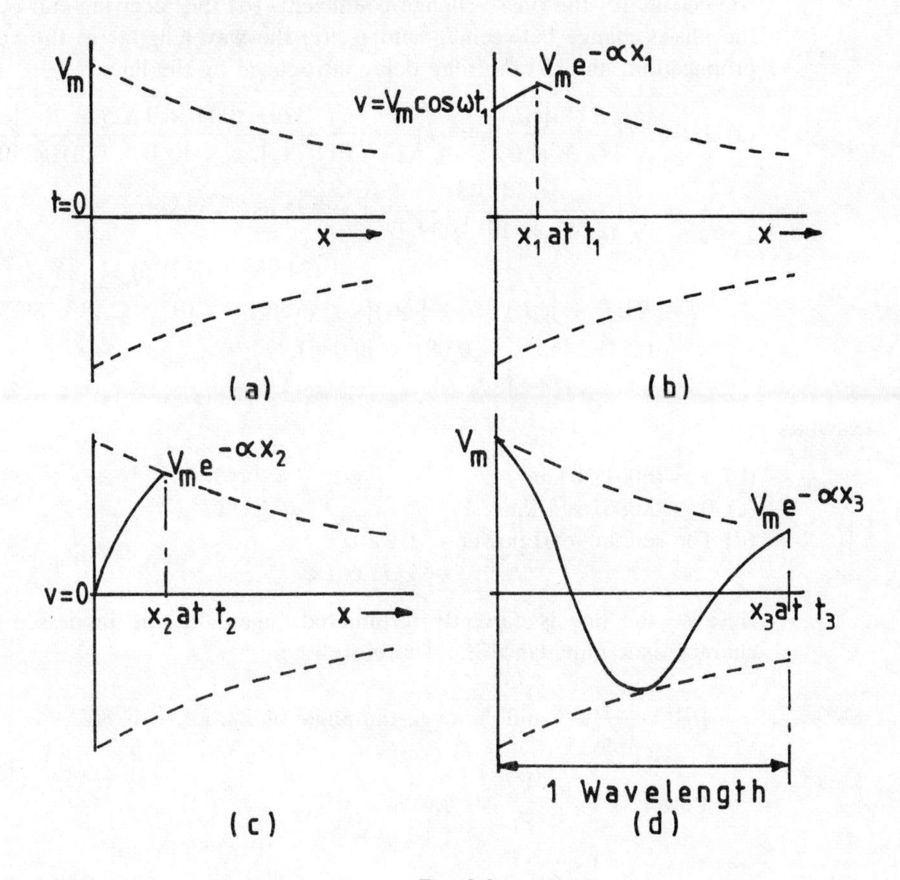

Fig. 9.9

$x_3 = \lambda$.

How long has it taken for the effect to be felt? It has taken the periodic time of the wave, namely $1/f$ seconds. Now

$$\text{velocity} = \frac{\text{distance}}{\text{time}} = \frac{\lambda}{1/f} = \lambda f \text{ s.} \qquad (9.14)$$

Equation (9.14) defines the phase velocity or velocity of propagation.

WORKED EXAMPLE 9.6

An underground cable is 20 km long and has the following primary constants per kilometre length

$$R = 56\ \Omega,\ L = 0.625\ \text{mH},\ C = 0.04\ \mu\text{F and } G = 1\ \mu\text{S}.$$

The line is terminated in its own characteristic impedance and has an input power of 1 watt at a frequency of 1 kHz.

Determine (a) the characteristic impedance of the line, (b) the attenuation coefficient, (c) the phase-change coefficient, (d) the receiving-end power, (e) the phase change between I_R and I_S, (f) the wavelength, (g) the velocity of propagation, and (h) the time delay introduced by the line.

$$\text{(a) } Z_0 = \sqrt{\left(\frac{R + j\omega L}{G + j\omega C}\right)} = \sqrt{\left(\frac{56 + j2\pi x\ 1000 \times 0.625 \times 10^{-3}}{1 \times 10^{-6} + j2\pi \times 1000 \times 0.04 \times 10^{-6}}\right)}$$

$$= \sqrt{\left(\frac{56.14\angle 4.01^\circ}{2.513 \times 10^{-4}\angle 89.77^\circ}\right)} = 472.65\angle -42.88^\circ$$

$$= (346.35 - j321.6)\ \Omega.$$

$$\gamma = \sqrt{[(R + j\omega L)\ (G + j\omega C)]} = \sqrt{(56.14\angle 4.01^\circ \times 2.513\angle 89.77^\circ)}$$

$$= 0.119\angle 46.89^\circ = 0.081 + j0.0867.$$

ANSWERS

(b) $\alpha = 0.081$ N/km.
(c) $\beta = 0.0867$ rad/km.
(d) The sending-end power = 1 watt

$$= V_S I_S \cos \phi_S.$$

Because the line is correctly terminated, signals at the input see only the characteristic impedance Z_0. Therefore

$$|I_S| = \frac{|V_S|}{|Z_0|} \quad \text{and } \phi_S = \phi, \text{ the angle of } Z_0, \text{ i.e. } -42.88^\circ.$$

$$\text{Power} = V_S \times \frac{V_S}{Z_0} \cos \phi_0$$

$$1 = \frac{V_S^2}{Z_0} \cos \phi_0.$$

Substituting values and transposing

$$V_S^2 = \frac{472.65 \times 1}{\cos(-42.88^\circ)}$$

$$V_S = 25.4\ \text{V}.$$

Again, power $= V_S I_S \cos \phi$ watts. Hence

$$1 = 25.4 \times I_S \times \cos(-42.88^\circ).$$

Transposing

$$I_S = \frac{1}{25.4 \times 0.733} = 0.0537 \text{ A.}$$

$$\text{Now } V_R = V_S e^{-\alpha \ell} \qquad \text{(equation 9.10)}$$

$$V_R = 25.4e^{-0.081 \times 20} = 5.027 \text{ V.}$$

$$\text{And } I_R = I_S e^{-\alpha \ell} \qquad \text{(equation 9.12)}$$

$$I_R = 0.0537e^{-0.081 \times 20} = 0.0106 \text{ A.}$$

The phase angle between V_R and I_R is that of Z_0 since the termination is in Z_0. Therefore

$$\text{receiving-end power} = 5.027 \times 0.0106 \times \cos(-42.88°) = 0.039 \text{ W.}$$

Once the method is understood, it will be seen that the receiving-end power can more easily be calculated as

$$P_R = \text{sending power} \times (e^{-\alpha \ell})^2 \text{ watts.}$$

(e) Phase change = 0.0867 rad/km. Therefore over 20 km, phase change is $20 \times 0.0867 = 1.734$ rad (99.35°).

$$\text{(f) Wavelength } \lambda = \frac{2\pi}{\beta} \qquad \text{(equation 9.13)}$$

$$= \frac{2\pi}{0.0867} = 72.47 \text{ km.}$$

$$\text{(g) Phase velocity } V_p = f\lambda \qquad \text{(equation 9.14)}$$

$$= 1000 \times 72.47 = 72\,470 \text{ km/s.}$$

(Since the speed of light is 3×10^8 m/s, the phase velocity in this line is 0.24 times that of light.)

(h) To travel 20 km at 72 470 km/s takes

$$\frac{20}{72\,470} = 276\ \mu\text{s.}$$

This is the time delay introduced by the line.

SELF-ASSESSMENT EXAMPLE 9.7

A cable 8 km in length has the following primary constants per 100 m length

$$R = 4.5\ \Omega,\ L = 0.035 \text{ mH},\ C = 0.013\ \mu\text{F};\ G \text{ may be ignored.}$$

The line is correctly terminated in its characteristic impedance and is supplied at 50 V and 10 kHz at the sending end.

Determine (a) the value of Z_0, (b) the receiving-end current, (c) the wavelength, and (d) the velocity of propagation.

9.2.7 The incorrectly terminated line

It is not the intention in this book to deal in depth with the incorrectly terminated line since how to deal with the problems involved and artificial matching methods is a complete study in itself. The topic is introduced so that the desirability of correct matching is appreciated.

From equation (9.7)

$$V = V_S \cosh \gamma x - I_S Z_0 \sinh \gamma x \text{ volts} \qquad \text{(i)}$$

where V is the line voltage at any distance x from the sending end. For the correctly terminated line this reduces to

$$V = V_S e^{-\alpha x} \angle -\beta x \text{ V.}$$

For any other terminating impedance than Z_0, $V_S/I_S \neq Z_0$, and equation (i) above becomes

$$V = V_S \left(\frac{e^{\gamma x} + e^{-\gamma x}}{2}\right) - I_S Z_0 \left(\frac{e^{\gamma x} - e^{-\gamma x}}{2}\right)$$

$$= \frac{e^{\gamma x}}{2}(V_S - I_S Z_0) + \frac{e^{-\gamma x}}{2}(V_S + I_S Z_0).$$

In the work on the logarithmic spiral, an expression containing $e^{-\gamma x}$ reduces in size as it travels in the x direction, i.e. from the sending end towards the receiving end. An expression containing $e^{\gamma x}$ is one which increases in size as it travels in the x direction. This appears to be impossible since taken to the limit at $x = \infty$, $V = \infty$.

In reality, what is happening is that the original wave travels along the line until it reaches the termination where, since this is not correct, not all the power is absorbed. Some of it is retransmitted towards the sending end, being attenuated as it travels along the line. The returning wave is decreasing in magnitude as it moves in the $-x$ direction.

$$e^{-\gamma x}\left(\frac{V_S + I_S Z_0}{2}\right)$$

is a wave travelling from the sending end towards the receiving end, being attenuated as it progresses along the line.

$$e^{\gamma x}\left(\frac{V_S - I_S Z_0}{2}\right)$$

is a wave travelling from the receiving end towards the sending end, being attenuated as it progresses along the line. If it could be measured separately it would be increasing in size at increasing distances from the sending end.

Although the line is not terminated in Z_0 the line still has its characteristic impedance and it is in this that the current flows.

In a low-loss line there will be several reflections before the magnitude is reduced to zero. This is one way in which several images successively to the right can be produced in a television receiver. The receiver needs to be matched to the co-axial down-lead which in turn must be matched to the aerial. Failure to do so results in reflections from the receiver to the aerial and then back to the receiver possibly several times.

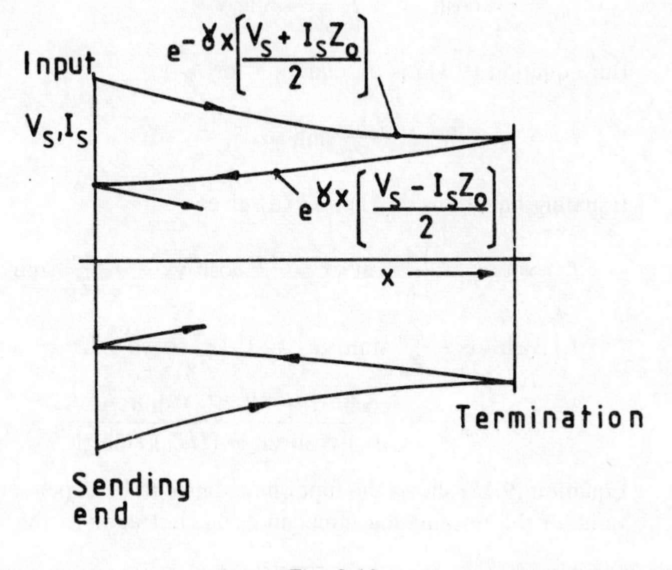

Fig. 9.10

The same phenomenon may be demonstrated by considering the current equation (9.11).

One further effect of incorrect termination is that it does not simply change the input impedance to another (constant) value but makes this a function of the length of line. Adjusting the length of line, while keeping the terminating impedance constant can have a considerable effect on the input impedance.

Consider a line terminated in an impedance Z_R as shown in Fig. 9.11. From equation (9.7)

$$V_R = V_S \cosh \gamma x - I_S Z_0 \sinh \gamma x.$$

$$\text{Now } I_R = V_R/Z_R \quad \text{or } V_R = I_R Z_R. \qquad \text{(i)}$$

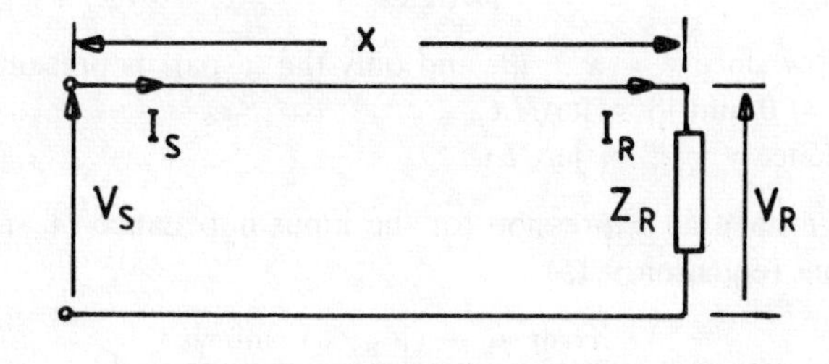

Fig. 9.11

Substituting the value for V_R from equation (i) into equation (9.7) gives

$$I_R Z_R = V_S \cosh \gamma x - I_S Z_0 \sinh \gamma x.$$

Dividing throughout by Z_R gives

$$I_R = \frac{V_S}{Z_R} \cosh \gamma x - I_S \frac{Z_0}{Z_R} \sinh \gamma x. \qquad \text{(ii)}$$

But equation (9.11) is an equation for I_R

$$I_R = I_S \cosh \gamma x - \frac{V_S}{Z_0} \sinh \gamma x.$$

Equating equations (9.11) and (ii) gives

$$I_S \cosh \gamma x - \frac{V_S}{Z_0} \sinh \gamma x = \frac{V_S}{Z_R} \cosh \gamma x - I_S \frac{Z_0}{Z_R} \sinh \gamma x$$

$$I_S\left(\cosh \gamma x + \frac{Z_0}{Z_R} \sinh \gamma x\right) = V_S\left(\frac{1}{Z_R} \cosh \gamma x + \frac{1}{Z_0} \sinh \gamma x\right)$$

$$\frac{V_S}{I_S} = Z_{input} = \frac{\cosh \gamma x + (Z_0/Z_R) \sinh \gamma x}{(1/Z_R) \cosh \gamma x + (1/Z_0) \sinh \gamma x}. \qquad (9.15)$$

Equation (9.15) shows the input impedance to be dependent not only upon the value of the terminating impedance Z_R, but upon x, the length of the line.

9.3 The loss-free line

Equation (9.2) gives $Z_0 = \sqrt{\dfrac{R + j\omega L}{G + j\omega C}}\ \Omega$.

At very high frequencies $j\omega L \gg R$ and $j\omega C \gg G$ so that equation (9.2) reduces to

$Z_0 = \sqrt{\dfrac{j\omega L}{j\omega C}} = \sqrt{\dfrac{L}{C}}\ \Omega.$ The units of this are resistive ohms and it is generally denoted by the symbol R_0. (9.16)

The equation for $\gamma = \sqrt{(R + j\omega L)\,(G + j\omega C)}$ reduces to $\sqrt{(j\omega L)\,(j\omega C)}$

$$\gamma = \sqrt{(j\omega)^2 LC}$$
$$= j\omega\sqrt{LC}.$$

Now since $\gamma = \alpha + j\beta$, and only the 'j' part is present
$\alpha = 0$ and $j\beta = j\omega\sqrt{LC}$.
Hence $\gamma = j\beta = j\omega\sqrt{LC}$. (9.17)

We have an expression for the input impedance of an incorrectly terminated line (equation 9.15)

$$Z_{input} = \frac{\cosh \gamma x + (Z_0/Z_R) \sinh \gamma x}{(1/Z_R) \cosh \gamma x + (1/Z_0) \sinh \gamma x}\ \Omega.$$

Using this equation in association with a loss-free line, Z_0 becomes R_0 (equation 9.16). Multiply through by $Z_R/\cosh \gamma x$.

$$Z_{\text{in}} = \frac{Z_R + R_0 (\sinh \gamma x/\cosh \gamma x)}{1 + (Z_R/R_0)(\sinh \gamma x/\cosh \gamma x)} = \frac{Z_R + R_0 \tanh \gamma x}{1 + (Z_R/R_0) \tanh \gamma x}.$$

Substituting $j\beta$ for γ (equation 9.17) and finally simplifying the bottom line by multiplying throughout by R_0

$$Z_{\text{in}} = \frac{R_0(Z_R + R_0 \tanh j\beta x)}{R_0 + Z_R \tanh j\beta x} = \frac{R_0(Z_R + jR_0 \tan \beta x)}{R_0 + jZ_R \tan \beta x} \quad (\tanh j\theta = j \tan \theta). \tag{9.18}$$

9.3.1 *Input impedance of an open-circuited loss-free line*

For the line open circuited at its receiving end, Z_R is infinite. Rewriting equation (9.18) with $Z_R = \infty$

$$Z_{\text{in}} = \frac{R_0(\infty + jR_0 \tan \beta x)}{R_0 + j\infty \tan \beta x} = \frac{R_0\infty + jR_0^2 \tan \beta x}{R_0 + j\infty \tan \beta x}.$$

Now $R_0\infty \gg jR_0^2 \tan \beta x$, and $j\infty \tan \beta x \gg R_0$ so that the value for the input impedance approaches

$$Z_{\text{in}} = \frac{R_0\infty}{j\infty \tan \beta x} = \frac{R_0}{j \tan \beta x}\ \Omega \tag{9.19}$$

9.3.2 *Input impedance of a short-circuited loss-free line*

For the line short circuited, the value of Z_R becomes zero. Substituting this value in equation (9.18)

$$Z_{\text{in}} = \frac{R_0(0 + jR_0 \tan \beta x)}{R_0 + j0 \tan \beta x} = \frac{jR_0^2 \tan \beta x}{R_0} = jR_0 \tan \beta x\ \Omega. \tag{9.20}$$

9.3.3 *Relationships between wavelength and input impedance*

From the results of sections 9.3.1 and 9.3.2 we see that the input impedances of both open-circuited and short-circuited, loss-free lines is purely imaginary; it has no real component.

Since $\beta = 2\pi/\lambda$ (equation 9.13), we can evaluate $\tan \beta x$ in terms of wavelength.

At $x = 0$, $\tan \beta x = 0$

At x = one quarter wavelength, i.e. $\lambda/4$, $\beta x = \beta \times \lambda/4$

$$= \frac{2\pi}{\lambda}\frac{\lambda}{4} = \frac{\pi}{2}$$

$\tan \beta x = \tan \pi/2 = \infty$ and $1/\tan \pi/2 = 0$.

At x = one half wavelength, i.e. $\lambda/2$, $\beta x = \beta \times \lambda/2 = 2\pi/\lambda \times \lambda/2 = \pi$

$\tan \beta x = \tan \pi = 0$ and $1/\tan \pi = \infty$.

For length $x = 3\lambda/4$, $\tan \beta x = \infty$ and $1/\tan \beta x = 0$.

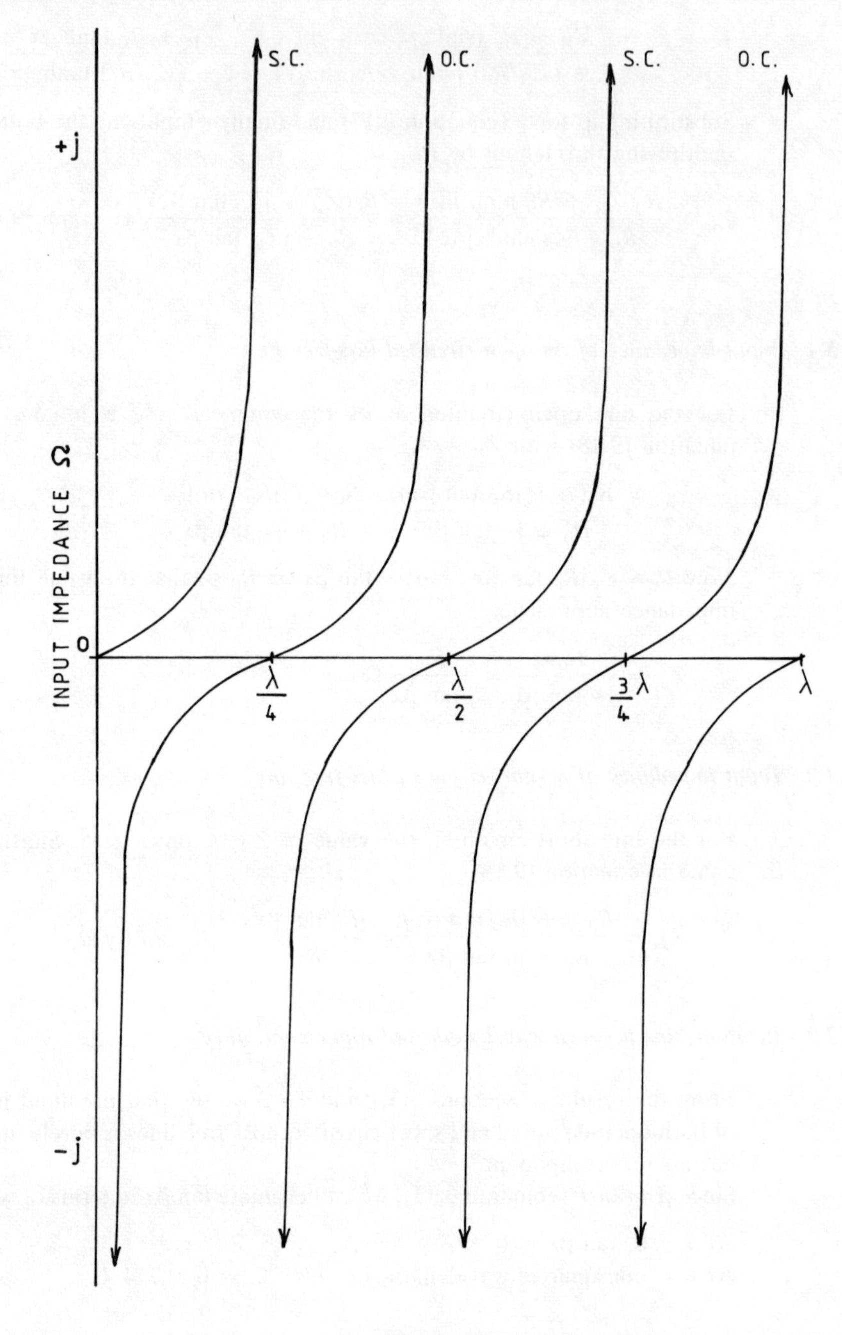

Fig. 9.12

We see that the input impedance of the open-circuited and short-circuited lines varies between 0 and infinity according to their lengths. The reader might like to evaluate the expressions for (say) λ/8, 7λ/8 and λ. Using these and similar results the curves shown in Fig. 9.12 may be drawn. Lengths of open-circuited or short-circuited line may be used to introduce into a system any value of impedance from $-j\infty$ to $+j\infty$. This is an important feature of stub matching of transmission lines.

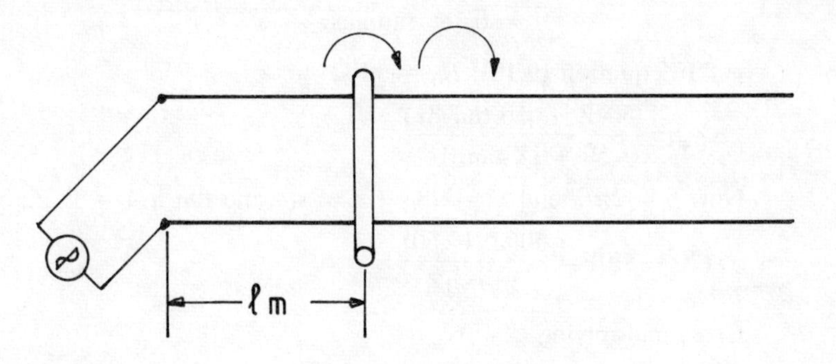

Fig. 9.13

It may also be used for the determination of high frequencies. In Fig. 9.13 we see two open copper wires being fed from a high-frequency generator. Starting at the generator end, a copper rod which short circuits the generator is rolled along until the input current to the wires falls to zero. The input impedance is now infinite. For a short-circuited line this must occur at a distance of $\lambda/4$ from the supply end. Being air insulated, the phase velocity is that of light, i.e. 3×10^8 m/s. From equation (9.14) we know that $V_p = \lambda f$ so that the frequency may be calculated.

WORKED EXAMPLE 9.8

An air-insulated loss-free line of length $\lambda/8$ has an inductance of 166.67 nH/m. It operates at a frequency of 50 MHz. It is terminated in an impedance of Z ohms. The measured input impedance = $(20 - j10)\ \Omega$.
Calculate (i) the value of β; (ii) the characteristic resistance R_0 of the line; (iii) the time delay introduced by the line; (iv) the phase change between input and output voltages; (v) the value of Z.

(i) $V_p = f\lambda$ (equation 9.14)
In air $V_p = 3 \times 10^8$ m/s $\quad \therefore 3 \times 10^8 = 50 \times 10^6\ \lambda$

$$\lambda = 6 \text{ m}$$

$\beta = 2\pi/\lambda$ (equation 9.13) $\quad \therefore \beta = 2\pi/6 = 1.0472$ rad/m

(ii) In a loss-free line $\beta = \omega\sqrt{LC}$ (equation 9.17)

$$\therefore 1.0472 = 2\pi \times 50 \times 10^6 \sqrt{166.67 \times 10^{-9}\, C}$$

$$C = \frac{1.0472^2}{(2\pi \times 50 \times 10^6)^2 \times 166.67 \times 10^{-9}} = 66.67 \times 10^{-12} \text{ F}$$

$$R_0 = \sqrt{L/C} = \sqrt{166.67\ 10^{-9}/66.67 \times 10^{-12}} = 50\ \Omega.$$

(iii) The length of the line = $\lambda/8 = 6/8 = 0.75$ m
At a velocity of 3×10^8 m/s the time taken for a signal to travel the length of the line = $0.75/3 \times 10^8 = 2.5$ ns.

The time delay of the line = 2.5×10^{-9} seconds.

(iv) Phase change = β rad/m × length of the line in metres

$$= 1.0472 \times 0.75$$
$$= 0.785 \text{ radians.}$$

(v) In equation (9.18), $R_0 = 50\ \Omega$

$$\therefore Z_{in} = \frac{50(Z + j50 \tan \beta x)}{50 + jZ \tan \beta x}$$

Now $\beta = 2\pi/\lambda$ and $x = \lambda/8$ $\therefore \beta x = \pi/4$ and $\tan \pi/4 = 1$. $Z_{in} = (20 - j10)\ \Omega$

$$\therefore (20 - j10) = \frac{50(Z + j50)}{50 + jZ}$$

Cross multiplying

$$1000 - j500 + j20Z + 10Z = 50Z + j2500$$

$$Z(-40 + j20) = -1000 + j3000$$

$$Z = \frac{-1000 + j3000}{-40 + j20} = \frac{(-1000 + j3000)(-40 - j20)}{1600 + 400}$$

from which it is found that $Z = (50 - j50)\ \Omega$.

9.3.4 *The quarter wavelength loss-free line*

Using equation (9.18), for a line of length $\lambda/4$, βx becomes $2\pi/\lambda \times \lambda/4 = \pi/2$ and $\tan \pi/2 = \infty$.

$$Z_{in} = \frac{R_0(Z_R + jR_0 \tan \pi/2)}{R_0 + jZ_R \tan \pi/2} = \frac{R_0 Z_R + jR_0^2 \infty}{R_0 + jZ_R \infty}$$

Now $R_0 \infty \gg R_0 Z_R$ and $Z_R \infty \gg R_0$ so that the value of Z_{in} approaches

$$Z_{in} = \frac{jR_0^2}{jZ_R} = \frac{R_0^2}{Z_R} \tag{9.21}$$

R_0 is the characteristic resistance of the line and Z_R is the terminating impedance.

The quarter wavelength line is sometimes referred to as an impedance transformer since Z_{in} may be given any desired value by varying Z_R.

WORKED EXAMPLE 9.9

Two loss-free lines A and B have characteristic resistances $R_{0(A)}$ and $R_{0(B)}$, respectively. These are terminated in their correct impedances. They are connected using a quarter wavelength of line with characteristic resistance $R_{0(C)}$ as shown in Fig. 9.14.

(a) Deduce the relationships between $R_{0(A)}$, $R_{0(B)}$ and $R_{0(C)}$ in order that there shall be perfect matching of the system.

(b) (i) Given that $R_{0(A)} = 75\ \Omega$ and $R_{0(B)} = 100\ \Omega$ determine the required value of $R_{0(C)}$. (ii) If the linking line C is 0.375 m in length and is air insulated, at what frequency must the system operate?

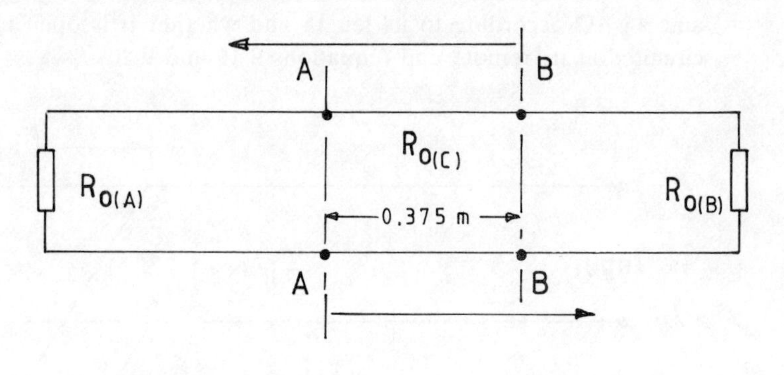

Fig. 9.14

Looking to the right from point A we see a quarter wavelength line which is connected to an impedance of $R_{0(B)}$.

From equation (9.21) $Z_{in} = \dfrac{R^2_{0(C)}}{R_{0(B)}}$. For matching, this will need to be equal to the impedance looking to the left, i.e. $R_{0(A)}$.

$$\therefore R_{0(A)} = \frac{R^2_{0(C)}}{R_{0(B)}}.$$

Transposing $R_{0(A)} \times R_{0(B)} = R^2_{0(C)}$.
Again, looking from point B to the left we see the same quarter wavelength line which is connected to an impedance $R_{0(A)}$. This impedance has to match that looking to the right, i.e. $R_{0(B)}$.

$$\therefore R_{0(B)} = \frac{R^2_{0(C)}}{R_{0(A)}}.$$

Transposing $R_{0(B)} \times R_{0(A)} = R^2_{0(C)}$.

ANSWERS

(a) The two lines will be matched if the joining quarter wavelength line has a characteristic resistance given by

$$R_{0(C)} = \sqrt{R_{0(A)} \times R_{0(B)}}$$

(b) (i) $R_{0(C)} = \sqrt{100 \times 75} = 86.6\ \Omega$
(ii) $\lambda = 4 \times 0.375$ m $\quad V_p = 3 \times 10^8$ m/s in air and $V_p = f\lambda$
$\therefore f = V_p/\lambda = 3 \times 10^8/4 \times 0.375 = 200 \times 10^6$ Hz.

9.4 Single stub matching

If a short section of loss-free line is connected across a main transmission line it will introduce, at the point of connection, a reactance varying between $-j\infty\ \Omega$ and $+j\infty\ \Omega$ according to its length and whether it is open circuited or short circuited at its remote end (equations 9.19 and 9.20).

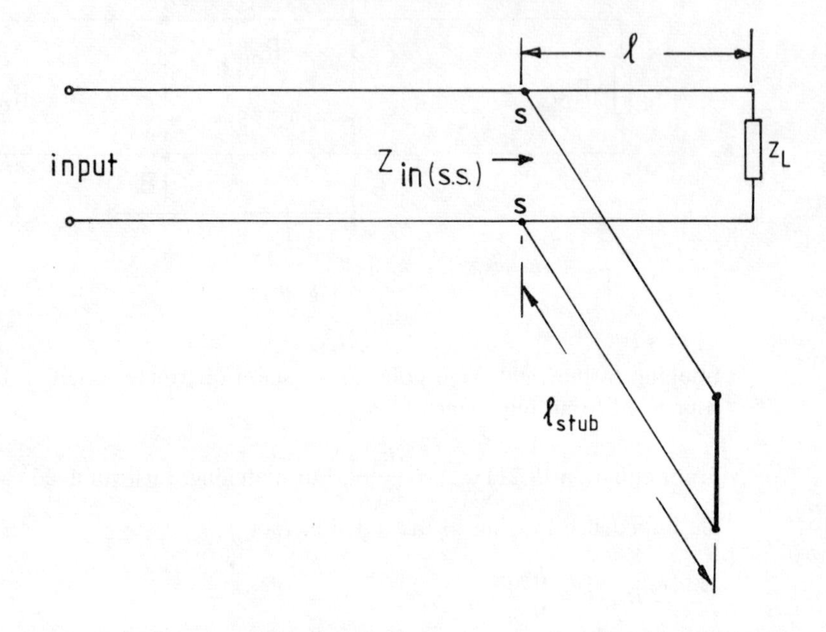

Fig. 9.15

Referring to Fig. 9.15, since the stub and line to the right of the attachment points S,S are in parallel it is convenient to consider admittance.

If for some unavoidable reason the main line is not terminated in its characteristic impedance, reflections are set up leading to power loss and distortion. The input impedance of the main line looking towards the load from S,S (disregard the line to the left of S,S) will be a function of the terminating impedance Z_L and the distance of S,S from the load, l m in Fig. 9.15. Equation (9.18) gives Z_{in} of an incorrectly terminated transmission line of total length x. In this case we are considering a length of line l m. $Z_{in(S,S)}$ is the value of equation (9.18) with x replaced by l. Imagine sliding the points S,S gradually further from the load. As l increases $Z_{in(S,S)}$ changes. At one point on the line the admittance $Y_{S,S}$ $(= 1/Z_{in(S,S)})$ will have a real component equal to the characteristic conductance of the line $1/R_0$. There will also be a reactive susceptance $1/\pm jX$. A stub is attached to the line at this point, the susceptance of which is made equal to that of the line but with the opposite j sign. We then have a condition in which the two susceptances sum to zero, the j terms disappear and to the left of S,S the line appears to be perfectly terminated in conductance $1/R_0$ (characteristic resistance R_0).

Having determined the point of attachment of the stub, the length is calculated from equations (9.19) or (9.20) to give the correct susceptance. There will be four positions per wavelength at which these conditions may be met, but the point nearest the load is chosen to give the maximum possible length of matched line.

WORKED EXAMPLE 9.10

It is required to match an open-wire loss-free feeder with a characteristic resistance of 600 Ω to a load of 100 Ω at a frequency of 100 MHz by the use of a short-circuited stub attached to the line as close to the load as possible. The stub has the same characteristic resistance as the main feeder. Determine the best position along the line at which to connect the stub and the length of the stub.

Since the line is open wire, the velocity of propagation is that of light = 3×10^8 m/s. Now $V_P = f\lambda$ and hence $\lambda = 3 \times 10^8/100 \times 10^6 = 3$ m.
Also $\beta = 2\pi/\lambda$. $\therefore \beta = 2\pi/3 = 2.0944$ rad/m.
$R_0 = 600\ \Omega$, $Z_R = 100\ \Omega$. Let $\beta l = \theta$ for the moment.
Inverting equation (9.18) to obtain admittance, and substituting values

$$Y_{in} = \frac{1}{R_0}\left(\frac{R_0 + jZ_R \tan \beta l}{Z_R + jR_0 \tan \beta l}\right) = \frac{1}{600}\left(\frac{600 + j100 \tan \theta}{100 + j600 \tan \theta}\right).$$

Rationalising

$$Y_{in} = \frac{1}{600}\left(\frac{(600 + j100 \tan \theta)(100 - j600 \tan \theta)}{100^2 + 600^2 \tan^2 \theta}\right)$$

$$= \frac{1}{600}\left(\frac{60\,000 - j360\,000 \tan \theta + j10\,000 \tan \theta + 60\,000 \tan^2 \theta}{100^2 + 600^2 \tan^2 \theta}\right). \quad \text{(i)}$$

The real part of equation (i) must equal the conductance of the line, i.e. be equal to 1/600.

$$\therefore \frac{1}{600} = \frac{1}{600}\left(\frac{60\,000 + 60\,000 \tan^2 \theta}{100^2 + 600^2 \tan^2 \theta}\right)$$

$$100^2 + 600^2 \tan^2 \theta = 60\,000 + 60\,000 \tan^2 \theta$$

$$300\,000 \tan^2 \theta = 50\,000$$

$$\tan^2 \theta = 0.1665 \quad \text{(ii)}$$

$$\tan \theta = 0.408. \quad \text{(iii)}$$

But $\beta l = \theta$. $\therefore 2.0944\ l = 0.408$ and $l = 0.185$ m.
The stub must be attached to the line 0.185 m from the load end.
The length of the stub is found by equating the imaginary (j) part of equation (i) to the admittance of a short-circuited stub with its j sign changed.
From equation (9.20) the stub has input impedance $Z_{in} = jR_0 \tan \beta l_{stub}$.

$\therefore\ Y_{in} = 1/\text{j}600 \tan\theta$ S. Change the sign of j when:

$$\frac{1}{-\text{j}600 \tan \beta l_{stub}} = \frac{1}{600}\left(\frac{-\text{j}360\,000 \tan\theta + 10\,000 \tan\theta}{100^2 + 600^2 \tan^2\theta}\right)$$

Substituting values for $\tan\theta$ and $\tan^2\theta$ from equations (ii) and (iii), and cancelling 600 on both sides

$$\tan \beta l_{stub} = -0.49$$

$$\tan 2.686 \text{ radians} = -0.49 \text{ so that } \beta l_{stub} = 2.686$$

$$l_{stub} = 2.686/2.0944 = 1.28 \text{ m.}$$

FURTHER PROBLEMS

9.11 The per-phase resistance, reactance and capacitive susceptance of a three-phase transmission line are 15 Ω, 45 Ω, and 4×10^{-4} S respectively. Calculate the values of the four-terminal network parameters for the line considering it as a T-equivalent.

9.12 The four-terminal network parameters for each phase of a 132 kV, 3-phase, 50 Hz overhead line are as follows

$$A = D = 0.9975\angle 0.6°,\ B = 47.3\angle 72° \text{ and } C = 4 \times 10^{-4}\angle 90°.$$

The line delivers a load of 50 MVA at 0.8 power factor lagging at 132 kV to a factory sub-station.

Calculate (a) the sending-end voltage, (b) the sending-end current, and (c) the transmission efficiency.

9.13 A 160 km length of three-phase, 132 kV, 50 Hz transmission line has conductor resistance 0.184 Ω/km/phase, inductance 1.32 mH/ km/phase and capacitance 9 nF/km/phase. Leakage current may be ignored. Using a T-representation, calculate for a receiving-end load of 40 MW at 0.8 power factor lagging and 132 kV (a) the sending-end voltage, (b) the sending-end current, and (c) the transmission efficiency.

9.14 A transmission line 16 km in length has the following primary constants per kilometre length

$$R = 31.25\ \Omega,\ L = 0.625 \text{ mH},\ C = 37.5 \text{ nF},\ G = 0.$$

(a) Calculate the value of Z_0 for $\omega = 5000$ rad/s.
(b) When the line is terminated in its characteristic impedance and the sending-end voltage equals 2.5 V at 5000 rad/s, calculate (i) the current in the terminating impedance, (ii) the wavelength, (iii) the phase velocity V_P.

9.15 A transmission line 5 km long has a characteristic impedance of $600\angle 0°$ Ω and a propagation coefficient (0.025 + j0.18) per km at a frequency of 1591.55 Hz.

Given that the line is terminated in its characteristic impedance and has a p.d. of $10\angle 0°$ V at the above frequency applied to the sending end, calculate (a) the magnitude of the receiving-end current, (b) the angle through which the receiving-end voltage has been retarded with respect to the sending end, and (c) the magnitude of the line voltage at a point 2.8 km from the sending end.

9.16 A co-axial cable has inductance 171 nH/m, resistance 0.1 Ω/m and a capacitance of 65 pF/m. Zero conductance may be assumed. A signal of 0.2 mV at 2 MHz is applied to a length of the cable terminated in its characteristic impedance Z_0.
Calculate (a) the value of Z_0, (b) the length of the cable if the time delay introduced by it is 0.05 s, and (c) the voltage at the termination.

9.17 A co-axial cable 15 m in length has inductance 250 nH/m, resistance 0.3 Ω/m and capacitance 51.02 pF/m. It has an input signal of 5 mV at 28 MHz and is terminated in an impedance designed to prevent reflections.
Calculate (a) the characteristic impedance Z_0, (b) the phase shift in the line, (c) the time delay introduced by the line, and (d) the power dissipated in the line termination (Z_0).

9.18 The primary constants of a telephone pair per metre length are

$$R = 0.01875\ \Omega,\ L = 0.0125\ \text{mH},\ C = 37.5\ \mu\mu\text{F},\ G = 5 \times 10^{-9}.$$

The line is 10 km in length and the frequency is 3400 Hz. An oscillator with an internal resistance of 600 Ω which generates an e.m.f. of $6\angle 0°$ V at 3400 Hz is connected to the sending-end of the line. Assuming correct termination, calculate (a) the value of Z_0, (b) the input voltage to the line, (c) the power input to the line, (d) the power at the termination, and (e) the time delay introduced by the line.

9.19 An open-wire loss-free line with characteristic resistance = 600 Ω is matched at a single frequency of 100 MHz to a load of 100 Ω using a single short-circuited stub. (See also worked example 9.10) The best position for the stub is a distance of 0.185 m from the termination, when the stub needs to be 1.28 m long.
(a) Verify that the next possible position for the stub is at a distance of 1.315 m from the termination.
(b) Determine the required length of short-circuited stub to be attached at this point to give matching over the rest of the line.

9.20 (a) A length of loss-free line with inductance 1.32 mH/km and capacitance 0.0089 μF/km is to be matched to a length of cable with inductance 200 μH/km and capacitance 0.166 μF/km using a quarter wavelength section of line. Determine the necessary characteristic resistance of the quarter wavelength section.
(b) If the matching section of line is air insulated and the matching is to be correct for a frequency of 10 MHz, what is the length of this section?

9.21 A pure-resistance load with value 90 Ω is used to terminate a transmission line which has characteristic resistance R_0 = 75 Ω and a wavelength λ = 6 m.
(a) Determine the distance from the load at which an open-circuited stub should be attached to optimise the length of matched line. (b) Determine the minimum length of stub required.
The stub is formed using the same cable as the main line.

9.22 A transmission line with a characteristic resistance R_0 = 400 Ω is terminated in an impedance of (108.6 − j121.5) Ω. One wavelength = 0.75 m.
(a) Determine the distance of the point on the line at which an open-circuited

stub should be connected to give matching over the greatest length of line.
(b) Determine the length of the stub.
(*Hint*: Using equation (9.18) the term $jZ_R \tan\theta = j(108.6 - j121.5)\tan\theta$, giving a real and j term. As in worked example 9.11 collect real terms and equate to 1/400.)

Answers

9.2 $34.18\angle{+4.23}°$ kV

9.4 $A = D = 1 + YZ/2 = 0.986\angle 0.39°$; $B = Z = 72.75\angle 66.07°$;
$C = Y + (Y/2)^2 Z = 4.49 \times 10^{-4}\angle 90.2°$; $I_R = 218.7\angle{-36.9}°$ A
(Observe: 50 MVA not MW)
$V_S = 89\,400\angle 5.3°$ V (phase) 154.8 kV line (values vary slightly with degrees of accuracy in using A, B, C and D)

9.7 (a) $Z_0 = 78.3\angle{-31.98}°\ \Omega$ (b) $0.638e^{-2.71} = 0.0423$ A (c) 11.6 km
(d) 115 600 km/s

9.11 $A = D = 0.991 + j0.003$; $B = 14.9 + j44.8$; $C = j0.0004$

9.12 $I_R = 218\angle{-36.8}°$ A (a) $V_S = 85\,000\angle 4.6°$ V (phase)
(b) $I_S = 199.4\angle{-28.7}°$ A (c) 0.938 p.u.

9.13 (a) $89\,200\angle 5.3°$ V (phase) (b) $196.9\angle{-28.4}°$ A (c) 0.91 (this is the same line as in self-assessment example 9.4 for comparison of π with T)

9.14 (a) $409.3\angle{-42.15}°\ \Omega$ (b) (i) 2.68×10^{-3} A (ii) 110.5 km
(iii) 87 933 km/s ($\gamma = (0.0515 + j0.0569)$ per km)

9.15 (a) 14.72 mA (b) 0.9 rad (51.6°) (c) 9.32 V

9.16 (a) $51.3\angle{-1.33}°\ \Omega$; $\gamma = (9.798 \times 10^{-4} + j4.189 \times 10^{-2})$ per metre;
$V_p = 3 \times 10^8$ m/s (b) 15 m (c) 0.197 mV

9.17 (a) 70 Ω (b) 9.42 rad (539.7°) (c) 53.57 ns (d) 0.33 μW

9.18 (a) $578\angle{-1.83}°\ \Omega$ (b) 2.944 V (c) 14.98 mW (d) 12.56 mW
(e) 216.6 μs

9.19 In part (a) $\tan\theta = -0.408$. Use in (b) to find $\beta l_{stub} = 0.455$. Length of stub = 0.2175 m

9.20 (a) $R_0 = 115.6\ \Omega$ (b) 7.5 m

9.21 Attach 0.993 m from load, length of stub 2.826 m

9.22 $\mathrm{Tan}^2\,\beta l - 0.8336\tan\beta l = 0.144$; $\tan\beta l = -0.1472$ or 0.9808; $\beta l = 6.137$ rads or 0.7757 rads. Best $l = 0.092$ m. Length of stub = 0.118 m

Chapter 10
Electron Ballistics

In general the analysis of the performance of an electron beam in an external field can be performed satisfactorily if the following conditions are fulfilled. (i) The electrons are fairly widely spaced so that the external field has more influence on the electron than does each electron on its neighbour. This would be the case if a substantial electron cloud were allowed to form, since like charges repel one another. (ii) Where acceleration and velocity are involved, the final velocity does not become a large proportion of that of light since, under these circumstances, adding energy increases the mass of the electron and not its velocity.

10.1 The electron beam

10.1.1 The electron gun

In a situation of near-perfect vacuum a heated cathode is situated a short distance from a flat anode which is maintained at a high positive direct potential with respect to the cathode. Since unlike charges attract one another, the electrons emitted from the cathode are attracted towards the positive anode. The force on an electron is given by the formula

Force = Ee Newtons

where E is the electric field strength between cathode and anode in volts/metre and e is the charge on the electron (equation 4.10, chapter 4, with charge q replaced by the charge on a single electron e)

Also force = mass of the electron × acceleration.

The mass of the electron is very small (9.1×10^{-31} kg), so that the large potential gives rise to considerable acceleration and the velocity of the electron as it approaches the anode is also very large.

The anode has a small hole in it and some of the electrons pass through this hole and travel on until they impact with something in their path. More electrons may be caused to pass through the anode hole by placing a metal cylinder called the grid between cathode and anode. The cylinder is maintained at a small negative potential with respect to the cathode. Since like charges repel one another, electrons are repelled by the grid cylinder which tends to cause them to travel on a central path, so more of them pass through the anode hole.

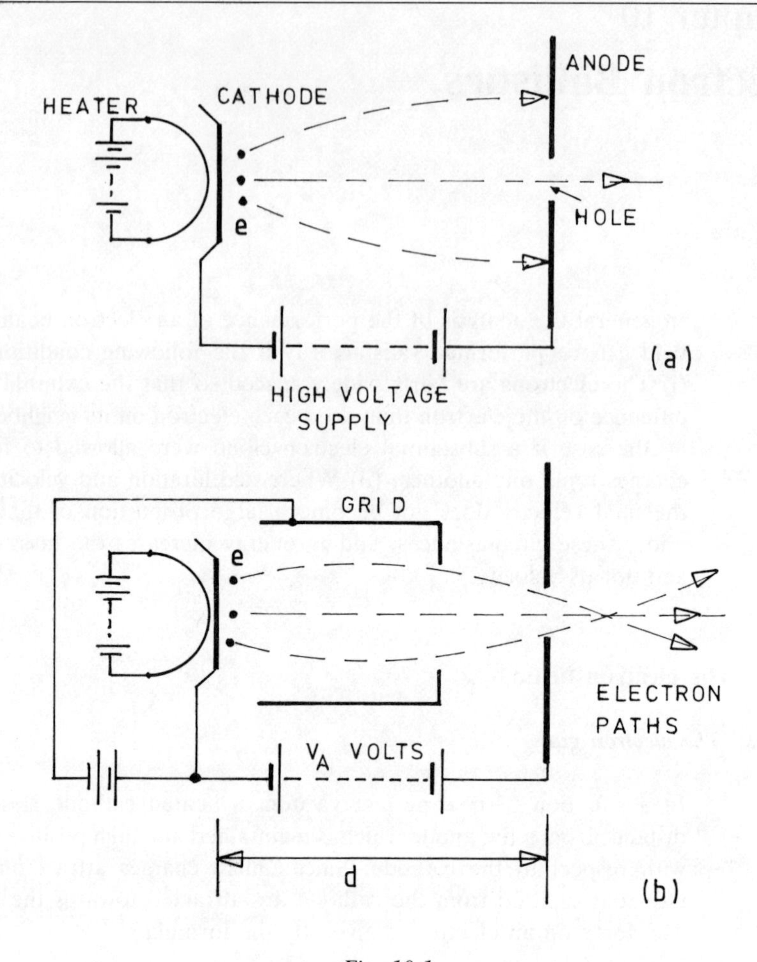

Fig. 10.1

Possible electron paths without and with the grid cylinder are shown in Fig. 10.1. If the grid is made very strongly negative, the electron flow is cut off since it repels electrons back to the cathode. Varying the grid voltage therefore varies the number of electrons leaving the gun and hence, for example, the brightness of the trace on a cathode ray tube.

10.1.2 Electron velocity leaving the gun

Given the anode voltage = V_A volts and distance between cathode and anode = d metres as shown in Fig. 10.1(b)

$$\text{Force on an electron} = Ee = \frac{V_A}{d} e \text{ N}$$

$$\text{Acceleration} = \frac{\text{force}}{\text{mass}} = \frac{V_A}{d} \times \frac{e}{m} \quad \text{m/s/s}$$

Final velocity v m/s = acceleration × time taken to travel between cathode and anode and

$$\text{Time } t = \frac{\text{distance}}{\text{average velocity}} = \frac{\text{distance}}{\text{final velocity}/2} = \frac{d}{v/2} = \frac{2d}{v} \text{ s.}$$

$$\therefore v = \frac{V_A}{d} \times \frac{e}{m} \times \frac{2d}{v} \text{ m/s}$$

$$v = \sqrt{\frac{2V_A e}{m}} \text{ m/s.} \qquad (10.1)$$

The same result may be obtained by equating the energy of the moving mass of the electron to the energy given to it by the accelerating anode potential.

Energy of a moving mass = $\frac{1}{2}mv^2$ J.

Energy given to the charge when raised through a potential V_A volts = $V_A e$ J. (equation 4.7)

$$\therefore \tfrac{1}{2}mv^2 = V_A e \qquad (10.2)$$

Transposing equation (10.2) gives the same result as equation (10.1).

10.2 Deflection of an electron beam

10.2.1 Electrostatic deflection

In Fig. 10.2 an electron travelling at a velocity v_x m/s travels between two parallel plates a distance d_d m apart. The bottom plate is held at a positive potential of V_d volts with respect to the top plate. As the electron enters the space between the plates it will suffer a force given by

$$\text{force} = Ee \text{ N} = \frac{V_d}{d_d} e \qquad \text{acceleration} = \text{force/mass} = \frac{V_d}{d_d}\frac{e}{m}$$

$$\text{Time taken for the electron to pass between the plates} = \frac{l}{v_x} \text{ s.}$$

Owing to the positive potential on the bottom plate, the electron will be attracted downwards. Let v_y be the velocity imparted to the electron in the vertical direction.

$$v_y = \text{acceleration} \times \text{time}$$

$$\therefore v_y = \frac{V_d}{d_d}\frac{e}{m}\frac{l}{v_x} \text{ m/s.} \qquad (10.3)$$

The original velocity v_x is maintained since there has been no force in a horizontal direction to change it. The addition of a downward velocity has served to increase the final velocity v_F.

In Fig. 10.2(b), using Pythagoras' theorem $v_F = \sqrt{v_x^2 + v_y^2}$ m/s.

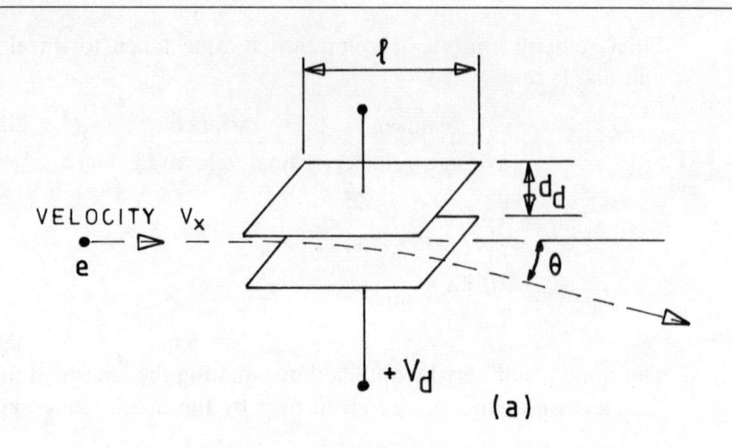

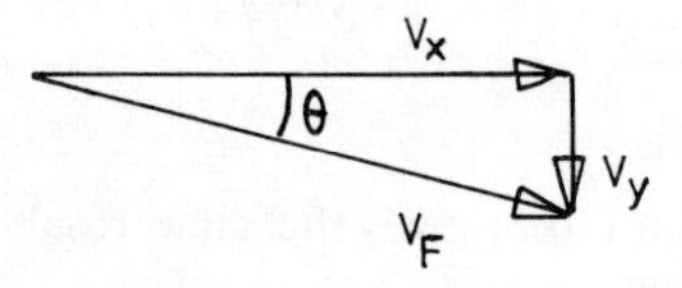

Fig. 10.2

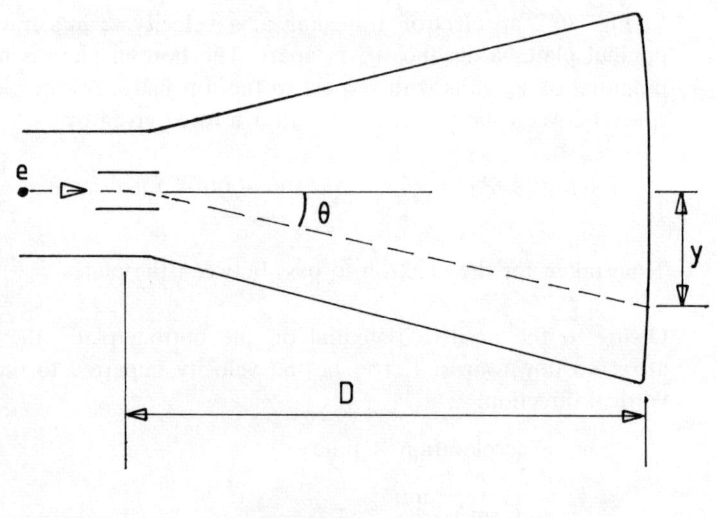

Fig. 10.3

In Fig. 10.3, the parallel-plate deflection system is shown as part of a cathode ray oscilloscope (CRO). With the electron travelling with a horizontal velocity of v_x m/s, the time taken to strike the tube end = D/v_x seconds. With vertical velocity v_y m/s, in this time it will travel downwards a distance y metres.

$$y = v_y \times \frac{D}{v_x} \text{ m.} \tag{10.4}$$

Now v_x is the exit velocity of the electron from the gun in equation (10.1). Substituting this value and that for v_y from equation (10.2) in equation (10.4)

$$y = \frac{D \times \dfrac{V_d e}{d_d m} \times l}{2 \times \dfrac{e}{m} \times V_A} = \frac{V_d D l}{V_A 2 d_d} \text{ m.} \tag{10.5}$$

Horizontal deflection of the beam is achieved in the same manner using a pair of vertically mounted plates.

The angle θ through which the beam can be deflected is small, being determined by the ratios V_d/V_A and l/d_d. In equation (10.5), using a small value of d_d would appear to make the deflection y large. However, in practice with d_d small the angle of deflection can only be small before the electron beam would hit the deflection plates. With d_d large the beam could be deflected more before hitting the plates but the value of V_d would need to be very much greater to give the same electric field strength between the plates. Increasing V_A to increase the speed of the electrons and hence the brightness of the trace on the screen has the effect of reducing the deflection. CROs using this method of deflection employ long tubes to produce a useful sized trace on the screen. The deflection may be increased by shaping the deflector plates as shown in Fig. 10.4. The plates are close together at the entry end giving a high electric field strength V_d/d_d, and further apart at the exit end to avoid making contact with the beam.

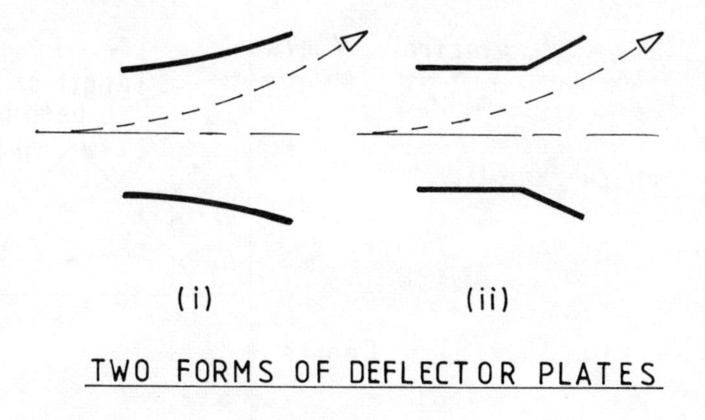

Fig. 10.4

10.2.2 *Magnetic deflection*

An electric current situated in a magnetic field suffers a force. We are often concerned with current in a wire and the direction of the force can be determined by the use of Flemming's left-hand rule as illustrated in Fig. 10.5(a).

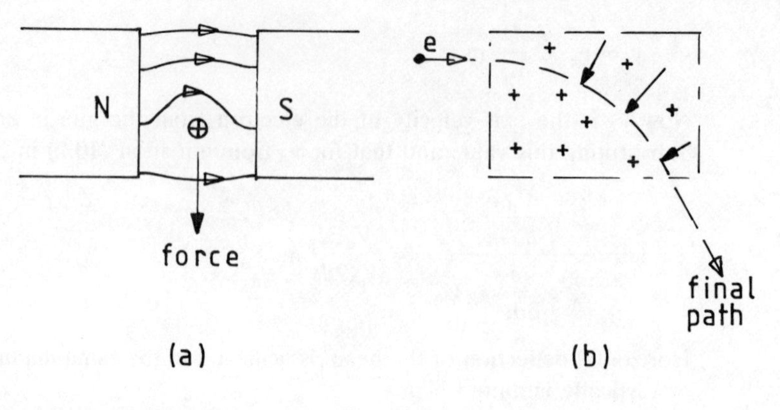

Fig. 10.5

This rule involves consideration of the direction of conventional current flow. Electrons move in the opposite direction, however. With electrostatic deflection the direction of the deflecting force was always at right angles to the face of the plates. With electromagnetic deflection the force is at right angles to the direction of current (electron) motion. This will cause it to take up a circular orbit. The velocity of the electron does not change.

The centrifugal force on a rotating mass m is $\dfrac{mv^2}{R}$ (10.6)

where R = radius of circle in metres, m = mass in kilograms and v = tangential velocity in metres/second.

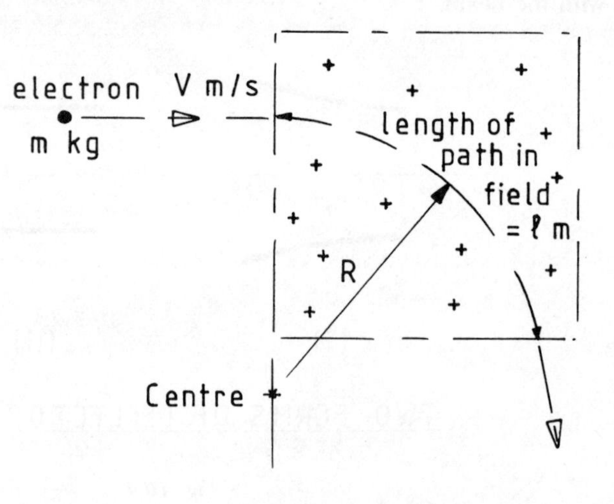

Fig. 10.6

With reference to Fig. 10.6. At velocity v m/s, the time taken for the electron to travel through the magnetic field = distance travelled/velocity

$$t = \frac{l}{v} \text{ s.}$$

Considering a single electron being deflected:
The charge on 1 electron = e coulombs. Now, the ampere = charge movement in coulombs/second.

$$\therefore \text{ 1 electron travelling at } v \text{ m/s is a current of } e/t = e \div l/v = ve/l \text{ A.}$$

The force on a current I amperes in a magnetic field of strength B tesla in a conductor of length l metres is given by

$$F = BIl \text{ N}$$

$$= B \times \frac{ve}{l} \times l \text{ N on a single electron}$$

$$F = Bev \text{ N.} \qquad (10.7)$$

For the orbit to remain circular, this force must be exactly balanced by the centrifugal force on the mass of the electron. From equations (10.6) and (10.7)

$$\frac{mv^2}{R} = Bev$$

Transposing

$$R = \frac{mv^2}{Bev} = \frac{mv}{Be} \text{ m.} \qquad (10.8)$$

m, v, B and e are all constants and determine the radius R. The length of time in the magnetic field will determine the angle through which the electron is deflected. Large angles are possible and this type of deflection is used in television receivers where a large picture is obtained on a very short tube by turning the electron beam through approximately 60°. However, quite large coil power is required to produce the magnetic field, and because of the inductance of these windings the range of operating frequencies is limited.

$$\text{The angular velocity of the electron } \omega = \frac{v}{R} = \frac{v}{mv/Be} = \frac{Be}{m} \text{ rad/s.} \qquad (10.9)$$

Time taken for one complete revolution in the field would be

$$\frac{2\pi}{\omega} = \frac{2\pi}{Be/m} = \frac{2\pi m}{Be} \text{ s.} \qquad (10.10)$$

WORKED EXAMPLE 10.1

An electron is accelerated using an electron gun when it enters a uniform magnetic field of flux density 1 milli-tesla in a direction perpendicular to that field. The electron leaves the field 0.005 μs after entry. Calculate the angle between the direction of entry and the direction of exit from the field.

$$e/m = 1.76 \times 10^{11} \text{ C/kg.}$$

From equation (10.10)

$$\text{time for one revolution} = \frac{2\pi m}{Be} = \frac{2\pi}{1 \times 10^{-3} \times 1.76 \times 10^{11}}$$
$$= 35.7 \times 10^{-9} \text{ s.}$$

In 0.005 μs it turns $0.005 \times 10^{-6}/35.7 \times 10^{-9} = 0.14$ revolutions
$= 50.4°$.

The electron leaves the field after having been turned through 50.4°.

10.2.3 Helical path of an electron in a magnetic field

If an electron enters a region in which a magnetic field exists whilst travelling along a line which is not at right angles to the magnetic force lines, then its path will take up the form of a helix.

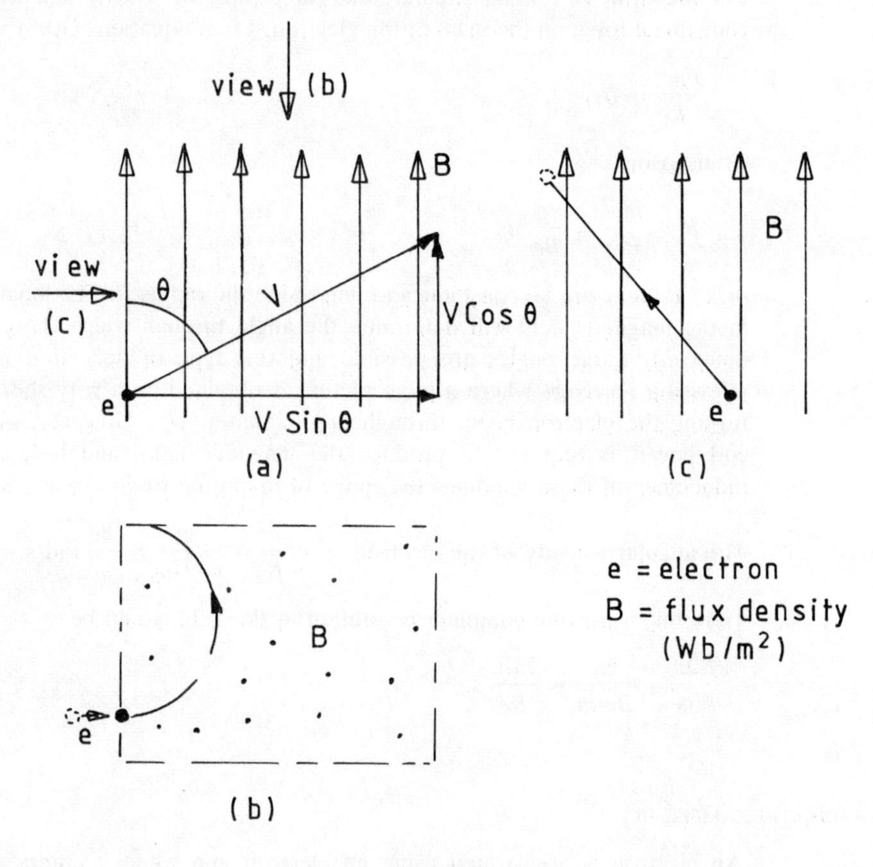

Fig. 10.7

Figure 10.7(a) shows a vertically oriented magnetic field with an electron entering at the left along a line making an angle θ to the direction of the field and travelling v m/s. This velocity may be resolved into two components, one at right angles to the field lines, $v \sin \theta$; the other along the field lines, $v \cos \theta$.

An electron travelling along magnetic field lines will be unaffected, whilst an electron travelling at right angles to the field lines will be caused to take up a circular path as described in the previous paragraph.

To make clear what is happening, two other views of the electron are presented in Fig. 10.7(b) and (c). Looking at Fig. 10.7(a) from the left we see view (c), and looking from the top we see view (b). The electron maintains its velocity $v \cos \theta$ in an upward direction (view a) whilst moving in circular orbit (view b). The total motion is that of a helix.

WORKED EXAMPLE 10.2

A solenoid of 2000 turns is 40 cm in length and it is assumed that the magnetic field produced within it is perfectly uniform. It is situated in a vacuum. An electron is accelerated using a potential of 750 V and it enters one end of the solenoid, crossing its centre line, at an angle of 3° to the magnetic force lines.

(a) Determine the value of coil current that will cause the electron to turn through one complete revolution and so exit from the other end of the coil, again crossing its centre line.

(b) Determine the radius of the turning circle of the electron.

$$e/m = 1.76 \times 10^{11} \text{ C/kg} \qquad \mu_o = 4\pi \times 10^{-7} \text{ H/m}.$$

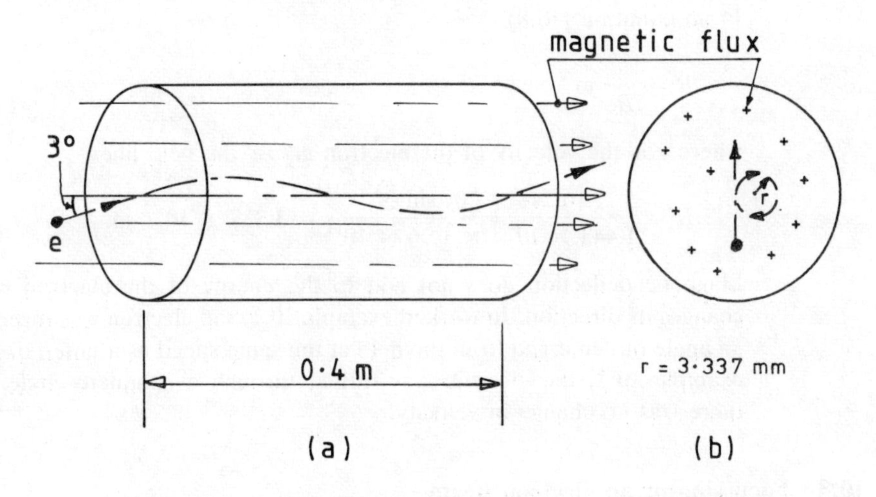

Fig. 10.8

From equation (10.1)

$$\text{exit velocity of the electron from the gun } v = \sqrt{2v_A \frac{e}{m}}$$

$$= \sqrt{2 \times 750 \times 1.76 \times 10^{11}}$$

$$v = 16.248 \times 10^6 \text{ m/s}.$$

The axial velocity (refer to Figs. 10.7 and 10.8) $= v \cos \theta$
$= 16.248 \times 10^6 \times \cos 3°$
$v_{axial} = 16.226 \times 10^6$ m/s.

The electron has to travel 40 cm in the axial direction to emerge at the other end of the coil.

Therefore, time in the coil $= 0.4 \text{ m}/v_{axial} = 0.4/16.226 \times 10^6 = 24.65 \times 10^{-9}$ s.

Owing to its velocity $v \sin 3°$ across the field lines, it will move with circular form. Equation (10.10) expresses the time taken for one revolution as

$$t = \frac{2\pi m}{Be} \text{ s.}$$

Substituting values $24.65 \times 10^{-9} = \dfrac{2\pi}{B \times 1.76 \times 10^{11}}$

Transposing $B = \dfrac{2\pi}{1.76 \times 10^{11} \times 24.65 \times 10^{-9}} = 1.448 \times 10^{-3}$ tesla.

For a coil in vacuum $H = NI/l$ and $B = \mu_0 H$

$$\therefore 1.448 \times 10^{-3} = 4\pi \times 10^{-7} \times \frac{2000\ I}{0.4}$$

$$I = 0.23 \text{ A.}$$

From equation (10.8)

$$R = \frac{mv}{Be} \text{ m}$$

where v is the velocity of the electron *across* the field lines

$$R = \frac{16.248 \times 10^6 \sin 3°}{1.448 \times 10^{-3} \times 1.76 \times 10^{11}} = 3.337 \times 10^{-3} \text{ m.}$$

Magnetic deflection does not add to the energy of the electron but merely changes its direction. In worked example 10.1, the electron was turned through an angle but emerged from the field at the same speed as it entered. In worked example 10.2, the electron was turned through a complete circle but again there was no change in velocity.

10.3 Focussing of an electron beam

10.3.1 Focussing using an electrostatic field

(See also chapter 4 for work field plotting.)

When an electron enters a region in which an electric field is present, that electric field influences the direction and velocity of the electron. We have used

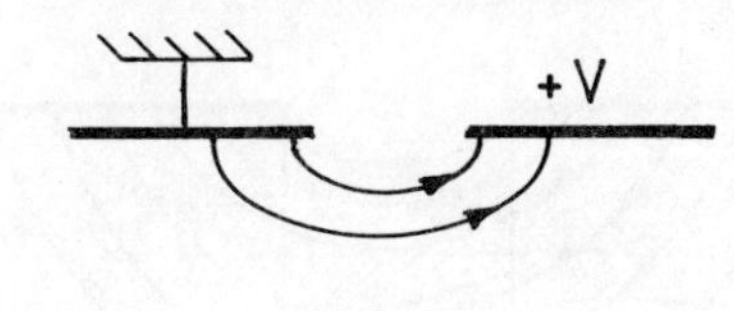

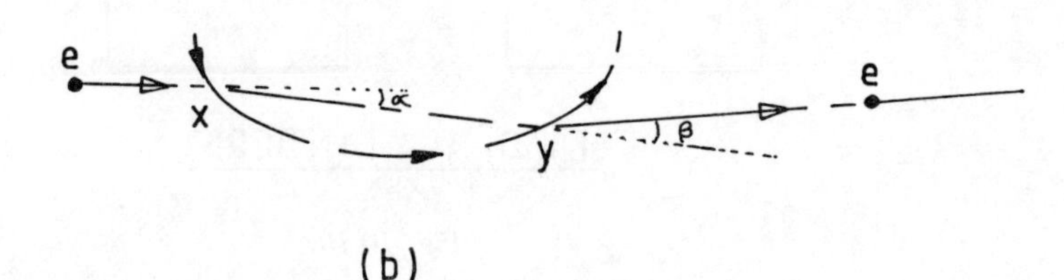

Fig. 10.9

the effect in this chapter to accelerate an electron in the electron gun and subsequently to deflect it. The same principles apply to focussing the electron beam to give a sharp display on the cathode ray tube.

In Fig. 10.9(a) two small plates are shown. The right-hand plate is at a positive potential with respect to the left-hand plate. The electric stress between them is shown in part using two lines only. The arrows indicate the directions of forces on an electron. In Fig. 10.9(b) an electron is travelling to the right and meets one of the stress lines shown in enlarged form. When it reaches point x it will suffer a force in the direction of the stress line which will cause it to be deflected in a downward direction through an angle α. It will carry on until it re-crosses the line at y when again it will be deflected, upwards this time, through an angle β. The deflection angles will depend upon the electron velocity (the faster it is travelling the less will be the deflection), the angle at which the electron meets the stress line, and upon the potential difference between the two plates.

Figure 10.10 shows two cylinders, the right-hand one being at $+V$ volts with respect to the left-hand one. Stress lines are drawn for the top half only, the bottom half will be symmetrical. The approaching electron beam, coming from the electron gun, is shown to be divergent. As the electrons enter the electric field they are at first deflected towards the central axis and then, as they leave, away from the axis. By varying the voltage on the second cylinder the effect is varied and the electrons may be brought to a single point on the screen. Using two cylinders of different sizes the effect is the same. This arrangement is shown in Fig. 10.11.

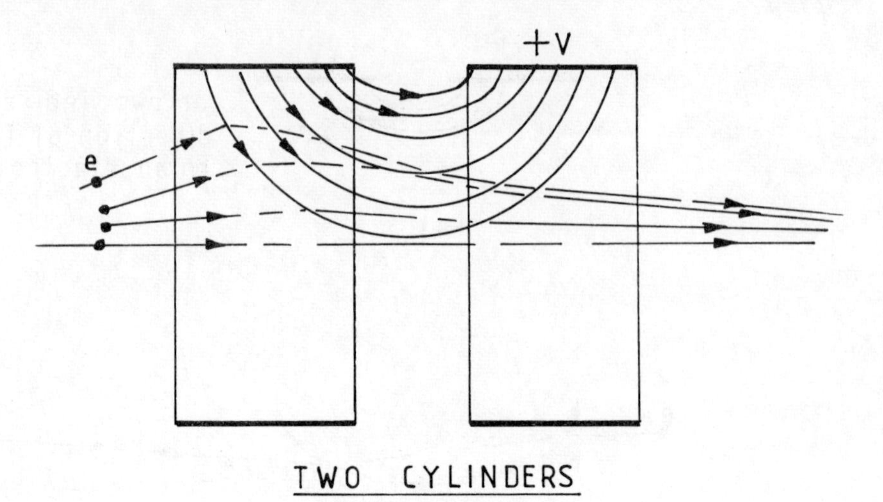

Fig. 10.10

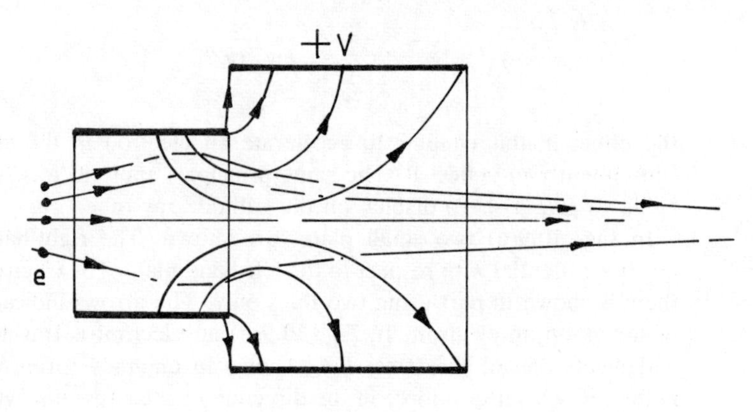

Fig. 10.11

10.3.2 *Focussing using a magnetic field*

A uniform magnetic field, as demonstrated in worked example 10.2, causes the electron beam to rotate but the electrons leave at the end of the solenoid in a variety of directions dependent upon entry angle and the number of revolutions or part revolutions completed. A short coil (as shown in Fig. 10.12), however, produces a non-uniform field along its length and this may be used to bring an electron beam to a focus point. The top half of the field only is considered; the bottom half is symmetrical.

Consider the typical section of magnetic field XX. The electron beam enters the field making an angle θ to the lines on the plane XX. There is a component of velocity along the field lines v_1, which will result in zero effect, and a component at right angles to the lines, v_2, which is in an upward direction in the section view on XX (Fig. 10.12(b)). The resultant force is in a clockwise

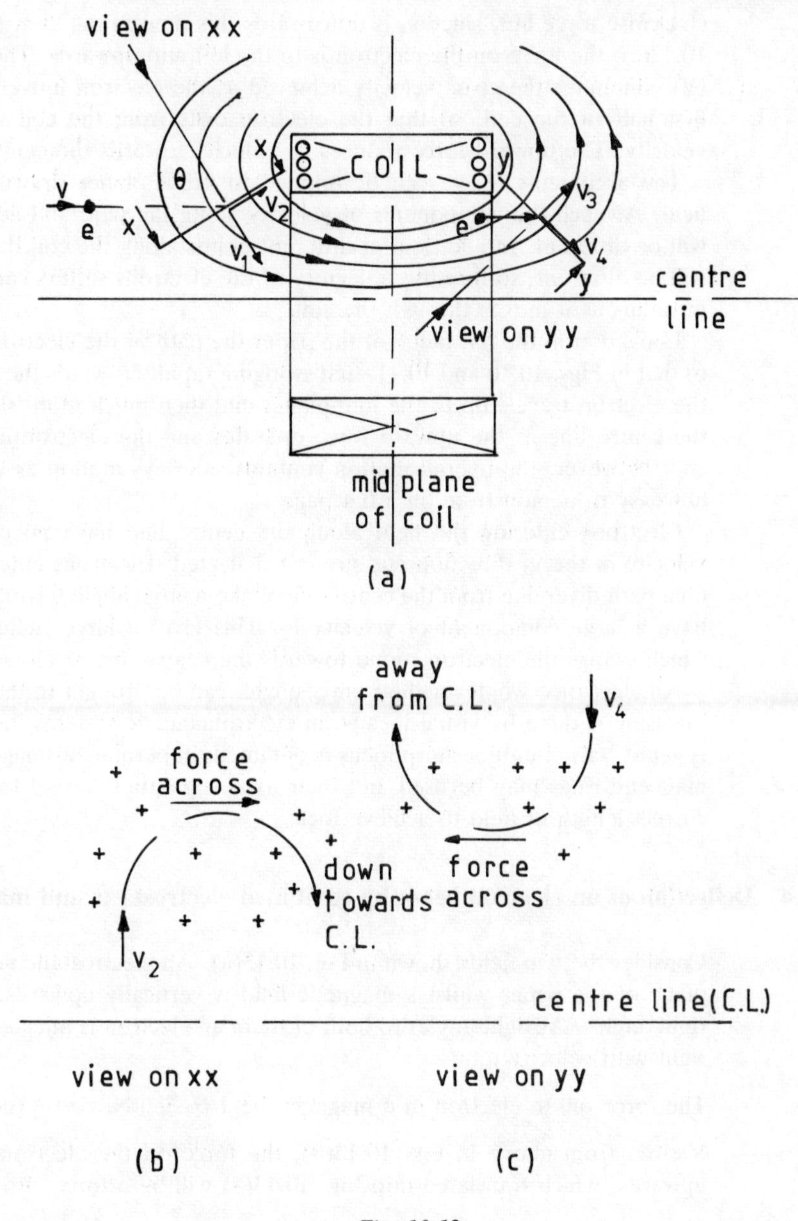

Fig. 10.12

direction, which pushes the electron to the right as viewed and downwards towards the centre line of the coil. As the electron crosses the mid-plane it is travelling sideways to the right, and downwards towards the centre line of the coil.

After the mid-plane a further typical plane is that at YY. As the beam crosses YY, the component of velocity along the flux lines is v_3 and this has no effect. The component across the lines is v_4 and this will again cause a clockwise force but, since v_4 is downwards this time in the view shown in Fig. 10.12(c), the force on the electron is to the left and upwards. The force to the left eliminates the cross velocity achieved as the electron moved through the first half of the coil, so that the electron exits from the coil with no cross velocity. The upward force reduces its velocity towards the centre line.

The arguments above can be applied to other planes drawn through the field. At each the components of velocity along the field and across the field will be different. In addition, at different points along the coil the flux density will be different, so that the trajectory of the electrons suffers constant change in radius as it moves through the coil.

Looked at in the flat plane of the paper the path of the electrons is identical to that in Figs. 10.10 and 10.11, first swinging rapidly towards the centre line as the electron travels up to the mid-plane, and then much more slowly towards the centre line as the upward force operates and the electron exits from the coil. However, the overall motion contains sideways motion as well, which is not easy to demonstrate on a flat page.

Electrons entering the field along the centre line have no component of velocity in the v_2 direction and are not deflected. Electrons entering the field on a path diverging from the centre line make a large angle θ with the field and have a large component of velocity v_2. This gives a large radius of rotation which swings the electron round towards the centre line as already described.

All electrons, whatever their entry angle, can be brought to the same point. Focusing is done by visual means: in electromagnetic systems the coil current is gently varied until a sharp focus is obtained; in permanent-magnet types two magnetic rings may be used and their axial separation varied to produce the correct length of field to achieve focus.

10.4 Deflection of an electron beam by combined electrostatic and magnetic fields

Consider the two fields shown in Fig. 10.13(a). An electrostatic field enters the plane of the paper whilst a magnetic field is vertically upwards. They are at right angles. At right angles to both of them an electron is injected from left to right with velocity v m/s.

The force on an electron in a magnetic field $= Bev$ N. (equation 10.7)

Viewed from above in Fig. 10.13(b), the force on the electron $F_{(m)}$ will be upwards, which translated into Fig. 10.13(a) will be a force into the paper.

The force on the electron due to the electrostatic field $= Ee$ N. (equation 4.10)

This force $F_{(e)}$ will be towards the positive potential producing the electrostatic field, i.e. downwards in Fig. 10.13(b) or out of the paper in Fig. 10.13(a).

When $Bev = Ee$, there will be no deflection. The electron will travel straight on.

Cancel e on both sides of the equation.

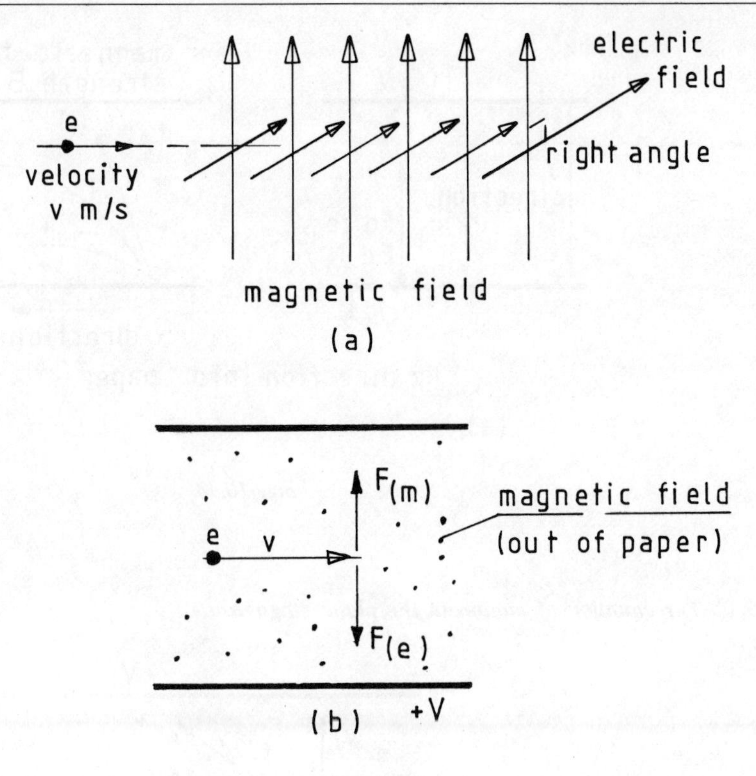

Fig. 10.13

Transposing gives $v = \dfrac{E}{B}$ m/s.

When the electron enters the combined fields at a velocity of E/B m/s, the electron will not be deflected.

10.5 The magnetron effect

In Fig. 10.14(a) we see an electron injected into the space between two plates between which there is an electrostatic field of strength E V/m. The positive potential of the top plate will attract the electron which will move in the y direction.

The force on the electron $= Ee$ N. (equation 4.10)

In addition to the electrostatic field, at right angles to the two plates there is a magnetic field entering the plane of the paper. This is shown in Fig. 10.14(b). Now, because the electron is moving upwards in the magnetic field it will be deflected to the right. The electron may never reach the top plate and be returned to the cathode further along to the right. Three possible paths for the electron are shown (i), (ii) and (iii).

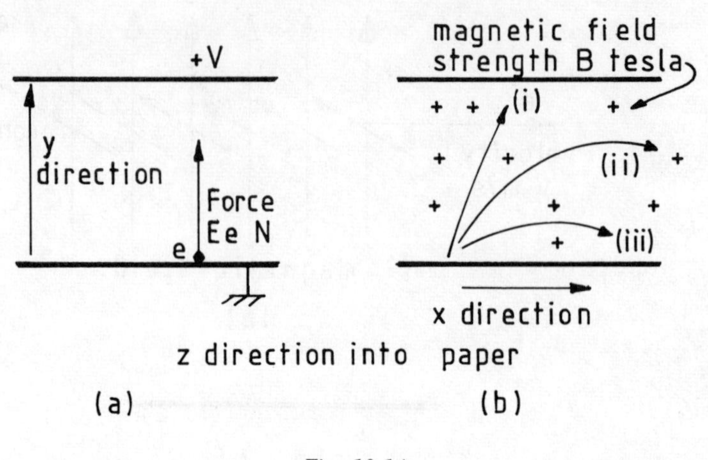

Fig. 10.14

10.5.1 The equation of motion in the plane magnetron

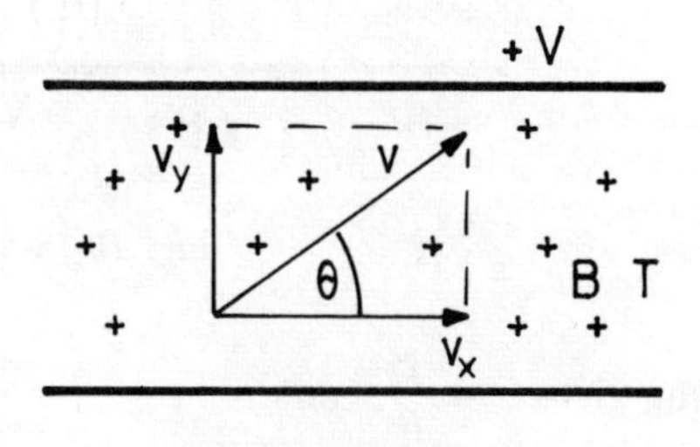

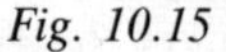

Fig. 10.15

After a short interval of time the electron considered in Fig. 10.15 will be travelling with velocity v m/s at an angle θ to the horizontal, as drawn. Resolving this velocity into two components at right angles

$$v_x = v \cos \theta \qquad \text{and } v_y = v \sin \theta.$$

In the y direction:

The upward force due to the electrostatic field = Ee N.
Since the electron has a horizontal velocity v_x through the magnetic field, there will be a downward force on it of Bev_x N.

The resultant force on the electron = $Ee - Bev_x$ N.

But force = mass × acceleration (in the y direction)

$$\therefore \text{ force} = m\frac{dv_y}{dt} = Ee - Bev_x. \qquad (10.11)$$

In the x direction:

Because of the vertical velocity through the magnetic field there will be a force in the x direction of Bev_y N.

Again, force = mass × acceleration (in the x direction)

$$\therefore m\frac{dv_x}{dt} = Bev_y. \tag{10.12}$$

Transposing equation (10.12) $\dfrac{dv_x}{dt} = \dfrac{Be}{m}v_y.$ (10.13)

Differentiating equation (10.11) with respect to t, $\quad m\dfrac{d^2v_y}{dt^2} = -Be\dfrac{dv_x}{dt}.$

Substituting for dv_x/dt from equation (10.13)

$$m\frac{d^2v_y}{dt^2} = -Be \times \frac{Be}{m}v_y$$

$$\frac{d^2v_y}{dt^2} = -\frac{B^2e^2}{m^2}v_y. \tag{10.14}$$

Using Laplace transforms (chapter 8)

$$\frac{d^2f(t)}{dt^2} = s^2F(s) - sf_{(0)} - f'_{(0)} \qquad v_y \text{ is a function of } t.$$

$$\therefore s^2\overline{v_y} - sv_{y(0)} - v'_{y(0)} = \frac{B^2e^2}{m^2}\overline{v_y}.$$

$$\overline{v_y}\left\{s^2 + \frac{B^2e^2}{m^2}\right\} = sv_{y(0)} + v'_{y(0)}.$$

At $t = 0$, $v_y = 0$ and $\dfrac{dv_y}{dt}$ is the acceleration due to $E = \dfrac{\text{force}}{\text{mass}} = \dfrac{Ee}{m}$.

Therefore $sv_{y(0)} = 0$ and $v'_{y(0)} = Ee/m$.

$$\overline{v_y}\left(s^2 + \frac{B^2e^2}{m^2}\right) = \frac{Ee}{m}$$

$$\overline{v_y} = \frac{Ee}{m}\left(\frac{1}{s^2 + [Be/m]^2}\right) \quad \text{and } v_y = \frac{E}{B}\sin\frac{Be}{m}t \text{ m/s.} \tag{10.15}$$

From equation (10.12)

$$m\frac{dv_x}{dt} = Bev_y.$$

Substituting for v_y from equation (10.15)

$$\frac{dv_x}{dt} = \frac{Be\ E}{m}\sin\frac{Be}{m}t.$$

Integrating both sides of the equation with respect to t gives

$$v_x = -\frac{E}{B} \cos \frac{Be}{m} t + C \text{ m/s.}$$

To evaluate the constant of integration C, we know that at $t = 0$ there is no velocity in the x direction. Therefore at $t = 0$, $v_x = 0$ and $C = E/B$.

$$\text{Therefore } v_x = \frac{E}{B}\left(1 - \cos \frac{Be}{m} t\right) \text{ m/s.} \tag{10.16}$$

The distance travelled by the electron in the x direction in time t seconds is found by integrating equation (10.16) between the limits 0 and t.

$$\text{Distance travelled in the } x \text{ direction} = \frac{E}{B}\left(t - \frac{m}{Be} \sin \frac{Be}{m} t\right) \text{ m.} \tag{10.17}$$

Similarly, integrating equation (10.14) for v_y between limits of 0 and t gives

$$\text{distance travelled in the } y \text{ direction} = \frac{Em}{B^2 e}\left(1 - \cos \frac{Be}{m} t\right) \text{ m.} \tag{10.18}$$

Equations (10.17) and (10.18) are those of a cycloid. This is the name given to the curve described by a point on the circumference of a circle as it rolls along a flat plane.

The maximum value of equation (10.18) occurs when $(Be/m)\, t = \pi$ radians

$$\text{when } y_{\max} = \frac{Em}{B^2 e}\{1 - \cos \pi\} = \frac{2Em}{B^2 e} \text{ m.} \tag{10.19}$$

If this is less than the distance between the plates the electron will return to the cathode. If it is greater than the distance between the plates, the electron will collide with the anode.

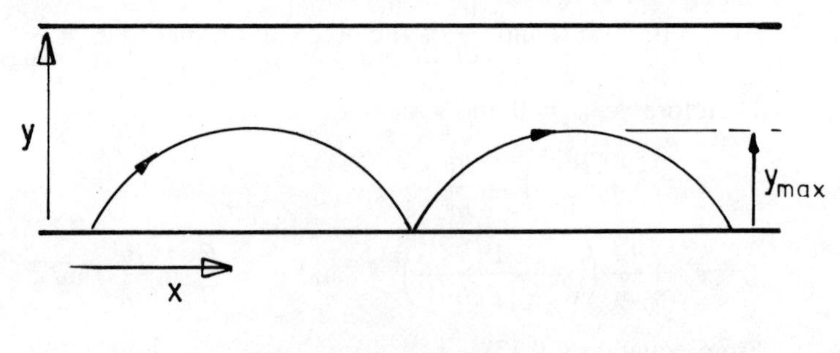

Fig. 10.16

The curve and $y_{\max}$ are shown in Fig. 10.16.

WORKED EXAMPLE 10.3

Two parallel flat plates as shown in Fig. 10.14 are situated in vacuo and are 2 cm apart. The top plate is at a potential of +5000 V with respect to the

bottom plate. An electron is liberated from the bottom plate with zero initial velocity.
(a) What is the maximum value of magnetic flux density B in order that the electron shall just "brush" the top plate and return to the cathode?
(b) What is the horizontal distance between the point of entry of the electron and the point where it strikes the cathode?

$$e/m = 1.76 \times 10^{11} \text{ C/kg.}$$

(a) From equation (10.19) $y_{max} = \dfrac{2Em}{B^2e}$ m and in this case $y_{max} = 2$ cm $= 0.02$ m

$$E = 5000/0.02 = 250\,000 \text{ V/m}$$

$$\therefore 0.02 = \frac{2 \times 250\,000}{B^2 \times 1.76 \times 10^{11}} \qquad B = \sqrt{\frac{2 \times 250\,000}{1.76 \times 10^{11} \times 0.02}} = 11.92 \times 10^{-3} \text{ T.}$$

(b) To give maximum movement in the y direction, from equation (10.18) we know that

$$\frac{Be}{m}t = \pi \qquad \therefore t = \frac{\pi m}{Be} = \frac{\pi}{11.92 \times 10^{-3} \times 1.76 \times 10^{11}}$$

$$= 1.4975 \times 10^{-9} \text{ s.}$$

Now, this is the time for the electron to reach the maximum height of its travel, i.e. to just reach the top plate. It will take a similar time to return to the bottom plate.
Total travelling time in the horizontal direction $= 2 \times 1.4975 \times 10^{-9}$ s.

From equation (10.17)

$$\text{Distance in the } x \text{ direction} = \frac{E}{B}\left(t - \frac{m}{Be}\sin\frac{Be}{m}t\right)$$

in which $t = 2 \times 1.4975 \times 10^{-9}$ s when $\dfrac{Be}{m} \times 2 \times 1.4975 \times 10^{-9} = 2\pi$

and since $\sin 2\pi = 0$, distance $= \dfrac{250\,000}{11.92 \times 10^{-3}} \times 2 \times 1.4975 \times 10^{-9}$

$$= 0.0628 \text{ m.}$$

This result may be deduced directly from a consideration of the size of the rotating circle forming the cycloid. The diameter of the circle is such that it just touches both plates, i.e. it has a diameter of 0.02 m. In rotating completely once, a spot on the circumference will move a distance equal to its circumference along the bottom plate, i.e. $2\pi \times 0.02 = 0.0628$ m.

SELF-ASSESSMENT EXAMPLE 10.4

Using the information from worked example 10.3, determine the velocity and direction of the electron 2 ns after leaving the bottom plate.

10.6 Devices employing resonant cavities

10.6.1 Inductance and capacitance of a wire-wound coil at high frequencies

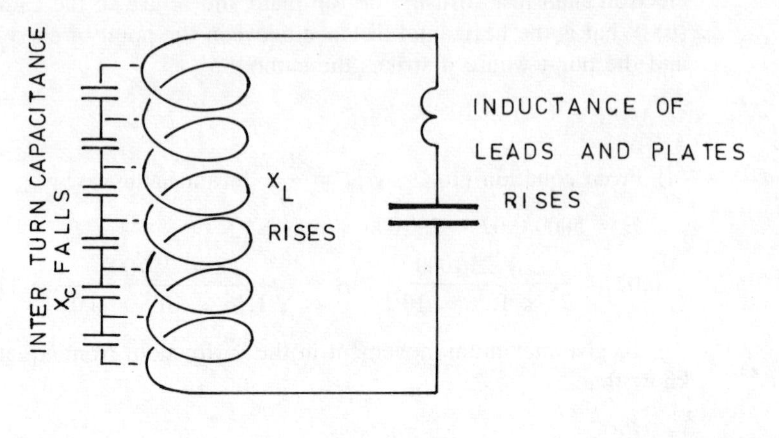

Fig. 10.17

Figure 10.17 shows a coil of wire connected to a capacitor. There are small capacitances present between the turns of the coil since they are closely situated and the leads and plates of the capacitor will have some inductance since they are linked by a self-produced magnetic field. At one particular frequency this circuit will exhibit the characteristics of resonance. Up to frequencies in the hundreds of kilohertz, normal construction of winding wire on a former to form the coil and using metal plates separated by a dielectric is satisfactory. However, at higher frequencies the inter-turn capacitance becomes significant since as capacitive reactance falls, more current takes the capacitive route rather than flow in a rapidly increasing inductive reactance of the wire

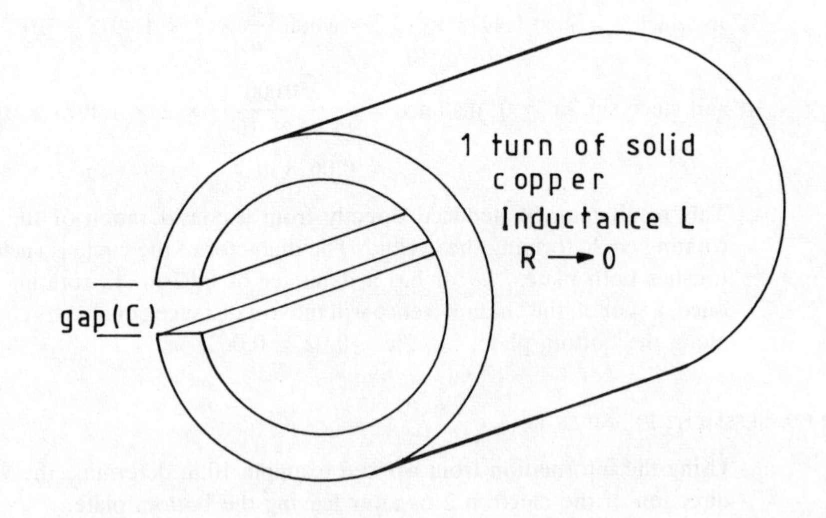

Fig. 10.18

circuit. At the same time the inductance of the capacitor becomes significant limiting current flow. At very high frequencies coils become capacitive and capacitors become inductive.

The problem is overcome by reducing the coil to a single turn with a gap creating the necessary capacitance. One form of this arrangement is shown in Fig. 10.18. The 'turn' is of solid copper and so has very little resistance. The inductance and capacitance are both small so that the resonant frequency is high, often in the thousands of megahertz range.

10.6.2 The cylindrical cavity magnetron

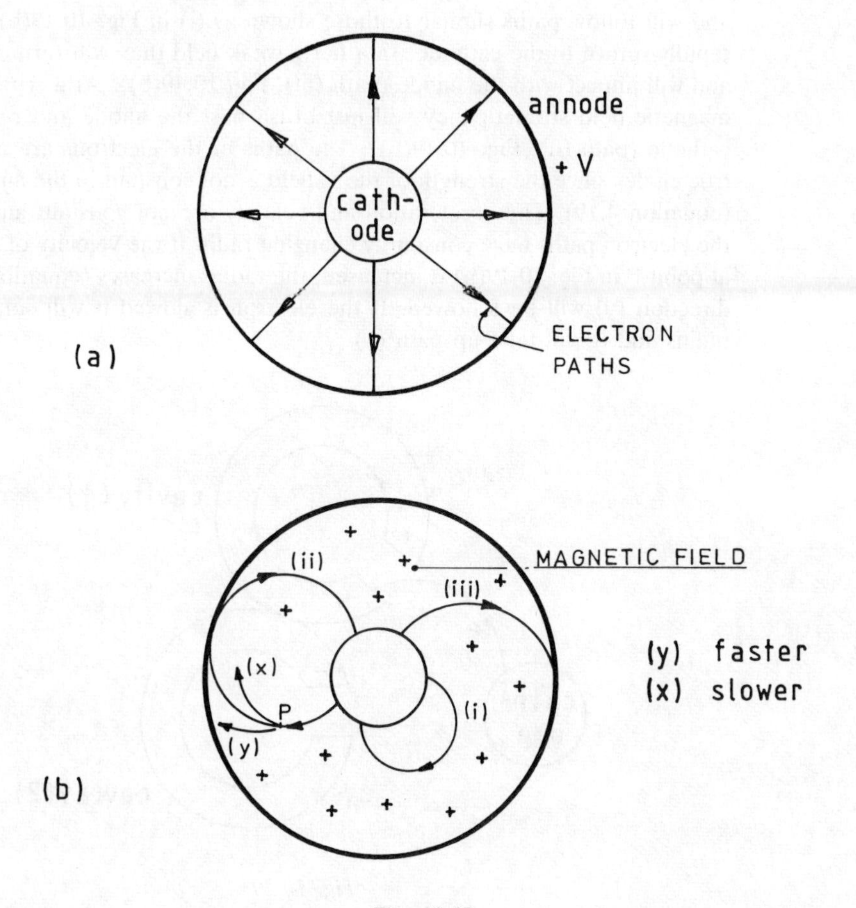

Fig. 10.19

The cylindrical magnetron is a high-power microwave oscillator which is at the heart of most modern radars operating with wavelengths of a few centimetres, for example a 3 cm wavelength which is equivalent to a frequency of 10 000 MHz. The magnetron is capable of delivering considerable power and it is therefore the last component in the radar transmission system, its output being taken via a waveguide or coaxial cable directly to the aerial system.

The magnetron consists of a heated cylindrical cathode situated within a concentric anode, the structure being in a vacuum. The cathode is maintained at a strong negative potential with respect to the anode which is earthed. This produces an electrostatic force between cathode and anode in exactly the same manner as in the plane magnetron. The cathode emits electrons into the anode/cathode space. Under the influence of the E field they will be accelerated towards the anode along paths shown in Fig. 10.19(a).

Across the anode/cathode space along the axis of the cathode, at right angles to the E field, there is a magnetic field with density B tesla referred to as the H field which will cause the high-velocity electrons to be deflected in a clockwise direction. In a very strong magnetic field the electrons will be rapidly turned and will follow paths similar to those shown as (i) in Fig. 10.19(b). They will rapidly return to the cathode. In a fairly weak field they will turn only slightly and will impact with the anode (path (iii), Fig. 10.19(b)). At a critical value of magnetic field strength they will just brush past the anode and return to the cathode (path (ii), Fig. 10.19(b)). The paths of the electrons are not, in fact, true circles since the strength of the E field is not constant in the annular space (equation 4.19). The acceleration and velocity are not constant and therefore the electron paths have constantly changing radii. If the velocity of an electron at point P in Fig. 10.19(b) is increased, the radius increases (equation 10.8) and direction (y) will be followed. If the electron is slowed it will suffer reduced radius and it will take up path (x).

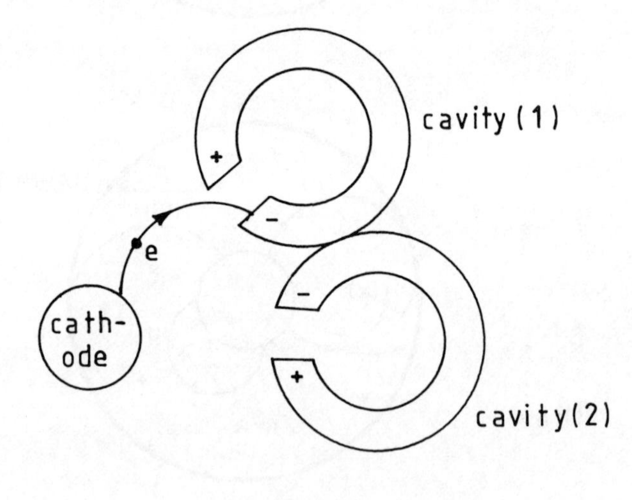

Fig. 10.20

In Fig. 10.20, the cathode is shown with two resonant cavities, forming part of the anode, disposed around it. Under the conditions envisaged in Fig. 10.19, with critical magnetic field in use, countless electrons will circulate around the cathode but none will ever come into contact with the anode and no power can be drawn from the magnetron. However, the operation of the magnetron depends upon the cavities carrying currents at resonant frequency; currents at this frequency have to be started. This requires a stray electron to impact with one of the faces of a cavity which, since the device does in fact work, must be assumed to occur at some stage. Such an impact is shown in Fig. 10.20 on one face of cavity (1). This face is now more negative than the rest of the structure and has the same effect as momentarily connecting a battery between the two faces. The capacitor is now charged and oscillation commences, charge moves backwards and forwards round each of the cavities at its natural resonant frequency, each pole reversing its polarity at twice the resonant frequency. An oscillatory electrostatic field is set up in the inter-electrode space. Since the resistance of each cavity is very small, once started, the oscillations continue without further electron arrivals for some time.

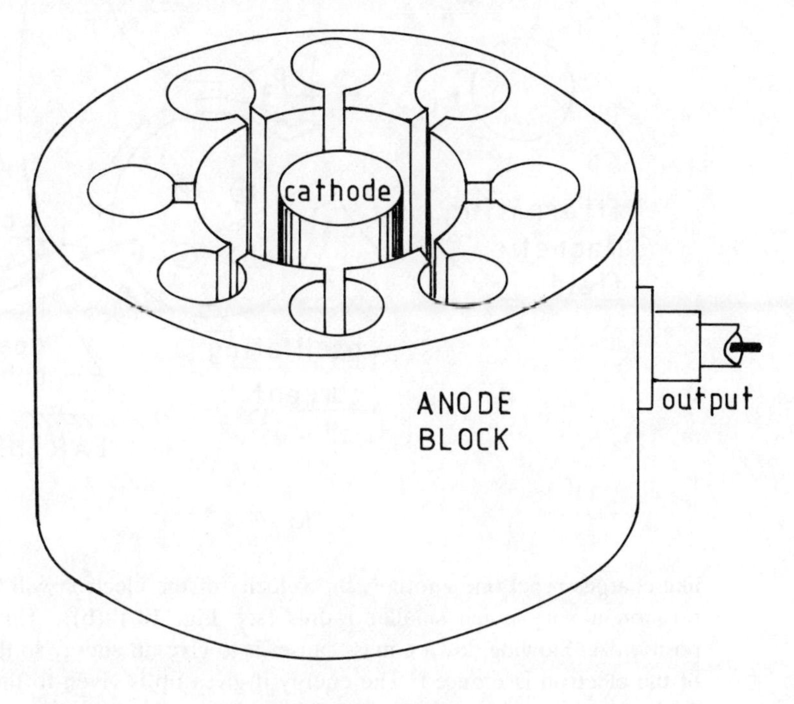

Fig. 10.21

Now consider the multi-cavity magnetron as shown in Fig. 10.21, part of which is enlarged in Fig. 10.22. Observe that since the external gaps left by forming the anode from separate resonant cavities serve no purpose, the actual construction is by drilling cavities in a solid block of copper. As already discussed, the steady E and H fields cause electrons to swirl around the cathode but do not cause other than chance electrons to collide with the anode. We have to look at the oscillatory field for an explanation of its operation. At one instant in time the anode poles will have the polarities shown in Fig. 10.22. An electron leaving the cathode at (a) will rapidly accelerate towards positive Pole 1. As it increases in velocity it will turn with ever increasing radius. It will turn clockwise which will bring it under the influence of negative Pole 2. Since

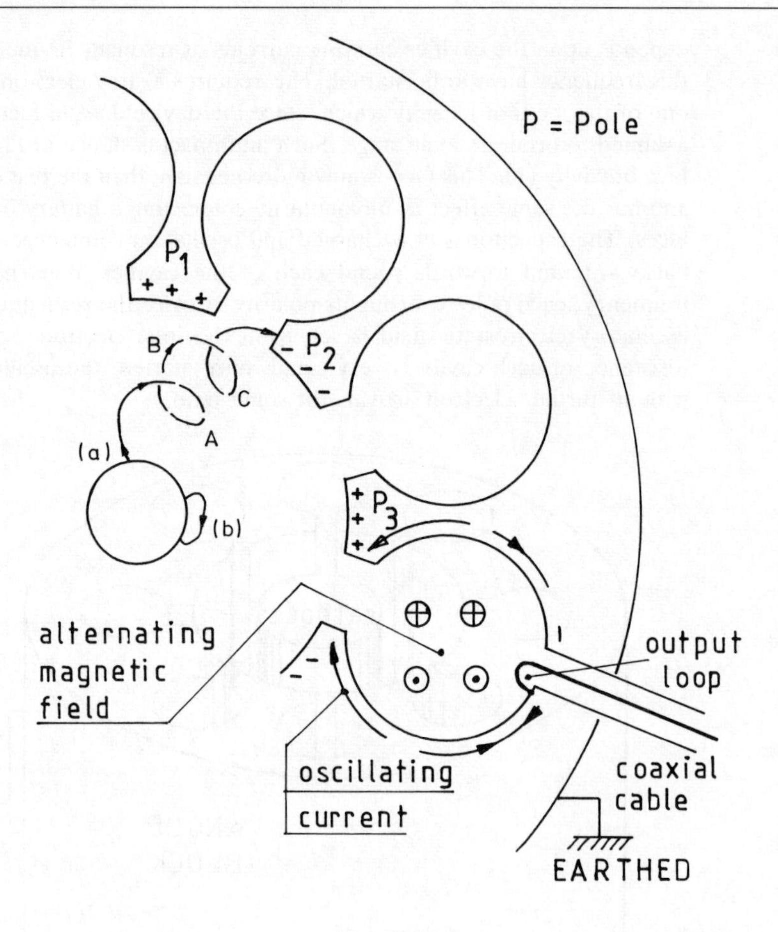

Fig. 10.22

like charges repel one another, the velocity of the electron will fall and it will turn on a very much smaller radius (see Fig. 10.19(b)). This is shown at position A. Slowing down a mass causes it to give up energy so that the energy of the electron is reduced. The energy it gives up is given to the electrostatic field which caused it to be retarded and each time an electron is slowed down in this manner the electrostatic field is reinforced. As the electron swings round to position B it regains some of its lost energy, gains velocity, increases its radius somewhat and again is slowed down at C where again it loses energy to the field. A final turn and as the electron approaches negative Pole 2 it is further retarded giving up most of its remaining energy to the field. This electron then impacts with the anode which heats it a little and in so doing the electron gives up the last of its energy. Clouds of electrons follow routes similar to that described. Other electrons may make several turns as electron (a) giving energy to the electrostatic field, and finally impact with a positive pole at higher velocity (unlike charges attract one another), so giving up more of their energy as heat. The anode is finned externally to facilitate the removal of the heat produced. An electron (b) returns quickly to the cathode. A similar electron pattern exists between the cathode and each of the pairs of anode poles. As the poles reverse and for example Pole 3 (Fig. 10.22) becomes negative the whole pattern of electron movements turns clockwise to make a similar pattern advanced by one pole. For an eight-pole magnetron the whole electron/field pattern will have made one complete revolution round the cathode in four cycles of output.

Electrons leaving the cathode and arriving at the anode represent an anode current being driven by the cathode/anode voltage. This requires an energy input. Energy given to the anode by the final impact in the form of heat is a power loss and must be got rid of. Electrons which circle round and impact immediately with the cathode give their energy to it in the form of heat and once the magnetron is operating the heater power to the cathode is either turned down or off to prevent overheating. The difference between the input power and the heat loss is energy given to the field. This is removed by the output loop and is the useful output from the magnetron.

The large oscillating E field produced causes currents to flow backwards and forwards round the cavities as the poles are alternately positive and then negative. They are all linked physically and by the fields. Power may therefore be taken from any one cavity which is effectively drawn from all the cavities. The oscillating currents round the cavities set up large alternating magnetic fields. This field, in one cavity, links with a small loop formed by folding back the core of a piece of coaxial cable and fixing it to the cavity wall. A voltage is induced in the loop which drives current into the external circuit at the resonant frequency of the cavity. Alternatively the output may be directed straight into a waveguide.

10.6.3 The multi-cavity klystron

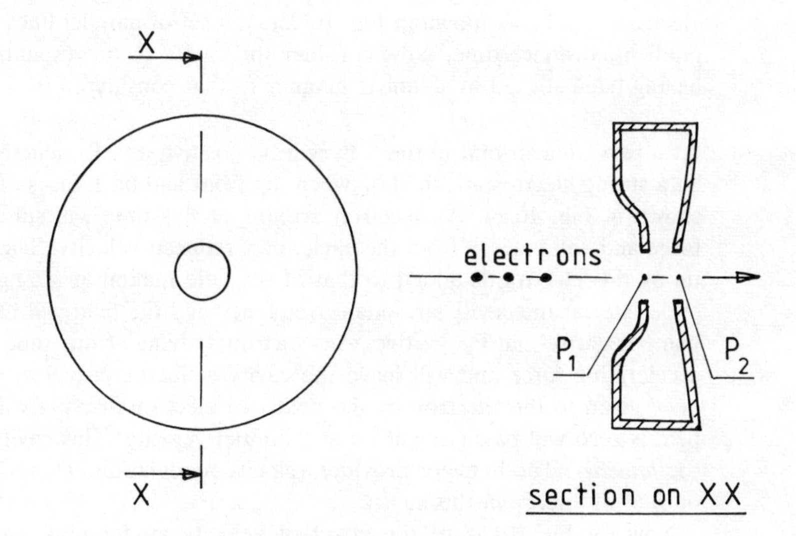

Fig. 10.23

Another form of microwave generator is the klystron which employs another type of resonant cavity as shown in Fig. 10.23. Situated in a vacuum, electrons enter on the left through port P_1 and leave on the right through port P_2. A single cavity in vacuum acts as a velocity modulator to a steam of electrons injected along its axis.

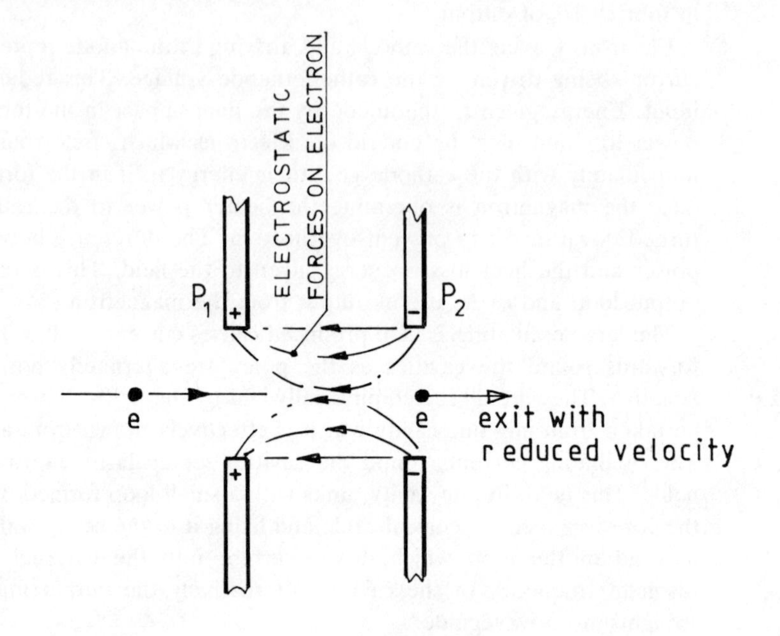

Fig. 10.24

Consider the steam of electrons from an electron gun moving with uniform velocity from left to right as shown in Fig. 10.23. The distance/time graph for electrons will be as shown in Fig. 10.25(a), a set of parallel lines with constant gradient, distance/time. Now consider the cavity to be resonating, this state having been started in a similar manner to that considered in the magnetron.

At a particular instant in time, P_1 will be positive and P_2 negative. There will be a strong electrostatic field between the front and back faces of the orifice as shown in Fig. 10.24. An electron arriving at this time will suffer a retarding force and will emerge from the cavity at a reduced velocity. The energy given up by this electron is added to that of the field making it stronger. One half-cycle later at the cavity's resonant frequency and the field will be reversed, P_1 being negative and P_2 positive. An electron arriving at this time will suffer an accelerating force and will leave the cavity at increased velocity. Energy has been given to this electron by the field. An electron arriving whilst the cavity field is zero will pass straight on at its initial velocity. This cavity is known as the *buncher*. The buncher provides velocity modulation. There is no nett loss or gain of energy in this cavity.

Now, in Fig. 10.25(b) the effect of velocity modulation is demonstrated. Electrons (1) have been retarded and hence are travelling slowly. Notice that

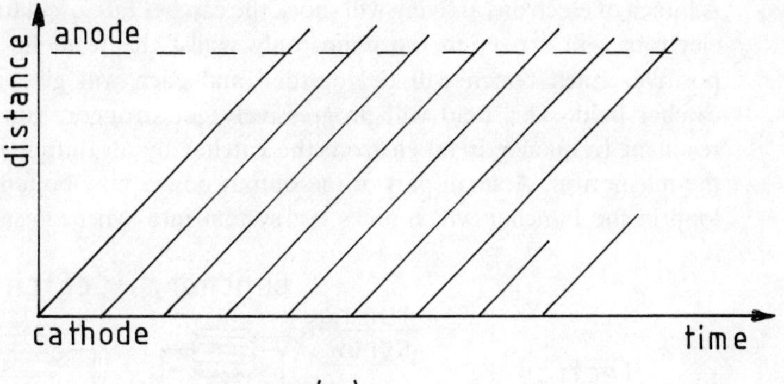

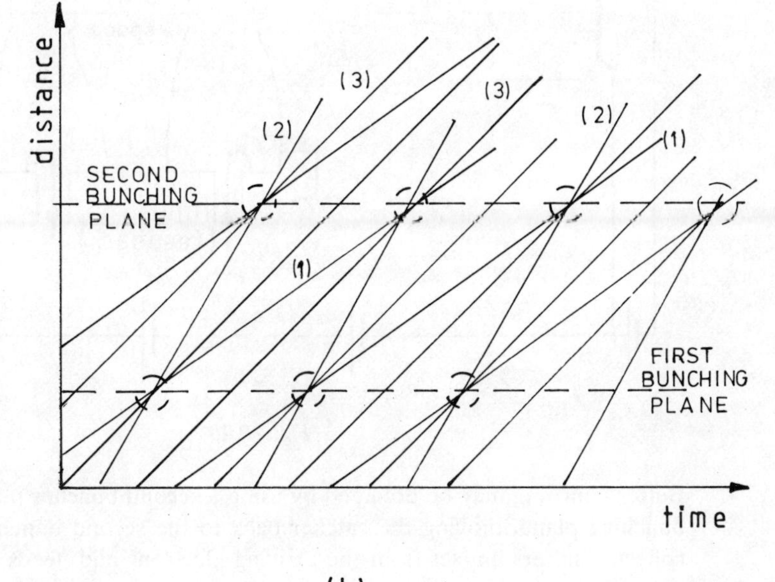

Fig. 10.25

they take a longer time to travel a given distance. Electrons (2) have been accelerated and electrons (3) are travelling at their original velocity. (Slope equal to that in Fig. 10.25(a).) The steeper the slope of the line, the greater is the velocity. There will, of course, be many other electrons arriving at various times when P_1 and P_2 are not at their maximum values and these will all be accelerated or retarded to a lesser degree. However a significant number of electrons will cross what are known as the bunching planes at the same time, the fast electrons catching up on the slower electrons to pass these lines together.

A second cavity known as the catcher is situated on the first bunching plane.

A bunch of electrons arriving will shock the catcher into oscillation and thereafter electrons will arrive, in the main, only whilst the front lip of the catcher is positive. Each bunch will be retarded and each will give up energy to the catcher field. This field will progressively get stronger. Energy at the cavity resonant frequency is taken from the catcher by an output loop exactly as in the magnetron. A small part of the output power may be fed back to an input loop in the buncher which locks the system into synchronism.

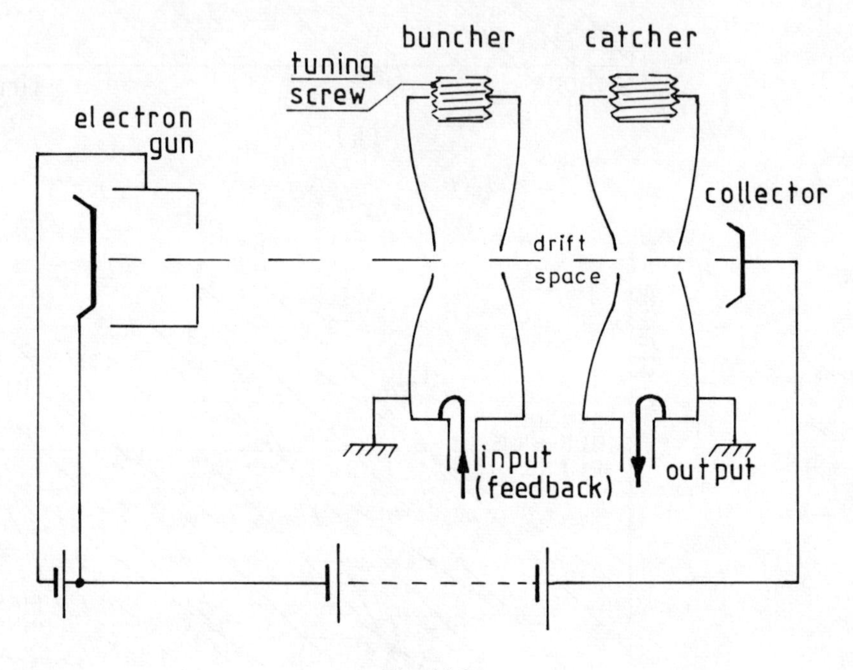

Fig. 10.26

Better bunching may be obtained by using a second buncher placed at the first bunching plane, moving the catcher back to the second bunching plane. The collector suffers impact from the arriving electrons and needs cooling in high power klystrons.

10.7 Linear particle accelerator

The linear particle accelerator consists of a number of cylinders with small entry and outlet ports. The odd-numbered cylinders are connected to the top end of the winding of a centre-tapped transformer and the even-numbered cylinders are connected to the opposite end of the winding. The transformer is fed from a sinusoidal supply. During the positive half-cycles of alternating input the odd cylinders are positive with respect to the centre point. During the negative half cycles the even cylinders are positive with respect to the centre point.

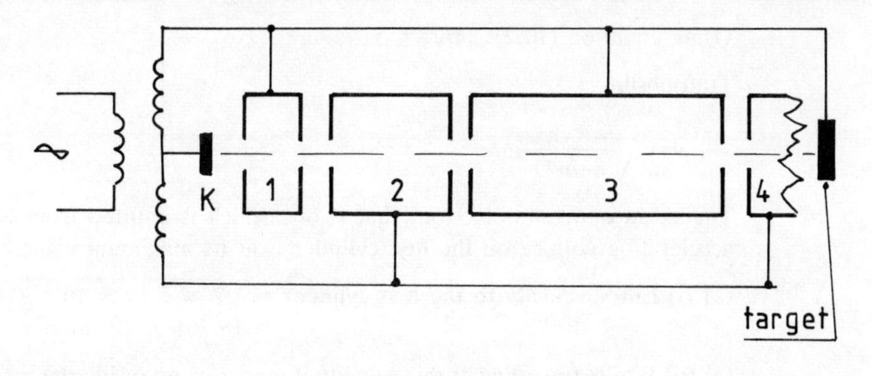

Fig. 10.27

Electrons (or ions) are produced by cathode K which is connected to the centre tapping of the transformer and are accelerated towards the front face of the first cylinder which acts as did the anode in the electron gun in Fig. 10.1. They pass through the anode hole and enter the cylinder where, since they are surrounded by an equipotential surface, they drift on until upon emerging from the cylinder they 'see' the front of the next cylinder which ideally is at its maximum potential at this time. This acts as another anode so that they are further accelerated and drift on through cylinder 2. Emerging from this at the correct time to 'see' the front of cylinder 3 at its highest potential, yet more acceleration takes place. Each cylinder needs to be longer than the previous so that it takes one half of the periodic time of the supply voltage for the electron to pass through the cylinder and to emerge at the correct instant. Electrons that are injected into cylinder 1 whilst the voltage is less than its maximum will travel too slowly and after passing through several cylinders will find that the cylinder behind them is positive rather than the one in front, they will be pulled back, will strike the cylinder and return to the supply transformer and K. Only electrons travelling at the ideal speed, that is to say those which were accelerated by the peak alternating voltage on cylinder 1 and subsequently emerge from each of the cylinders at the instant of peak voltage, arrive at the target.

WORKED EXAMPLE 10.5

A linear particle accelerator as shown in Fig. 10.27 employs an accelerating voltage described by

$$V_A = 1000 \sin(2\pi \times 10^6)\, t \text{ V}.$$

It accelerates electrons for which the ratio $e/m = 1.76 \times 10^{11}$ C/kg.

(a) Determine (i) the velocity of an electron starting from rest at K as it enters the first cylinder, (ii) the length of the first cylinder, (iii) the velocity of the electron as it enters the third cylinder, and (iv) the length of the third cylinder.

(b) Why would it not be safe to continue the analysis to determine the velocities and cylinder lengths after a large number of accelerations?

Using equation (10.2): $\frac{1}{2}mv^2 = V_A e$.

Transposing

$$v = \sqrt{\frac{2\ e\ V_A}{m}}\ \text{m/s}.$$

Taking an oportune electron, that is one which is emitted from K whilst the accelerating voltage on the first cylinder is at its maximum value of 1000 V

(a) (i) Entry velocity to the first cylinder $= \sqrt{2 \times 1.76 \times 10^{11} \times 1000}$
$= 18.761 \times 10^6$ m/s.

(a) (ii) Whilst travelling at this velocity it must remain inside the cylinder whilst the voltage wave goes through one half cycle, i.e. until the second cylinder potential rises to +1000 V.

One half of the periodic time $= 1/2f = 1/2 \times 10^6 = 0.5 \times 10^{-6}$.

Distance = velocity × time $= 18.761 \times 10^6 \times 0.5 \times 10^{-6} = 9.38$ m.
Required length of cylinder = 9.38 m.

(a) (iii) As it leaves cylinder 1 and is accelerated towards cylinder 2, energy is added to the electron.
Added energy $= V_A e$ J.

This occurs again between cylinders 2 and 3. Therefore, as the electron enters cylinder 3 its total energy $= 3V_A e$ J.

Equating energy of motion to energy supplied $\frac{1}{2}mv^2 = 3V_A e$

$$v = \sqrt{2 \times 3 \times 1.76 \times 10^{11} \times 1000}$$
$$= 32.496 \times 10^6 \text{ m/s}.$$

(a) (iv) Length of cylinder $= 32.496 \times 10^6 \times 0.5 \times 10^6 = 16.248$ m.

(b) Carrying on with this analysis, we might argue that after 15 accelerations (say), i.e. the electron is entering cylinder 15, total energy supplied = 15 × $V_A \times e$ and velocity $= \sqrt{2 \times 15 \times 1.76 \times 10^{11} \times 1000} = 72.663 \times 10^6$ m/s.

The speed of light $= 3 \times 10^8$ m/s so that the velocity of the electron represents 0.24 times that of light. At velocities of this order and above, adding energy to the electron increases its mass to a significant degree and hence the ratio of e/m changes. The calculations fall into increasing error. It has never been found possible to accelerate an electron to the speed of light no matter how many cylinders are used and however much energy is put in.

10.8 The cyclotron

Ions are liberated between two hollow, D-shaped accelerating electrodes which are mounted in an evacuated envelope between the poles of a very powerful magnet. The electrodes are connected to a high-frequency generator. The ions

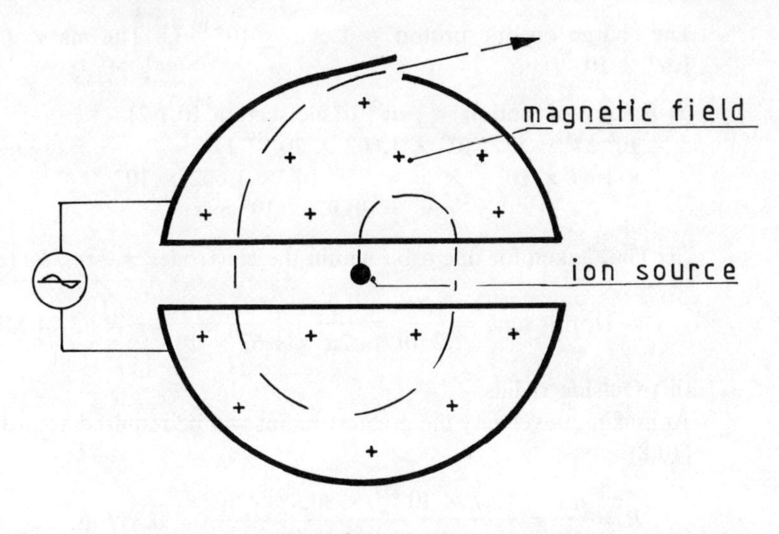

Fig. 10.28

are accelerated towards the opposite-polarity electrode and under the influence of the magnetic field follow a circular orbit within the electrode.

Using equation (10.8), with the charge e on a single electron replaced by q, the charge on an ion and m = mass of the ion

the radius on which the ions turn $= \dfrac{mv}{Bq}$ m

and from equation (10.10), time for one orbit $= \dfrac{2\pi m}{Bq}$ s.

If the source frequency is chosen such that its periodic time is equal to the time taken for one orbit, the ions, after one half of a revolution, will arrive back at the gap at the instant when the polarity has reversed and risen to its maximum value. The velocity will increase, giving a corresponding increase in radius and a longer trajectory in the field. The longer path at higher velocity means that the ion will re-emerge into the electrode gap at exactly the correct time for further acceleration. After several half revolutions the ion is ejected and travels on in a straight line until it impacts with a target.

Electrons are not used in the accelerator because of their small mass which would result in impossibly high frequencies being required.

WORKED EXAMPLE 10.6

A cyclotron is to be employed to accelerate protons so that they emerge with an energy of 5×10^6 eV. (1 eV = 1.602×10^{-19} J). The magnetic field across the electrodes of the cyclotron has a flux density of 0.9 T. Determine (i) the velocity at which the protons emerge (ii) the required frequency of the accelerating supply, and (iii) the outside radius of the D-shaped electrodes.

The charge on one proton = 1.602×10^{-19} C. The mass of one proton = 1.67×10^{-7} kg.

(i) Energy of motion = $\frac{1}{2}mv^2$ J (see section 10.1.2)
5×10^6 eV = $5 \times 10^6 \times 1.602 \times 10^{-19}$ J
$\therefore \frac{1}{2} \times 1.67 \times 10^{-27} \times v^2 = 5 \times 10^6 \times 1.602 \times 10^{-19}$
$v = 30.97 \times 10^6$ m/s.

(ii) Time taken for one orbit within the electrodes = $2\pi m/Be$ (equation 10.10)

$$\therefore f = 1/\text{orbit time} = \frac{Be}{2\pi m} = \frac{0.9 \times 1.602 \times 10^{-19}}{2\pi \times 1.67 \times 10^{-27}} = 13.74 \text{ MHz.}$$

(iii) Outside radius:
At maximum velocity the greatest radius will be required according to equation (10.8).

$$R = \frac{mv}{Be} = \frac{1.67 \times 10^{-27} \times 30.97 \times 10^6}{0.9 \times 1.602 \times 10^{-19}} = 0.359 \text{ m.}$$

FURTHER PROBLEMS

10.7 Calculate the time taken for an electron to travel a distance of 10 mm given that the potential difference over that distance is 100 V. The electron commences with zero energy.

10.8 An electron is liberated from a flat plate at earth potential with zero initial velocity. A second plate, parallel to the first, situated 10 cm away is at a potential of +5000 V. Calculate (i) the force on the electron, (ii) the time taken for the electron to strike the top plate, (iii) the final velocity of the electron (immediately before striking the top plate), and (iv) the energy in joules given to the top plate by the impact of the electron. $e = 1.602 \times 10^{-19}$C, $m = 9.1 \times 10^{-31}$ kg.

10.9 (a) An electron is accelerated by a potential of V volts and injected into a uniform magnetic field of strength B tesla in a direction at right angles to the field lines. Demonstrate that the motion of the electron will be circular and derive an expression for the radius.
(b) It is required to cause an electron beam to be turned through exactly 60° on a radius of 5 cm. Given that the accelerating voltage = 20 kV, determine (i) the time the electron must remain in the field, (ii) the required magnetic field strength, and (iii) the velocity of the electron as it leaves the electron gun.

10.10 A cathode ray tube employs an accelerating potential of 600 V. Its vertical deflection system consists of two parallel plates 1 cm apart and 1.5 cm in length. The tube face is 25 cm from the deflection plates. It is required to achieve a 1 cm deflection of the trace on the screen for an input of 0.1 V. Determine (i) the velocity of the electron beam as it leaves the electron gun, (ii) the deflection voltage required between the plates to give the required deflection, (iii) the gain of the amplifier feeding the plates to give the required sensitivity, and (iv) the vertical velocity given to the electron by the deflection system.

10.11 (a) What problems are encountered when using wire-wound coils at very high frequencies?
(b) How are these problems overcome in the cavity magnetron?

10.12 A cavity magnetron using an accelerating voltage of 16 kV is supplied with a direct current of 20 A during an 'on' pulse. The power taken from the high-frequency output lead is 150 kW.
(a) What is the efficiency of the magnetron?
(b) Explain the mechanism whereby the direct power input is partly transformed into an alternating output.
(c) Account for the difference between input and output powers.

10.13 (a) Derive an expression for the radius and periodic time characterising the motion of a charged particle moving in a magnetic field. (b) Determine the required frequency and magnet radius in order that a cyclotron employing a magnet with field strength 1 tesla shall deliver a beam of 2 MeV protons. For a proton: charge = 1.602×10^{-19} C, $m = 1.67 \times 10^{-27}$ kg.

10.14 (a) A linear particle accelerator employs a sinusoidal voltage described by the equation $v = V_m \sin 2\pi ft$ volts. Assuming that a particle with charge e coulombs and mass m kg is liberated from the cathode with zero initial velocity deduce the formula

$$l_n = \frac{\sqrt{(2\,neV_m/m)}}{2f}$$

where l_n = length of the nth cylinder in the accelerator.
(b) Why do particles leaving the cathode at non-opportune times never reach the target at the end of the accelerator?

10.15 Electrons are liberated from a cathode with zero energy and are accelerated by a potential of V volts to a velocity v m/s. They travel in a beam l m in length to reach a target.
(a) Deduce that the beam current I A is given by the formula:

$$I = \frac{nve}{l}\ \text{A}$$

where n = number of electrons in the beam at any instant.
(b) Given that $V = 5$ kV, $l = 25$ cm and the beam current = 2 mA, deduce (i) the number of electrons leaving the cathode per second, (ii) the velocity of the electrons, (iii) the time taken for the electrons to travel from the cathode to the target, and (iv) the number of electrons in transit at any instant.

Answers

10.4 $v_x = 3.133 \times 10^6$ m/s, $v_y = -1.824 \times 10^6$ m/s
velocity = $3.625 \times 10^6\angle -30.2°$m/s

10.7 3.37×10^{-9} s
10.8 (i) 16.02×10^{-15} N, (ii) 2.383×10^{-9} s, (iii) 41.96×10^{6} m/s
(iv) 801×10^{-18} J
10.9 (b) (i) 0.641×10^{-9} s (ii) 9.788×10^{-3} T, (iii) 81.73×10^{6} m/s
10.10 (i) 14.156×10^{6} m/s, (ii) 32 V, (iii) 320, (iv) 566250 m/s
10.12 (a) 46.875% (c) heat
10.13 $f = 15.27$ MHz, $R = 0.204$ m
10.15 (i) 1.248×10^{16}, (ii) 41.96×10^{6} m/s, (iii) 5.96×10^{-9} s
(iv) 74.37×10^{6}

Appendix 1
The Star/Delta Transformation

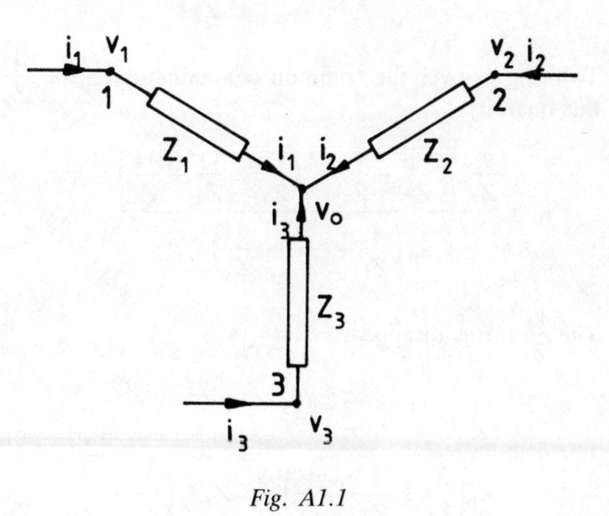

Fig. A1.1

Firstly consider the star shown in Fig. A1.1.

Currents i_1, i_2 and i_3 flow towards the star point. Voltages v_1, v_2, v_3 and v_0 are measured with respect to a datum point, usually earth.

From Kirchhoff's first law $i_1 + i_2 + i_3 = 0$

$$i_1 = \frac{v_1 - v_0}{Z_1}; \quad i_2 = \frac{v_2 - v_0}{Z_2}; \quad i_3 = \frac{v_3 - v_0}{Z_3}$$

each being the potential difference across an impedance divided by the value of that impedance. Adding the three currents

$$\frac{v_1 - v_0}{Z_1} + \frac{v_2 - v_0}{Z_2} + \frac{v_3 - v_0}{Z_3} = 0.$$

Transposing to get all the v_0 terms on the right-hand side

$$\frac{v_1}{Z_1} + \frac{v_2}{Z_2} + \frac{v_3}{Z_3} = \frac{v_0}{Z_1} + \frac{v_0}{Z_2} + \frac{v_0}{Z_3}$$

$$\frac{v_1}{Z_1} + \frac{v_2}{Z_2} + \frac{v_3}{Z_3} = v_0\left(\frac{1}{Z_1} + \frac{1}{Z_2} + \frac{1}{Z_3}\right)$$

$$v_0 = \frac{\dfrac{v_1}{Z_1} + \dfrac{v_2}{Z_2} + \dfrac{v_3}{Z_3}}{\dfrac{1}{Z_1} + \dfrac{1}{Z_2} + \dfrac{1}{Z_3}}.$$

Substituting for v_0 in any of the current equations (selecting the equation for i_1)

$$i_1 = \frac{v_1 - v_0}{Z_1} = \frac{v_1 - \left[\dfrac{\dfrac{v_1}{Z_1} + \dfrac{v_2}{Z_2} + \dfrac{v_3}{Z_3}}{\dfrac{1}{Z_1} + \dfrac{1}{Z_2} + \dfrac{1}{Z_3}}\right]}{Z_1}.$$

To bring v_1 over the common denominator $\left(\dfrac{1}{Z_1} + \dfrac{1}{Z_2} + \dfrac{1}{Z_3}\right)$, multiply it by this quantity.

$$i_1 = \frac{\dfrac{v_1}{Z_1} + \dfrac{v_1}{Z_2} + \dfrac{v_1}{Z_3} - \left(\dfrac{v_1}{Z_1} + \dfrac{v_2}{Z_2} + \dfrac{v_3}{Z_3}\right)}{Z_1\left(\dfrac{1}{Z_1} + \dfrac{1}{Z_2} + \dfrac{1}{Z_3}\right)}.$$

The $\dfrac{v_1}{Z_1}$ terms disappear.

$$i_1 = \frac{\dfrac{v_1}{Z_2} + \dfrac{v_1}{Z_3} - \dfrac{v_2}{Z_2} - \dfrac{v_3}{Z_3}}{Z_1\left(\dfrac{1}{Z_1} + \dfrac{1}{Z_2} + \dfrac{1}{Z_3}\right)} = \frac{v_1 - v_2}{Z_1 Z_2\left(\dfrac{1}{Z_1} + \dfrac{1}{Z_2} + \dfrac{1}{Z_3}\right)} +$$

$$\frac{v_1 - v_3}{Z_1 Z_3\left(\dfrac{1}{Z_1} + \dfrac{1}{Z_2} + \dfrac{1}{Z_3}\right)}. \qquad \text{(i)}$$

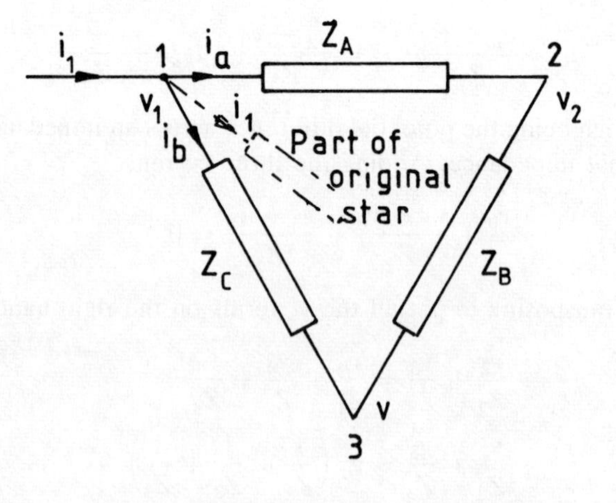

Fig. A1.2

Now consider the equivalent delta as in Fig. A1.2.
The input current from terminal 1 was i_1 A in the original star. If the delta is

to be exactly equivalent, the current entering terminal 1 must still be i_1 A. In Fig. 2.30, $i_1 = i_a + i_b$

$$i_a = \frac{v_1 - v_2}{Z_A} \quad \text{and } i_b = \frac{v_1 - v_3}{Z_C}.$$

$$\text{Hence } i_1 = \frac{v_1 - v_2}{Z_A} + \frac{v_1 - v_3}{Z_C}. \qquad \text{(ii)}$$

Comparing equations (i) and (ii) it is seen that for equivalence

$$Z_A = Z_1 Z_2\left(\frac{1}{Z_1} + \frac{1}{Z_2} + \frac{1}{Z_3}\right) \quad \text{and } Z_C = Z_1 Z_3\left(\frac{1}{Z_1} + \frac{1}{Z_2} + \frac{1}{Z_3}\right).$$

By a similar process it may be deduced that $Z_B = Z_2 Z_3\left(\frac{1}{Z_1} + \frac{1}{Z_2} + \frac{1}{Z_3}\right)$.

Appendix 2
The Delta/Star Transformation

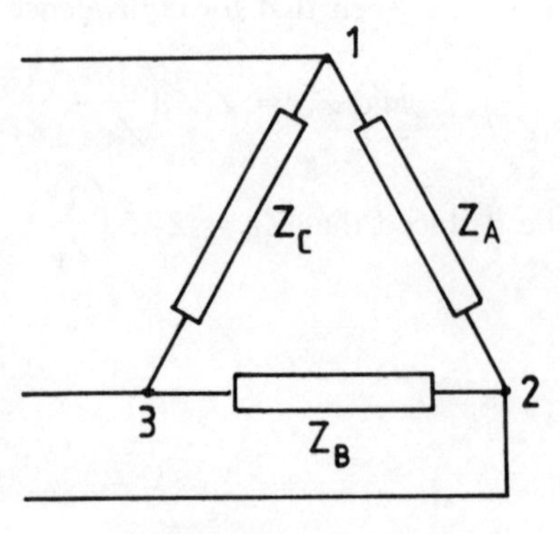

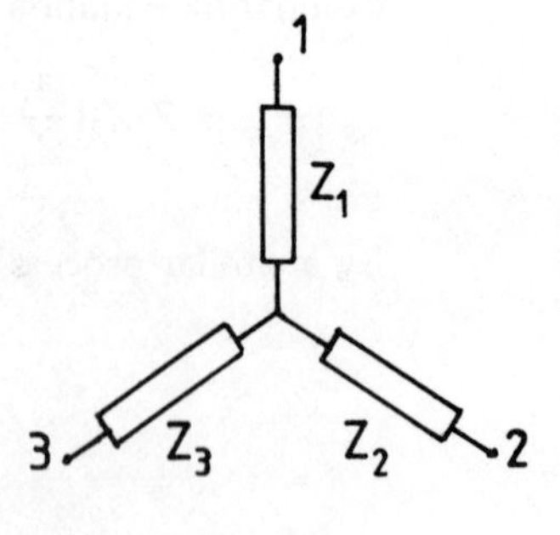

Fig. A2.1

The values of the impedances in the equivalent star are determined by considering the impedance between each pair of terminals in both networks. Consider the impedance from terminal 1 to terminal 2 in the delta network. An electrical path exists through Z_A direct. Another parallel path exists through Z_C and Z_B in series.

$$\text{Equivalent impedance of these two parallel paths} = \frac{\text{product of impedances}}{\text{sum of impedances}}$$

$$= \frac{Z_A\,(Z_B + Z_C)}{Z_A + Z_B + Z_C}.$$

In the star, between terminals 1 and 2, impedance $= Z_1 + Z_2$. For equivalence

$$Z_1 + Z_2 = \frac{Z_A(Z_B + Z_C)}{Z_A + Z_B + Z_C}. \qquad \text{(i)}$$

Considering terminals 2 and 3, in the delta Z_B is in parallel with $(Z_A + Z_C)$. In the star between terminals 2 and 3 the impedance $= Z_2 + Z_3$. For equivalence

$$Z_2 + Z_3 = \frac{Z_B(Z_A + Z_C)}{Z_A + Z_B + Z_C}. \qquad \text{(ii)}$$

Finally, between terminals 3 and 1

$$Z_1 + Z_3 = \frac{Z_C(Z_B + Z_A)}{Z_A + Z_B + Z_C}. \qquad \text{(iii)}$$

Taking equation (iii) from equation (i) and then adding equation (ii) to the result yields a left-hand side of

$$(Z_1 + Z_2) - (Z_1 + Z_3) + (Z_2 + Z_3) = Z_1 + Z_2 - Z_1 - Z_3 + Z_2 + Z_3 = 2Z_2.$$

On the right-hand side

$$\frac{Z_A(Z_B + Z_C) - Z_C(Z_B + Z_A) + Z_B(Z_A + Z_C)}{Z_A + Z_B + Z_C}$$

$$= \frac{Z_A Z_B + Z_A Z_C - Z_C Z_B - Z_C Z_A + Z_B Z_A + Z_B Z_C}{Z_A + Z_B + Z_C}$$

$$= \frac{2Z_A Z_B}{Z_A + Z_B + Z_C}.$$

Equating left-hand and right-hand sides gives

$$2Z_2 = \frac{2Z_A Z_B}{Z_A + Z_B + Z_C}$$

$$Z_2 = \frac{Z_A Z_B}{Z_A + Z_B + Z_C}.$$

Note that this is similar in form to the equations for the star/delta transformation.

By a similar method it is found that

$$Z_1 = \frac{Z_A Z_C}{Z_A + Z_B + Z_C} \quad \text{and} \quad Z_3 = \frac{Z_B Z_C}{Z_A + Z_B + Z_C}.$$

Appendix 3
Mutual Inductance

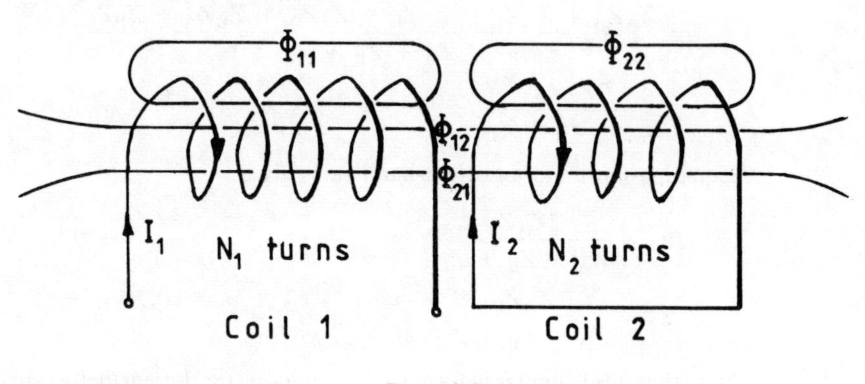

Fig. A3.1

Figure A3.1 shows two coils in close proximity. A current I_1 is supplied to coil 1 from an external source of e.m.f.

Φ_{11} is the flux set up in coil 1 which only links with coil 1.
Φ_{22} is the flux set up in coil 2 which only links with coil 2.
Φ_{12} is the flux set up in coil 1 which links with coil 2.
Φ_{21} is the flux set up in coil 2 which links with coil 1.

Assume that the flux produced is proportional to the current.
An expression for self-inductance from equation (5.13) is

$$L = \frac{N\,\mathrm{d}\Phi}{\mathrm{d}i}\ \mathrm{H}. \qquad \text{(i)}$$

For self-inductance the flux linking with the turns creating it is involved.
Assume that initially the circuit is not energised. All fluxes and currents are zero. Establishing a current in coil 1 creates a flux $\Phi_{12} + \Phi_{11}$, that is, the flux which links with coil 2 plus the flux which links only coil 1 or part of it.

$\mathrm{d}i$ is the change in current from zero to its final value I_1

$$\mathrm{d}i = I_1.$$

The change in flux is from zero to $(\Phi_{12} + \Phi_{11})$

$$\mathrm{d}\Phi = (\Phi_{12} + \Phi_{11}).$$

Therefore, using equation (5.13)

$$L_1 = \frac{N_1(\Phi_{12} + \Phi_{11})}{I_1} \text{ H.}$$

Multiplying out the parentheses gives two terms, each of which is a self-inductance.

$$L_1 = \frac{N_1\Phi_{12}}{I_1} + \frac{N_1\Phi_{11}}{I_1}. \qquad \text{(ii)}$$

Dealing with the second term, $N_1\Phi_{11}/I_1$ is an inductance present because of the imperfection in coupling between the coils. If perfect coupling were present all the flux from coil 1 would link with coil 2 and Φ_{11} would be zero.

$$\frac{N_1\Phi_{11}}{I_1} = L_{11}$$

where L_{11} is the leakage inductance of coil 1.

$$\text{Let } \frac{N_1\Phi_{12}}{I_1} = L_A$$

where L_A is an inductance due to the flux which does link the two coils.

Equation (ii) may therefore be written as $L_1 = L_{11} + L_A$.

The same reasoning applies to the secondary. Owing to the changing flux, an e.m.f. is induced which drives a current I_2 if the circuit is closed. The current produces two fluxes, Φ_{21} and Φ_{22}. Hence $L_2 = L_{22} + L_B$, where L_2 = total inductance, L_{22} = leakage inductance.

The two coils may be represented schematically as in Fig. A3.2.

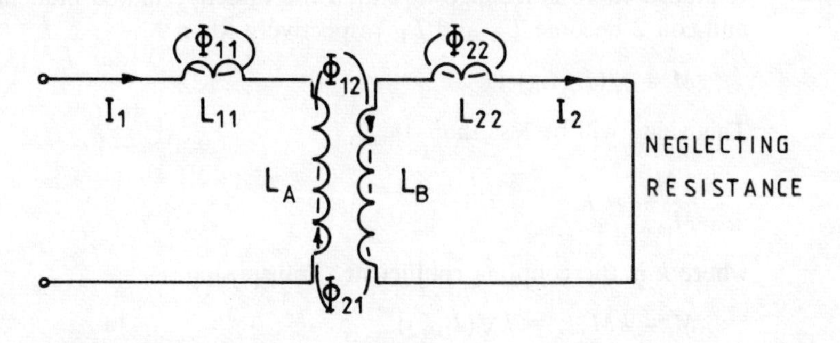

Fig. A3.2

Now if there were no leakage

$$L_1 = L_A = \frac{N_1\Phi_{12}}{I_1}$$

$$\text{and } L_2 = L_B = \frac{N_2\Phi_{21}}{I_2}.$$

$$\text{Hence } L_1L_2 = \frac{N_1\Phi_{12}}{I_1} \times \frac{N_2\Phi_{21}}{I_2}$$

or $$L_1L_2 = \frac{N_1\Phi_{21}}{I_2} \times \frac{N_2\Phi_{12}}{I_1}. \tag{iii}$$

Now $N_1\Phi_{21}$ are flux linkages of secondary-produced flux by I_2 with the primary coil of N_1 turns. Just as self-inductance is defined in equation (5.13) as flux linkages with a coil created by a current in that coil, mutual inductance may be defined as flux linkages with a second coil created by a current in the first coil or vice versa. Hence

$$\frac{N_1\Phi_{21}}{I_2} = M$$

where M is the mutual inductance between the coils. Similarly, $N_2\Phi_{12}$ are flux linkages of primary-produced flux by current I_1 with the secondary of N_2 turns. Hence

$$\frac{N_2\Phi_{12}}{I_1} = M.$$

Therefore equation (iii) becomes

$$L_1L_2 = M \times M = M^2. \tag{iv}$$

This result is achieved assuming perfect coupling, i.e. no leakage flux and so the mutual inductance is at its maximum possible value M_{max}.

Taking the square root of both sides of equation (iv) gives

$$M_{max} = \sqrt{(L_1L_2)}\ \text{H}. \tag{7.9}$$

If indeed there is leakage flux then the effective linked inductances of coil 1 and coil 2 become L_A and L_B respectively when

$$M = \sqrt{(L_AL_B)}\ \text{H}.$$

This value will be less than M_{max}.

$$\frac{M}{M_{max}} = k$$

where k is the coupling coefficient. Transposing

$$M = kM_{max} = k\sqrt{(L_1L_2)}. \tag{7.10}$$

Index